Strategic Management of Technology and Innovation

SECOND EDITION

Strategic Management of Technology and Innovation

Garry D. Bruton
Texas Christian University

Margaret A. White
Oklahoma State University

SOUTH-WESTERN
CENGAGE Learning

Australia • Brazil • Japan • Korea • Mexico • Singapore • Spain • United Kingdom • United States

SOUTH-WESTERN
CENGAGE Learning™

Strategic Management of Technology and Innovation, Second Edition
Garry D. Bruton, Margaret A. White

Vice President/Editorial Director: Jack Calhoun

Editor-in-Chief: Melissa Acuña

Senior Acquisitions Editor: Michele Rhoades

Developmental Editor: Erin Guendelsberger

Senior Editorial Assistant: Ruth Belanger

Marketing Manager: Nate Anderson

Senior Marketing Communications Manager: Jim Overly

Content Project Manager: PreMediaGlobal

Media Editor: Danny Bolan

Senior Frontlist Buyer: Sandee Milewski

Print Buyer: Arethea Thomas

Production House/Compositor: PreMediaGlobal

Senior Art Director: Tippy McIntosh

Senior Rights Acquisitions Account Manager, Text: Mardell Glinski Schultz

Internal Designer: Craig Ramsdell, Ramsdell Design

Cover Designer: Patti Hudepohl

Cover Image: Shutterstock

Photo Credits:

B/W Image: iStockphoto

Color Image: Shutterstock Images/Kentoh

For product information and technology assistance, contact us at
Cengage Learning Customer & Sales Support, 1-800-354-9706
For permission to use material from this text or product, submit all requests online at **www.cengage.com/permissions**
Further permissions questions can be emailed to
permissionrequest@cengage.com

Library of Congress Control Number: 2010930355

International Edition:

ISBN-13: 978-0-538-48173-1

ISBN-10: 0-538-48173-0

Cengage Learning International Offices

Asia
www.cengageasia.com
tel: (65) 6410 1200

Brazil
www.cengage.com.br
tel: (55) 11 3665 9900

Latin America
www.cengage.com.mx
tel: (52) 55 1500 6000

Represented in Canada by Nelson Education, Ltd.
tel: (416) 752 9100/(800) 668 0671
www.nelson.com

Australia/New Zealand
www.cengage.com.au
tel: (61) 3 9685 4111

India
www.cengage.co.in
tel: (91) 11 4364 1111

UK/Europe/Middle East/Africa
www.cengage.co.uk
tel: (44) 0 1264 332 424

Cengage Learning is a leading provider of customized learning solutions with office locations around the globe, including Singapore, the United Kingdom, Australia, Mexico, Brazil, and Japan. Locate your local office at: **cengage.com/global**

For product information: **www.cengage.com/international**
Visit your local office: **www.cengage.com/global**
Visit our corporate website: **www.cengage.com**

AVAILABILITY OF RESOURCES MAY DIFFER BY REGION. Check with your local Cengage Learning representative for details.

Printed in Canada
1 2 3 4 5 6 7 8 13 12 11 10

Dedicated to:

Our students and teachers
for sharing the learning voyage with us

and

Our families and friends
for inspiring us to pursue the voyage

BRIEF CONTENTS

CONTENTS

PREFACE

The management of technology and innovation (MTI) is an issue that faces all firms today. The waves of change in the business environment include new technologies and innovations that force industries and firms to find new ways to compete and to survive. Just as a storm of new products seems to be emerging rapidly, new ways of doing things (new processes) are emerging to help firms be more efficient and effective.

To meet these waves of change, business must find ways to manage technology and innovation. These changes include new ways to generate and implement strategic goals. These implementation issues include new ways to communicate needed information, organize tasks, and manage people. As a practical matter, these waves of change have resulted in an increasing number of engineers moving beyond technology concerns into management. Likewise, they have also pushed managers who never thought they would need to understand the intricacies of working with technology to seek knowledge about such issues. This text is the first to recognize that MTI is not strictly a technical concern or a business concern. Rather, MTI is a domain that needs an integrated approach for students and managers.

GOALS FOR THE TEXT

Prior texts have typically addressed only one or the other of these managerial groups, technical professionals (typically engineers) or those with business (or other) training. This text is designed to serve as an information link between the managerial needs of both domains. The authors believe that this type of approach and information is needed because in the reality of our MTI classes there are mixed groups of engineering and business students, and this is the situation that faces business practitioners. Therefore, the book was developed to meet several goals:

1. To integrate strategy and technology: Prior texts have assumed that the student has a strong engineering or business strategy background. Those texts then build on that knowledge, emphasizing the domain they believe the students already know. To the authors, however, this continues to stress functional silos and results in students not developing the integrated understanding of MTI that today's business requires. Therefore, this text develops an integrated approach with strategy and technology intertwined. This integrated approach is applicable no matter what the educational background of the student may be.

2. To provide insight into MTI that will be useful to students as they enter practice: While theory is important, MTI is an applied discipline that students need to be able to use. Throughout the text, the authors employ numerous realistic settings to ensure that students understand different concepts. Exercises and checklists at the end of each chapter are designed to help students apply their new knowledge in their careers. The net result is a useful set of tools to aid actual MTI decision making.

3. To help the instructor excel in the classroom: This view has led to the development of a full set of teaching supports for the text. Too often, MTI has been viewed as a minor domain and adequate teaching supports (including teachers' manuals with suggestions on how to use the chapter and additional material that instructors can use to supplement their lectures, test banks, sample answers on the exercises, and a complete set of PowerPoint slides) have not been provided. This text is fully supported and has a complete set of supplements that makes it the most user-friendly text on the market today. The authors have used the draft version of this text on multiple occasions. The result is a text that has been tested and further developed for maximum usability by the instructor.

ORGANIZATION OF THE BOOK

This second edition of the text is organized like the first with four sections. The first section introduces the concept of MTI and then establishes what is in the domain of management of technology and innovation. This section contains two chapters. There are two major strategic options that an organization can take in the development and maintenance of MTI—(1) internal innovation and (2) obtaining technology through external means. The process utilized in each of these two approaches to MTI involves (a) planning, (b) implementation, and (c) evaluation and control to ensure that the plans and goals of the firm are met. As a result of this view, the second section of the text examines internal innovation planning (Chapter 3), implementation (Chapter 4), and evaluation and control (Chapter 5). The third section repeats this pattern for external efforts to obtain technology planning (Chapter 6), implementation (Chapter 7), and evaluation and control (Chapter 8). The last section of the text examines the building of the capabilities necessary for MTI success (Chapter 9), and organizational learning and knowledge management (Chapter 10). These chapters are very rich and the authors found by using the book that the material for a course fits into both the quarter system and the fifteen-week semester.

Each of these four sections has a unifying real company case that introduces it. These are all new for this edition. In the first section, General Electric is the firm of interest. Chapter 1 uses the firm to illustrate the need and benefit of MTI to business and society. Chapter 2 then illustrates the strategic activities that impact the MTI process at GE. In a similar manner, GlaxoSmithKline is used in the section about internal innovation (Chapters 3–5) and Acer provides the example for the section about acquisition of technology (Chapters 6–8). Finally, Google illustrates how to build strategic MTI success (Chapters 9–10).

At the end of the text, an Appendices section addresses key topics that impact all chapters in each of the four sections. Appendix 1 focuses on ethics and corporate social responsibility. This is new for this edition. The recent wave of interest in "green" technology led to the inclusion of this topic. Appendix 2 centers on managing innovation projects. Appendix 3 addresses the issues of managing platforms and portfolios of technology. Forecasting is the focus of the Appendix 4.

The authors have sought to ensure that there are updated examples in the text to illustrate the concepts discussed. In addition, the authors have included examples from every continent except Antarctica—the global economy is a reality because of new technologies in communication, transportation, and information sharing. Within each chapter, there are examples that illustrate different aspects for that chapter. To further this emphasis, there are several mini-cases in each chapter. These cases have questions that can be used by instructors to generate analysis by and discussion with the students.

FOUNDATION FOR THE TEXT

This book was written by authors with over 50 years of combined teaching and consulting experience. Their primary goal was to provide a readable, useful text about strategic issues in MTI. This means that the book is focused on real contexts and organizational actions. The direct systematic planning, implementation, and evaluation and control approach to the two major strategic actions required to obtain technology is unique to this text. These activities are part of every strategic decision of successful organizations. However, this text is the first to lay a practical, understandable method for students, no matter their background, to see how to accomplish such activities.

The text is based on material that gives the student active learning opportunities from a variety of sources. For example, the exercises in each chapter include those that focus on using the World Wide Web, others that are intended to enable students to apply the knowledge learned in a chapter to a firm they know, whether it be one they research or one they work for, and still others that are intended to generate discussion in class. In addition, the supplements available to instructors and students provide other avenues for exploring MTI.

Third, the text is based on the recognition that MTI is a global phenomenon. To that end, the authors use numerous examples from around the world. Some of the names may be unfamiliar to some students, but the issues are the same. The hope is to allow students to learn more about businesses around the world and to help them see that MTI, in many ways, transcends national boundaries and reflects the global economy in which they will work.

FEATURES OF THE BOOK

- Integrated cases for each section that focus on high-profile companies such as General Electric, GlaxoSmithKline, Acer, and Google
- Exercises designed to engage the students at the end of each chapter

  - Real-life mini-cases
  - WWW exercises
  - Discussion questions
  - Exercises to help students apply the knowledge to their firms if they are currently employed or very knowledgeable about a given firm

- Managerial checklists and guidelines at the end of each chapter
- Four appendices at the end of the text discuss useful analytical tools and concepts for moving forward in MTI

OUTSTANDING INSTRUCTOR RESOURCES

- **Web Resources at www.cengage.com/international** include rich teaching and learning resources, including Key Terms with definitions in separate files by chapter, Flashcards, the Instructor's Manual, Test Bank, ExamView, PowerPoint slides, and a list of potential cases.
- **The Test Bank,** written by Garry Bruton and Margaret White, includes approximately 35 multiple-choice, 15 true/false, and 7 to 10 short-answer questions for each chapter. This has been updated.
- **ExamView** is an easy-to-use automated testing program that enables instructors to create exams by using provided questions, modifying questions, or adding new questions.
- **The Instructor's Manual** helps instructors streamline course preparation and get the most from the text. Included in the IM are answers to the questions at the end of each chapter. These include "Relating to Your World," questions that connect course material to students' own experiences, WWW Exercises, Technology Audit Exercises, Discussion Questions designed to generate broad-based classroom discussion, and

AVAILABILITY OF RESOURCES MAY DIFFER BY REGION. Check with your local Cengage Learning representative for details.

questions that connect the opening case for each section to the chapter.

- **PowerPoint Presentation Slides** let you incorporate images from your book right into your own lectures. Each chapter contains approximately twenty slides. In addition, the Instructor's Manual contains a guide on how to use the slides with each chapter.

ABOUT THE AUTHORS

GARRY D. BRUTON B.A., University of Oklahoma; M.B.A., George Washington University; Ph.D., Oklahoma State University. Professor Bruton has authored or coauthored over sixty articles in leading academic journals, including the *Academy of Management Journal, Strategic Management Journal, Journal of International Business Studies*, and the *Journal of Business Venturing*. His principal research interests include entrepreneurship and emerging economies. Professor Bruton is the editor of *Academy of Management Perspectives*, serves on the editorial boards of five journals, and is the president of the Asia Academy of Management.

MARGARET A. WHITE B.S., M.B.A., Sam Houston State University; Ph.D., Texas A&M University. Professor White is the coauthor of over 75 papers and articles. She has published in leading academic journals, such as *Academy of Management Review, Academy of Management Journal,* and *Strategic Management Journal*. Her current research interests include organizational structure and innovation and strategic management of technology. She has served as an ad hoc reviewer at the *Academy of Management Journal, Academy of Management Review, Administrative Science Quarterly, Journal of Management,* and the *Strategic Management Journal*. Moreover, she was an index editor of the *Academy of Management Journal*. Professor White is a member of the Academy of Management Association and the Strategic Management Society, where she was a board member for the Strategic Process Interest Group. She is also a member of the Project Management Institute.

ACKNOWLEDGMENTS

The book is dedicated to two groups of people in our lives: our professional support and our personal support groups. The first is our students and teachers—some individuals belong to both of these. We both believe we will stop teaching when we quit learning in the classroom. The beginning of this text was prompted by one of Margaret's students asking why she didn't write an MTI text that took the view she was teaching. Just as our formal teachers inspired us in the classroom, our students today inspire us and the text's development reflects that.

Our families and friends have provided both of us an anchor and sails throughout our lives. Right after signing the contract for the first edition, Luanne, Garry's wife, was diagnosed with breast cancer. The technological strides that have been made in fighting this disease are miraculous. We have seen the survival rates for this disease jump to over 70% because of new treatments and early detection. Luanne is doing well. She inspired and encouraged us both to complete this book and to undertake the second edition. Other family and friends have given us encouragement and support through the good times and bad. Thanks to our parents who encouraged us as we matured and to Garry's mother who continues to be a great source of strength. John, Stephanie, and Faith—your father wants you to understand the world you will work and live in. Stefani and Lisa—your aunt wants to inspire you to reach for goals you did not know you had.

Our editors have shown incredible patience. As we all know, learning should never conclude and it continues for us. We continue to learn, to seek new oceans of knowledge. We hope you learn much from this text and that it provides an impetus for your continued learning and growth.

Thanks to the following reviewers who supplied many helpful suggestions:

Terry R. Adler
New Mexico State University

Scott Droege
Western Kentucky University

Susanna Khavul
University of Texas at Arlington

Steven Tello
University of Massachusetts Lowell

Don Wicker
Brazosport College

Strategic Management of Technology and Innovation

PART 1

Strategic Foundation

GENERAL ELECTRIC: CHANGING WITH THE TIMES

The successful management of technology and innovation has become one of the most critical aspects of business today. To recognize the importance of the management of technology and innovation (MTI) one can simply look at the impact of those skills on many of the leading firms around the world. Here is a brief overview of one of the world's leading technology firms, General Electric (GE). This overview illustrates many of the topics included in this part of the book.

GE: The Firm's History

Today, GE is one of the world's leading firms. Their products range from the light bulb, invented by GE's founder, Thomas Edison, to the handheld ultrasound machines that are revolutionizing the practice of medicine around the world. GE is the only firm that was included in the original 1896 Dow Jones Industrial Average that still operates as an independent firm. This highly diversified firm has over 300,000 employees worldwide and in 2009 had revenues over $156 billion, with net income exceeding $11 billion. GE is composed of a number of business units—some of which would be in the *Fortune* 500 if they were independent firms. How GE grew from Thomas Edison's small research facility in Menlo Park to today's multi-product, global giant is through a process and pattern of strategic decisions that include acquisitions, divestitures, innovations, and reorganizations.

GE was founded in 1890 when Thomas Edison brought his business interests together and formed Edison General Electric. Edison GE then merged with Thomson-Houston Company in 1892. This set the pattern of innovation and acquisition that is part of GE today. Throughout its history, GE has been a leader in patent applications and has acquired and divested a number of businesses in its drive to remain a leading innovator with a competitive advantage.

GE throughout its history has caught each new wave of innovation as it has emerged. GE caught the initial wave of innovation that occurred around electricity. The firm then was able to catch the next wave of innovations that included trains and radios. GE began manufacturing diesel locomotives and founded Radio Corporation of America in 1919.

This pattern of innovation continued in World War I as GE used its engineering knowledge to move into the aircraft engine business. GE pioneered superchargers for airplane engines. This innovation made it possible for aircraft to go higher and faster. Throughout the decades between WWI and WWII, GE continued to develop turbo superchargers. When WWII started, GE was in the unique position of being the leader in the development and manufacturing of exhaust-driven supercharging engines. Superchargers later became a key element in jet engines. As a result, as airplanes changed from propeller powered to jet powered GE became a dominant player in the industry. The result is that today GE is one of the leading manufacturers of jet engines in the world. (The last appendix in this text will discuss how managers can analyze and use such waves of innovation.)

In recent years technology has opened up new entertainment domains. Seeing this trend, GE today is the third largest media conglomerate in the world. Technology has allowed many firms to enter financial business domains. Again, GE responded and now GE Financial is one of the world's largest financial firms in both the consumer and commercial markets. GE is also extensively involved in the green energy sector as a major manufacturer of wind turbines for electricity generation. And, of course, GE still makes light bulbs.

This brief history demonstrates how GE was able to take advantage of and advance with major technological changes in the general business environment. If the firm had remained focused solely on light bulbs, it is doubtful the company would still exist. Instead, the company has taken advantage of environmental changes through strategic management and implementation of new technologies.

How GE Changes

In reviewing GE, it is critical to realize that the firm did not simply state, "We want to take advantage of changes in technology." Instead, the firm made numerous changes in its organization, including structure, personnel policies, and leadership, to make taking advantage of changes in technology possible.

To illustrate, in 2002 GE introduced its compact ultrasound machine. This machine sold for $30,000 and combined a laptop computer with sophisticated diagnostic software. GE followed this innovation five years later with a model that was half the price. While this machine does not have the clarity of GE's large ultrasound machines, which cost seven to eight times more, the portable ultrasound is usable in areas of the world where hospitals do not exist. The portable ultrasound machine is a growth product in China, India, and other countries where medical facilities are scarce. But developing this product required that GE also make changes in its processes and structure. The norm in product lines is to have a single integrated business unit for the global market. However, GE recognized that even though the large conventional ultrasound and the portable machine were related, the placement and functionality were very different. Therefore, GE created a facility for portable ultrasounds in Wuxi, China. Thus, the firm created a new organizational unit to handle this product. Not only is the unit separate but the organizational processes and policies were implemented to meet this unique market. This involved an innovative way of building structures known as local growth teams (LGT) and processes that followed five critical principles:

- Shift power to where the growth is
- Build new offerings from the ground up
- Build local growth teams like new companies
- Customize objectives, targets, and metrics to fit the business you want to grow
- Have the local growth team report to someone high in the organization so that appropriate resources are available.

From the beginning with portable ultrasound machines, GE now has expanded its local growth process to include more than a dozen LGTs in China and India. This has powered much of GE's recent growth.[1]

To further push innovation GE recently introduced (2005) Ecomagination. GE CEO, Jeff Immelt, announced at the time

> GE will help build tomorrow's smart energy grid: help drive electric vehicles out of the labs and onto the world's roadways; and work to build advanced, cleaner energy production in the United States, India, China, and the middle East at mammoth scale. Nobody else can do this like GE can.[2]

These changes illustrate one of the foundations for the management of technology and innovation: Not a single change but an organization-wide effort is needed to succeed. A firm cannot simply decide to introduce a new technology or take advantage of a given opportunity. Instead, the company must ensure that the firm has strategies and processes in place that allow it to fit with new technologies and introduce innovations to both the organization and its people. This alignment is critical to the successful management of technology and innovation.

Strategic Perspective

GE is a company well into its second century of existence. The firm continues to invest in innovative initiatives, make acquisitions to enhance GE's position, while divesting businesses that no longer fit the firm's strategic choices. The firm's long-range strategy is based on four keys:

- Be global—connect locally, scale globally
- Drive innovation—lead with technology and content innovation
- Build relationships—grow customer and partner relationships worldwide
- Leverage strengths—use GE's size, expertise, financial capability, and brand[3]

Overview of Part One

Part One of the text lays the foundation for the examination of the topic of managing technology and innovation. Chapter 1 will help to ensure that there is a common language as this topic is discussed. The chapter will also help to lay the foundation by defining the topic and establishing the flow of this book. Chapter 2 will provide the model we will use in studying the management of technology and innovation. Specifically, in this text, a strategic perspective is employed similar to that of GE's discussed above. That is, for a firm to be successful in the management of technology and innovation, a firm needs to focus its resources and capabilities to promote its success. As a result this text places a strong emphasis both on the strategic understanding of MTI and the process to implement it successfully.

SOURCES

GE Press Release. 2008. GE's 2008 Ecomagination Revenues to Rise 21%, Cross $17 Billion, Oct. 21.
GE web page: http://www.GE.com/

Immelt, J., V. Govindarajan, and C. Trimble. 2009. How GE is Disrupting Itself. *Harvard Business Review* (Oct.): 56–66.

CHAPTER 1

Strategic Perspective

OVERVIEW

How do firms manage technology and innovation to realize benefits? This book focuses on answering that question. This chapter lays a foundation for that understanding. It does so by establishing some of the basic definitions and outlooks that shape the development of the text. The specific issues addressed in this chapter include:

- The significance of technology, innovation, and their management
- The meaning of technology
- The process of managing technology
- The meaning of innovation
- The process of managing innovation
- The structure of this examination of the management of technology and innovation

INTRODUCTION

The opening vignette for Part One of this book about GE illustrates that the management of technology and innovation is not a new concern for businesses. However today, new products, processes, and approaches are emerging faster than in the past. As a result, the management of technology and innovation has been pushed to the forefront as a major focus for both business and society.

Importance of Technology and Innovation to Business

To illustrate the importance of technology to business, consider the following statement by Alan Greenspan, former Chairman of the Federal Reserve.

> *When historians look back at the latter half of the 1990s a decade or two hence, I suspect that they will conclude we are now living through a pivotal period in American economic history. New technologies that evolved from the cumulative innovations of the past half-century have now begun to bring about dramatic changes in the way goods and services are produced and in the way they are distributed to final users. Those innovations, exemplified most recently by the multiplying uses of the Internet, have brought on a flood of startup firms, many of which claim to offer the chance to revolutionize and dominate large shares of the nation's production and distribution system.*[1]

Former Chairman Greenspan goes further in his speech saying that not only will the future of business be directed by technology but also that the root of business today is driven by technology and its application. His belief in the growth of technology is supported by the growth in patents worldwide. In the United States, for example, during the years 1970–1985, patent growth was relatively flat. However, from 1985 to 2000, the number of patents awarded grew by more than 100 percent[2], although they have been relatively flat since then. However, one change is that now for the first time there were more patents of foreign origin filed with the United States Patent Office.[3] At the same time the number of lawsuits over patent rights has now doubled. Not only are innovation and technology important to the economy, but today firms believe they are clearly worth arguing and fighting over in today's business environment.

The practical impact on business of this growth in technology is illustrated by the fact that as recently as ten years ago information, including pricing on many different types of machinery and commodity products, was highly inefficient. It was difficult to know exactly what each firm would charge for its product and what the price would be for other firms. A businessperson could call and ask the price for that product. Whether the price was the same if you called a different salesperson in a different part of the month was not predictable. The result was that widely different prices were charged for the same products. Purchasing agents spent a lot of time looking for the best price. However, changes in telecommunications have all but eliminated this inefficiency. Internet availability has resulted in more transparent and efficient pricing for both capital goods and commodity products today.

The impact of technology on business is seldom one-dimensional, but rather, new technology causes a cascading effect within firms. To illustrate, consider the information technology from the prior example. In economic theory, we learn that price is a function of supply and demand. But the technology has resulted in both more demand and lower prices. New technology has made more information available to consumers. As more information becomes available, potential buyers become more aware of opportunities to obtain and use products. This leads to greater demand. But more precise information also leads to pricing being more systematic. Thus, technology leads to better prices. A similar cycle has taken place in other markets. Today, people use the Internet to buy automobiles, books, and other products. This has resulted in more buyers while, in many cases, exerting pressures to lower prices. For a firm to make a profit in this environment, it must be more efficient. One of the key ways that a firm obtains such efficiency is through technology. Thus, the use of technology in one domain typically leads to greater need for changes in technology in other areas.

Retail is one of the oldest industries in the United States. Walmart is the world's largest retailer and is a good example of this cascading effect in practice. Today, when you purchase goods and check out at a Walmart store, you or the cashier scans the various products you are purchasing. This process is more than a way to speed your checkout from the store. There is information generated on the sale and on the product itself. This information is used for reordering products and tracking sales patterns.

Walmart wants to expand the information generated at the checkout by implementing nationwide **radio frequency identification (RFID)** technology. RFID technology requires a small tag be placed on each item at the manufacturer. This tag allows the product to be actively tracked from the time it leaves the manufacturer until it leaves the store. One result of this ability to track the product, Walmart will have better control of shrinkage or loss due to theft or misallocation. RFID will help the firm improve inventory control in the stores because Walmart will know instantly if there is a shortage of any given product in any given store or a surplus in another store. Walmart will be able to estimate whether the transfer of their products between stores is possible and profitable. Initially, Walmart wanted all of its suppliers to implement RFID technology. However, even Walmart found its plans to be too ambitious for the new technology—WalMart missed its 2007 goals for converting its distribution centers to RFID, but is continuing to pursue the use of this technology although in a longer timeframe than initially expected. Part of the reason Walmart did not move as aggressively with RFID was that smaller suppliers protested the cost of implementing and using the technology. However, other larger suppliers like Proctor & Gamble (P&G) supported its use and has realized cost savings in implementing RFID. P&G found they are able to serve a very large customer like Walmart better and more efficiently using the technology. RFID allows P&G to obtain instant data on its products that are sold at WalMart. This in turn allows a firm like P&G to adjust its production process so that it has the supplies needed by Walmart when they are needed. It is estimated that RFID implementation could save Walmart $8.4 billion a year in costs if it is fully deployed.[4]

Importance of Technology and Innovation to Society

The impact of technology is not simply on individual firms. It also has broader societal impact—both positive and negative. Consider the positive effect by examining the findings on the impact of technology in a single state, Washington. This state has aggressively developed its technological foundation. It has found that technology-based industries support a total of 3.55 jobs for each technology-focused job; this compares to an average of 2.86 jobs in all other industries in Washington. Labor income in technology-based industries averaged $61,330 in 2000 compared to a state average of $32,748, or 87 percent above the average. It has also been found that technology-based businesses contribute more to the state's international exports than other types of businesses.[5]

As noted before, technology helps push firms to lower costs. However, this has led to increased levels of outsourcing by a number of firms to lower cost settings; technology advances in communication and computers help ensure that such outsourcing can be successful. For example, the cost of a computer programmer in the United States can be $90,000 a year. This same job in Russia, China, or India will cost less than half as much for the same quality work.[6] Technology allows many job activities to be done as easily in one part of the world as another. Thus, technology has encouraged and permitted the outsourcing of jobs to these lower cost environments to a degree not seen before. In the past, manufacturing jobs were the only jobs principally outsourced. However, today, the jobs outsourced include not only computer programmers but also other technical jobs such as reading MRI images from medical tests and preparing tax returns. In fact, Princeton economist Alan Blinder predicts that over the next 20 years over 40 million such jobs will be lost.

However, outsourcing is not all negative. Outsourcing impacts the United States as well as other developed economies. Countries as diverse as Ireland and Korea have experienced some of the same negative impact from technology as jobs are outsourced to lower cost environments. But as one country outsources some jobs other jobs will be insourced. While individually an outsourced job may be very painful, studies indicate that there is a net 14 percent benefit to the outsourcing nation through new job creation and increased efficiencies.[7] In addition, the development of the economies of India, China, Russia, and other similar nations provides new markets for other businesses from developed economies. The interaction between society and technology can be viewed in terms of **pushing** and **pulling**. When we say that technology is pushing society, we mean that new innovations in technology lead to changes in society that were not expected. For example, society was not demanding the development of the Internet. However, when it became a reality, it was quickly adopted and employed. Business can also be pulled by society to create technology. For example, society demanded through their legal representatives that there be new innovations in automobiles such as more safety features and better gas mileage. The major American automakers headquartered in Detroit insisted that it would be impossible to meet those goals. However, when laws were passed demanding the innovations, business rose to the task and developed the technology necessary to meet the demands. Thus, the relationship between society and technology is rich and multidimensional. We will explore these different dimensions throughout this book.

Technology and Innovation Do Not Stand Still

Technology and innovation influence both the firm and society as a whole, and this impact is ongoing. Entire industries can be created or can disappear very quickly because of new technologies. To illustrate, consider what has happened to the recorded music industry. In the last 40 years, the dominant technology has changed from records (LPs), to eight-track tapes, to cassette tapes, to compact discs. Turntables are antiques, and eight-track players are collectibles. Now with the emergence of MP-3 and other types of new technologies (4G phones), CDs and MP-3 players may soon become obsolete.

Individual companies can similarly be created or can disappear quickly due to technological changes. For example, a classic American company Polaroid went into bankruptcy because of the development of the digital camera, which made many of Polaroid's products obsolete. Today, Polaroid has reinvented itself with its innovative line of Polaroid PoGo digital products, digital cameras, digital photo frames, etc. Today, the firm has become a consumer electronics company that employs a wide range of cutting edge technologies—not a camera manufacturer. Therefore, as we begin to look at technology, we hope you recognize that technology is a key part of most businesses. Technology is typically pervasive in ways that we may not realize until we begin to explore it in depth. It is clear that an industry, firm, or individual who ignores technology and its development does so at great risk. A recent McKinsey report sums up how technology and innovation are changing how business is done as follows:[8]

> ...as globalization tears down the geographic boundaries and market barriers that once kept businesses from achieving their potential, a company's ability to innovate—to tap the fresh value-creating ideas of its employees and those of its partners, customers, suppliers, and other parties beyond its own boundaries ... has become a core driver of growth, performance, and valuation.

THE STUDY OF TECHNOLOGY, INNOVATION, AND ITS MANAGEMENT

The preceding discussion illustrates that technology and innovation management are important to societies, countries, firms, and individuals. Next we will look more closely at the various aspects of technology and innovation.

The Technology and Innovation Imperative Is Organization-wide

Technology and innovation influence not only the technical aspects of business but also the behaviors and attitudes of individuals and groups within the organization. The result is that technology and innovation are an organization-wide concern. An organization cannot isolate one unit and say its concern is technology while the rest of the organization ignores such issues.

1.1 **REAL** **WORLD** **LENS**	**Dr Pepper Snapple Group**

Dr Pepper Snapple Group

There are many other firms from mature industries that are leading examples of success in managing technological change. For example, the Dr Pepper Snapple Group developed the Deja Blue product line. This is the most basic product imaginable: bottled water. The bottling group was able to employ technology to become the low-cost producer in the industry. Each step of the production process used all the technology possible to lower costs. None of the technological innovations is radical. For example, the production process was designed to take out curves in the production line. This reduces the number of products falling off the line. Similarly, the production process is such that the machinery that fills the bottles has ninety different heads that never have to stop. The conveyor technology when the bottle is full is at a slight angle so that the bottles stay upright as they are placed on pallets. Since 2007, Deja Blue bottles have used 35 percent less plastic than the industry standard and have begun using green ink technologies on the label. This reduces waste for society, but also reduces costs for the Dr Pepper Snapple Group.

1. In what ways is the management of the Dr Pepper Snapple Group seeking and supporting innovation processes within the firm?
2. Do you think that their approach is appropriate for their industry and market? Why or why not?

References

Bruss, J. 2002. All grown up. *Beverage Industry*, 93 (9): 60.
Deja Blue, A Pure History. 2008. http://www.drpeppersnapplegroup.com/brands/deja-blue/.
Dr Pepper/Seven Up: Splashing into water. 2003. *MarketWatch: Drinks*, 2 (9): 7.

To illustrate, the portable cell phone has become part of our everyday lives in the last decade. This technological innovation means that employees who are "out of the office" are not out of contact. This has made it easier to work from locations other than the office. In fact today, with applications such as wireless connection and video capability available for many cell phones and laptops, an employee may never need to be in the office. As a result, processes must be in place to ensure that the person in the field behaves as desired by the firm. This means that managers must learn how to integrate and manage these individuals differently from employees who are physically present each day. Thus, managers must not only manage changes in technology but also the structures and systems of the organization where the technology is used. This increasing complexity provides opportunities for developing innovative ways to accomplish work; it also creates the need for changes in how the firm operates.

As we examine the management of technology and innovation, we need to ensure that we not only understand how technology is developed and innovation occurs but also the processes that surround these activities in the organization. A firm needs to understand what technology it has and how to manage that technology in the organization and its context.

The Technological and Innovation Imperative Is Worldwide

Today, it is difficult to segment technology as being from one country or another. For example, many Taiwanese semiconductor firms have their headquarters in Taiwan, produce their chips in mainland China, but maintain their principal research facilities in the United States.[9] Thus, technology firms are truly international entities.

It is true that much of the theory, and principles, as well as the investigation of the management of innovation and technology came from the United States and other developed countries. However, this does not limit their relevance to these countries. A theoretical foundation relevant for technology should be applicable in a wide variety of settings in management just as theory for physics or chemistry applies anywhere in the world. For a theory to be sound, it cannot simply apply to a single nation.

However, when we say that the theoretical foundation is applicable, it does not mean that there will be no differences in practice among various nations. There are many centers of excellence in technology development and application around the world. For example, in the appliance industry, most of the world's technological developments come from Japan. Similarly, much of the new technology for portable ultrasound equipment is being developed in China by GE. Individual and firm behaviors in these centers of excellence can be expected to be somewhat different from those found in the United States; however, in most cases, the similarities with the United States are greater than the differences.

The reasons there would be differences at all are due to differing institutions, or those subtle but pervasive characteristics that shape behavior. The institutions in the various domains can be described as regulatory, normative, and cognitive.[10] **Regulatory institutions** are the laws and regulations in a given country. **Normative institutions** are the norms of the industry and profession. For example, the values of an accountant or a doctor are very similar around the world. **Cognitive institutions** are those that come from the broader society and shape the individual's behavior. Most commonly, this is viewed as the culture of the country.

Regulatory institutions clearly can change from location to location. However, today, the power of the World Trade Organization and regional economic alliances such as the European Union have served to help ensure that there are some similarities in issues such as patents and other key technological concerns.

The normative institutions have also developed strong similarities around the world. This strong set of normative values has emerged from a variety of sources, including the fact that many of the leading professors in technological domains around the world attended a set of key institutions, including

Stanford and MIT.[11] The result has been the sharing of values these individuals learned in college. They took these values with them to diverse locations. Now they teach their own students how technological firms and professionals should act. Similarly, the increasing interaction between technological firms in different parts of the world has acted to homogenize the values of these various firms. The exchange of ideas at professional meetings and the increasing number of joint research activities have also contributed to this uniformity.

The cognitive institutions are the most difficult to change and would be expected to cause the greatest difference in behavior in different organizations. These are shaped principally by the culture of a nation. Initially, as technological firms moved internationally, some cultural conflicts did arise. However, today, such conflicts are less pervasive, and greater cultural knowledge and understanding of other nations exists. Therefore, the broad substance of the management of technology and innovation presented here will be relevant no matter where the firm is located. This does not mean that there will not be subtle differences around the world because of regulatory, cognitive, and normative institutions. However, more similarities than differences can be expected.

Value Creation Is the Key

Whether in the United States or elsewhere in the world, technology and innovation must add "value" to the firm or to society to flourish. The goal of technology and innovation processes is to add value to the business but not just for the purpose of creation. This typically means that there is a profit motive for the business or an efficiency and effectiveness motive for nonprofits. Basic research, which focuses on the creation of knowledge for the sake of knowledge, can have value to society, but it is not a major concern here.

In focusing on value creation, the manager must also recognize that in today's environment there is a need for technology to provide a visible and timely creation of value for the firm. Following the dot-com business crash of the mid-1990s, the spending on new technology by businesses decreased. However, this decrease must be kept in perspective. For four decades prior to the technology investment decline in the late 1990s, spending on new technology increased 10 percent annually. In 2003, that level of spending growth had declined. However, spending growth in new technology was still approximately 4 percent per year.[12] Thus, businesses are no longer willing to invest in technology if the strategic and performance benefits of the technology are not clear. During the boom years of the 1990s, firms invested with hopes that there would be a positive result. The new competitive environment requires more in the management of technology. Now the value addition of the technology must be clear and based on sound analysis and forecasts to justify the investment. This makes the processes of the management of technology and innovation more difficult and complex.

The focus on value creation has clearly been true in the turbulent economy of recent years. In 2009, the prediction was for a 4 percent increase in technology spending. However, in 2010 as the economy recovered spending is returning to historical averages closer to 10 percent. Thus, the global economy is affected by technology development and affects technology usage.[13] Individual technology-focused firms may experience ups and downs, but the core importance of technology continues for businesses in developed economies such as the United States and European Union, as well as emerging economies around the world.

KEY DEFINITIONS OF TECHNOLOGY

From the prior section, it is clear that we need to understand technology and innovation broadly, looking at organization-wide aspects not only in the United States but around the world. The focus in studying this book will be on how we create value for the firm through the management of technology and innovation.

Before we can establish our framework to examine these aspects of the management of technology and innovation, there must be some agreement about basic definitions and terms. Technology and innovation are related concepts but do represent separate concerns. Figure 1.1 illustrates some of these differences relating to one issue: the discovery of the atom.

The definitions presented and developed below reflect widely accepted views of technology and innovation. However, it should be recognized that there are a number of different ways to define technology, innovation, and the management of each. This initial section of the chapter focuses on issues related to technology. The section on innovation will follow.

Definition of Technology

Technology has been defined in a variety of ways. It is important to recognize these various approaches to the definition before we build one to focus on in this text. This range of definitions demonstrates that a variety of different

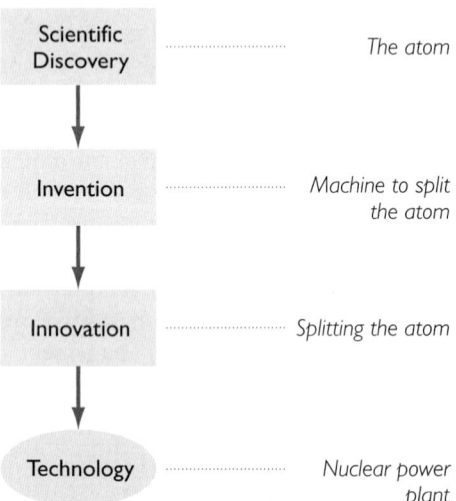

FIGURE **1.1** Process to Be Managed—Discovery to Application

perspectives on technology exist. A few of the major definitions of technology include:

- The processes used to change inputs into outputs
- The application of knowledge to perform work
- The theoretical and practical knowledge, skills, and artifacts that can be used to develop products as well as their production and delivery system
- The technical means people use to improve their surroundings
- The application of science, especially to industrial or commercial objectives; the entire body of methods and materials used to achieve such objectives

Although there is wide variety in the prior definitions of technology, there are also some common elements in each of the definitions. Each definition implies that there is a process involved in technology, that change is an outcome of technology, and that technology involves a systematic approach to deliver the desired outcomes (improvements, objectives, and outputs).

For the purposes of this text, we integrate these various definitions to define **technology** as *the practical implementation of learning and knowledge by individuals and organizations to aid human endeavor. Technology is the knowledge, products, processes, tools, and systems used in the creation of goods or in the provision of services.*

This definition has a strong systems view, as illustrated in Figure 1.2. A **systems view** presents the firm as an association of interrelated and interdependent parts. The systems approach to implementing technology involves a framework of inputs, transformations, outputs, and feedback along the entire process. It also involves individuals, groups, and departments that form the organization and the external environment that impacts the firm.

Definition of Management of Technology

The definition of technology also implies a process that involves the elements of strategic management. Therefore, the definition of the management of technology should also reflect this systematic, strategic approach. Such an approach requires an integration of different disciplines to the management of

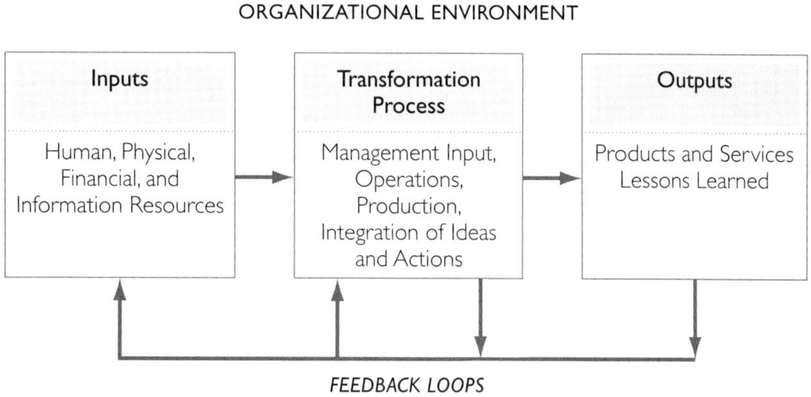

ORGANIZATIONAL ENVIRONMENT

FEEDBACK LOOPS

FIGURE **1.2** Systems View of Organizations

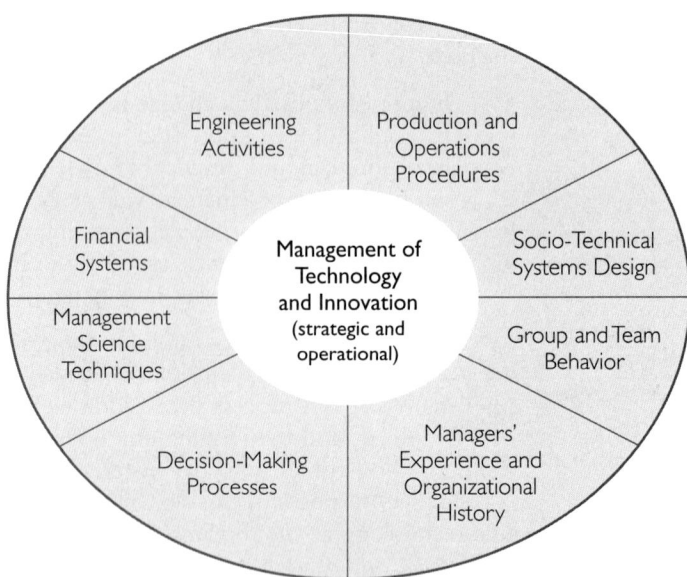

FIGURE **1.3** Areas Influencing the Management of Technology and Innovation

technology. Figure 1.3 illustrates the various disciplines that can influence the management of technology and innovation.

One of the most commonly cited definitions of the management of technology is consistent with this integration view.

> *Management of technology is defined as linking "engineering, science, and management disciplines to plan, develop, and implement technological capabilities to shape and accomplish the strategic and operational objectives of an organization."*[14]

The major shortcoming of this definition is its lack of attention to evaluation and control, which are required for a strategic approach to the management of technology. Evaluation and control involve monitoring technology to ensure that it meets the desired outcomes. It is necessary that after a technology is implemented, the firm monitors changes that may render the technology obsolete, dangerous, replaceable, or competitively weak. A prime example of the need for such evaluation and control is the National Cash Register Company, which was the leading manufacturer of mechanical adding and calculating machines. In the 1960s, the company embarked on a project to build a state-of-the-art manufacturing facility for these mechanical calculators. Just as the facility was being finished, the silicon chip and LED displays were becoming the technology of choice for these products. The technology for silicon chips and LED displays had existed for several years. However, National Cash Register had determined that they would still be the leaders in adding and calculating machines, and the new technology would not replace the need for its products for at least ten years. This turned out to be incorrect; the first hand-held calculators using the new technologies entered the market about the same time as the new facility began production.

The company had some very trying years as it adjusted. It did recover and emerged as NCR, but it was a difficult process. Better technology-auditing processes may have prevented those difficulties.

Therefore, we define the **management of technology** as follows: *The management of technology is the linking of different disciplines to plan, develop, implement, monitor, and control technological capabilities to shape and accomplish the strategic objectives of an organization.* This definition clearly recognizes the planning and implementation processes while recognizing the role of evaluation and control that many other definitions have omitted.

THE IMPORTANCE OF MANAGING TECHNOLOGY

Now that we have defined technology and its management, what will actually be needed to build an understanding of how to do these activities? The National Task Force on Technology has listed five specific reasons individuals and organizations should be concerned about the management of technology.[15] These reasons are as follows:

1. The rapid pace of technological change demands a cross-discipline approach if economic development is to occur in an effective and efficient manner to take advantage of technological opportunities.
2. The rapid pace of technological development and the increasing sophistication of consumers have shortened product life cycles. The result of these factors is a need for organizations to be more proactive in the management of technology.
3. There is a need to cut product development times as well as to develop more flexibility in organizations. The lead-time from idea to market is being reduced by the emergence of new or altered technologies.
4. Increasing international competition demands that organizations must maximize competitiveness by effectively using new technologies.
5. As technology changes, the tools of management must change, but the process of determining what those new tools should be is in its infancy.

Each of these issues will be dealt with in this book as we develop an understanding of how to manage technology. Although focusing on a single dimension of the management of technology may be interesting, it does not provide a usable basis to actually manage the firm. As a result, this text will address a wide range of issues and integrate those issues into a usable whole. At the heart of the various issues examined is the belief that the management of technology is the central strategic concern for the firm. If the business approaches the management of technology from this perspective, it will then have the foundation and insight to be successful.

THE PROCESS OF MANAGING TECHNOLOGY

The range of tools and issues that a firm must examine can be broad. To illustrate, consider the example of the iBOT, a new type of wheelchair that has been developed. The wheelchair has been in existence for more than

100 years, with very little change in its fundamental design. Wheelchair designs have historically confined their use to relatively flat and smooth surfaces. However, Dean Kamen, the inventor of the iBOT, saw how difficult it was for someone to handle a wheelchair in settings that were not flat, such as on stairs. So he went looking for a new solution. However, rather than thinking of a wheelchair traditionally, he sought to build a chair that could stand up and balance like a human. The end result would be a wheelchair that could carry a person up and down stairs.[16]

The development of the iBOT illustrates the role of various elements in the firm needing to work together for success. For example, the iBOT shows the need for a new approach and philosophy so that the problem could be attacked in a different way. Thus, it allows individuals in wheelchairs to roll across sand or "stand" to get products off the top shelf in their home or the grocery store.

This case demonstrates the need not only for engineers to design the product but also for financial experts to underwrite the costs and marketing personnel to test the product. The development of this product took substantial funds and investment. Marketing was also critical to the actual acceptance of the product. While the $29,000 cost per unit is high and its cost could be offset by the normal cost associated with modifying a house to meet the needs of a person who uses a wheelchair, it requires marketing to educate individuals about this benefit. Thus, it not only took the vision of one person to see a different solution, but it took an entire organization to develop the product. A full range of tools needs to be considered when examining the management of technology.

Making Decisions for Managing Technology

There are key decisions that need to be made as businesses and managers seek to manage technology. These decisions initially focus on the strategic posture the firm wants to assume. For example, the firm must determine if it wants to be a leader or follower in its industry. There are benefits to both, but the choice will result in the firm taking radically different steps and developing different processes and structures. The firm must also determine whether it will develop its own new technology or buy the technology. Again, each of these strategic approaches has benefits and drawbacks that will be detailed later, but the firm needs to weigh these pluses and minuses for itself.

The strategic decisions do not stop there. The firm will also have to determine the scope of products it wants to offer. A key element in this determination is how it can leverage its technology and innovations to create a total **platform** of products and processes. The firm must also determine the scale of products, how it will price the products, where it will market the products, and where it will manufacture the products.

The process that the firm needs to address each of these issues is critical. If the business responds in a reactive, piecemeal manner to the competition rather than actively determining its direction, the performance of the business will suffer. This book will examine the full range of issues with questions and key concerns for each provided throughout this book to help build an understanding about the many decisions managers must consider. The answers to the questions and the review of relevant concerns will help identify the tools that need to be employed in the decision-making processes associated with the management of technology.

Tools for Managing Technology

This perspective on the role of technology in the firm means that the specific tools necessary to properly manage technology can be very broad. Too often, managers of technology assume that, because the technology is interesting or attractive to them, it will be demanded by the consumer. However, for success, the manager does more than rely on his or her own judgment about the viability of the product. Instead, the manager needs to do things such as:

- Analyze the industry structure both domestically and internationally
- Understand the firm's capabilities and those of its competitors
- Conduct a financial analysis of the product and firm
- Forecast future changes

KEY DEFINITIONS OF INNOVATION

Now that the definition of technology and its management plus the nature of the tools and decisions related to those issues have been detailed, it is important to define innovation and its management. Innovation is part of technology management, but because of its characterization of having "newness," it is unique in how it is managed and developed within a business. The management of innovation requires technology; but the management of technology does not necessarily require innovation. If the processes, products, and structure of the organization are fairly stable and the environment is mature, innovation may not be appropriate. However, managers should be alert for the opportunity to be innovative. Therefore, innovation will be treated as a separate area.

Definition of Innovation

Defining innovation is not as easy as it would seem. Most of us think we know what innovation is, but we have our own frames of how to define it. Some have defined innovation as invention plus exploitation. In other words, it is not only the act of creation but the inventor or someone actually taking that product to market and selling it to people. This text goes beyond this definition to argue that innovation is more encompassing and includes the process of developing and implementing the invention. We believe that this broader definition is needed because the process elements of innovation are so critical. Thus, we prefer the definition of **innovation** by Rubenstein who defined innovation as *"the process whereby new and improved products, processes, materials, and services are developed and transferred to a plant and/or market where they are appropriate."*[17]

It is important to note that from this definition there are different types of innovations. Figure 1.4 summarizes these different types of innovation. There

PRODUCT/PROCESS

	New	Old
	Category 1	**Category 2**
Old	New solution to an old problem Product: new medications Process: just-in-time	No innovation
	Category 3	**Category 4**
New	Most innovative: new product that leads to new opportunities Product: PC in 1980 Process: bar codes for inventory control	Old product/process used in a new way Product: other uses for a paper clip; DVDs Process: testing soils with satellite imagery

USAGE/PROBLEM

FIGURE **1.4** Innovation Categories

can be newness of the product or process, newness of the usage, or a combination of both. The difficulty in managing these different types of innovations varies. For example, the most innovative approach is the development of a new product or process to solve a new problem or usage. These types of innovations are usually radical in their influence in change processes. For example, think about how the Internet changed how we work. Another example is the DVD, which illustrates an old process with a new usage. DVDs employ the same basic technology as CDs; however, the means of compression and reading hardware are more advanced.

The examples in the prior paragraph are all product-oriented, but there are also process innovations. **Just-in-time (JIT) inventory management** is a process innovation that ensures the inputs into a production process are there just as they are needed for the process. Such a process innovation allows firms to save on storage and capital costs. Frequently, product and process innovations are connected. For example, e-mail security that involves virus protection software is a product innovation. But many organizations also deal with the problems of e-mail security by building firewalls to protect company information—a process innovation. It is interesting to note, however, that almost as quickly as new software and processes are developed for protecting a firm's information, new problems emerge. It is a constant war of innovation.

Definition of Management of Innovation

With innovation defined, how do we manage it? Successful innovation management depends on the top management of the organization committing resources to empower individuals and groups to act on new concepts. This commitment by the top management to innovation, in turn, requires their recognition of several realities.[18] These realities are as follows:

1. Management of technology encompasses the management of innovation.
2. It requires fostering an environment where innovative thought and work are encouraged.
3. It involves leading a firm from existing processes and products to something that is "better" and more valuable.
4. It is proactive and encourages creativity and risk taking.

Therefore, we define the **management of innovation** as

a comprehensive approach to managerial problem solving and action based on an integrative problem-solving framework, and an understanding of the linkages among innovation streams, organizational teams, and organization evolution. It is about implementation—managing politics, control, and individual resistance to change. The manager is an architect/ engineer, politician/network builder, and artist/scientist.

THE PROCESS OF MANAGING INNOVATION

Just as for technology, there are special tools and decisions within the organization that must occur if innovation is to succeed.

Making Decisions for Managing Innovation

Fostering creativity is essential to managing innovation. However, it is more than encouraging individuals to think outside the proverbial box. It is a process that includes developing an environment of discovery in the organization. Delbecq and Mills[19] described the characteristics of firms that manage the innovation process well. These firms are characterized by:

1. Separate funds for innovation
2. Periodic reviews of informal proposals by a group outside line management
3. Clear direction on studies to be done and follow-ups that are expected
4. Extensive boundary-spanning activities to learn from others and to gain an understanding of what others are doing
5. Sets of realistic expectations
6. Supportive atmosphere for "debugging" and exploring variations as well as appropriate resources for maintenance and service

Pixar Animation Studios illustrates the way to build a supportive environment for innovation. This studio has created the movies *Toy Story*, *Wall-E*, *Cars*, and *Up* among others. It has pioneered the development of new computerized animation technologies, including Marionette, a software for animation, and Ringmaster, a software system for modeling, animating, and lighting. The studio has very creative individuals heading the firm (Steve Jobs, founder of Apple Computer) and others working throughout the firm. To ensure that individuals in the firm have the range of skills necessary, the business started Pixar University, which allows individuals to study for three months on a variety of topics related to Pixar's work. The company seeks to further encourage creativity by limiting its bureaucracy. Thus, the business has sought to create a total environment for creativity.

The management of innovation requires that the firm encourage creativity and risk taking by individuals. The firm must employ processes that allow failure and exploration. There are four key individual characteristics that enhance the initiative that sparks innovation.[20] If an organization manages the work environment in such a way as to encourage these behaviors, then innovation is more likely. The four behaviors are:

1. Asking questions to identify problems and opportunities
2. Learning new skills
3. Taking risks and being proactive
4. Aligning strong personal beliefs and values with the organization's values and goals

As you consider this innovation process, what becomes clear is that it should be a continuous process in the organization. It is not a process that occurs once and brings the firm all of the innovation it needs. Figure 1.5 summarizes the cyclical nature of the innovation process. The various aspects of this process will be examined in greater depth throughout the book.

To illustrate this process, consider Koch Industries. The firm is one of the largest privately held companies in the United States. Koch rewards individuals for developing new ideas like many firms. But Koch also actively seeks to

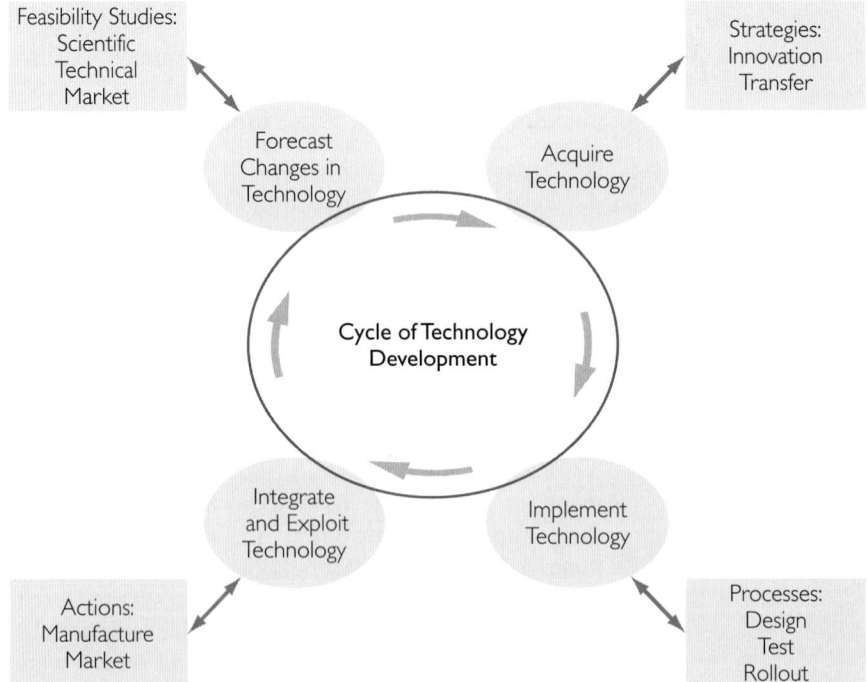

FIGURE **1.5** Cyclical Innovation Process Model

cross-train individuals in different areas of the firm so that they understand how the entire firm works. Additionally, the firm consciously seeks not to punish individuals if they try something new that does not work. The culture at Koch encourages risk taking. The end result is a firm that has been able to diversify from an oil and gas company into one that continually finds new markets into which it can expand.

Tools for Managing Innovation

The management of technology involves a much broader scope of continuing and nurturing existing technology than does innovation. Innovation directly involves the discovery and development of new products and/or processes.

1.2

REAL WORLD LENS

X-Rite

The management of innovation does not need a radical shift in a product; instead, it can be as simple as a fresh insight on how a product can be used. X-Rite traditionally produced equipment to match colors in industrial settings. Thus, an auto manufacturer would test to ensure that their sky blue was consistent on all cars. The innovation that occurred for the firm was that matching paint had similarities to matching the color of an individual's teeth in a dentist's office. Prior to this insight by X-Rite, the typical method of matching the color of teeth was for the dentist to hold a card with different variations of white and seek to judge which one matched the color of the patient's teeth. The nature of the X-Rite equipment had to be changed to a form that was more accessible and user friendly than the industrial version. Issues such as hygiene related to the product also became part of design for the first time. However, the technology was the same as before. This innovation insight allowed X-Rite to move into a profitable new market and become a major competitor in this area. X-Rite has become the leader in color matching in digital photography, dentistry, as well as paint color in industrial applications.

1. What other uses can you think of for X-Rite's technology?
2. How do you think the expansion of their customer base changed their organizational structure? Did this change of customer base change their organizational processes?
3. What does this tell you about what happens in organizations when new uses for old technology emerge?

References

Babyak, Richard J. 2003. Visionary design. *Appliance Manufacturer*, 51 (1): 24.
DaCosta, Victoria. 2004. The color of white. *RDH*, 24 (11): 34.
Jeppsson, Jessica. 2008. Emerging technologies: Innovative tools of the trade. *Industrial Engineer*, 40 (9): 56–57.

Most often, when we think of innovation, we think of radically new and inventive products and/or processes. For example, the innovation of the lean manufacturing system pioneered by Toyota has reshaped how manufacturers do business, with techniques such as JIT inventory now becoming the norm worldwide. However, innovation does not have to be so radical; it may be as simple as using an old product in a new way. For example, Scotch Masking Tape has been an innovation that has served 3M well since its invention in 1923. The original problem to be solved was making waterproof sandpaper. However, Richard Drew went to an auto body shop to test his ideas and discovered a need for tape that adhered to a painted surface and stripped off easily. This innovation and the resultant technology have led to over 900 other varieties of Scotch brand tape. Successful technology and innovation management have made 3M the international corporation it is today.[21]

In this book, we will differentiate the management of technology (MOT) and the management of innovation (MOI), but remember that they are interconnected within the organization. This differentiation helps us better analyze the firm's actions, but in reality, they are intertwined at a number of levels.

STRUCTURING THE EXAMINATION OF MTI

In this chapter, we have presented a broad overview of the issues and definitions central to the study of management of technology and innovation. These issues shaped the development of this book.

Strategy Perspective

Strategic management is a firm's effort to analyze its environment and its own strengths and weaknesses and then consciously choose the competitive path it wants to follow. On that path, the firm will seek to build up its strengths and address its weaknesses. As employed here the strategic perspective involves a strong process view. The strategic perspective is typically segmented into three distinct steps: planning, implementation, and evaluation and control. These steps are process activities that the firm must develop. The authors will use this three-step strategic framework throughout the text, with particular concern on how these steps are enacted.

Making Strategic Decisions

It was highlighted previously that the organization must make key decisions as it begins to examine technology and innovation. We believe from our experiences that the key element in these decisions is whether those processes are focused internally or externally. For example, if a firm chooses to purchase technology, it must focus on issues such as the integration of the technology and the nature of the firm that produced the technology. In contrast, if the focus is on the creation of technology, then how the firm encourages innovation internally through structure and compensation becomes more important.

As noted earlier, a strategic model will be employed to analyze the topics. Therefore, planning, implementation, and evaluation and control will be used to examine internal innovation and external acquisition of technology. Thus, this book will be separated by whether technology is internally developed (Part Two—Chapters 3, 4, and 5) or acquired (Part Three—Chapters 6, 7, and 8). To illustrate, Chapter 3 will examine the planning for internal innovation, Chapter 4 will examine implementation of internal innovation, and Chapter 5 will examine the evaluation and control of internal innovation. The third part of the book on external acquisition of technology will be organized similarly. The last part of the text will examine building capabilities and knowledge management in the business (Part Four—Chapters 9 and 10).

Strategic Tools for Managers

As noted previously, there is an emphasis on managerial tools that can be used in practice. This emphasis has led to the development of specific tools that will appear in the appendices at the end of the text. The first appendix will present information concerning social responsibility and ethics in the management of technology. The second will focus on tools concerned with project management, the third will discuss managing platforms and portfolios of technology, and the last appendix will discuss waves of innovation.

Additionally, at the end of each chapter, there will be an Audit Exercise focused on technology to help students understand how to apply the knowledge they learn in the chapter. This audit tool considers what a firm is actually doing with its technology and innovation capabilities in comparison to what the firm wants to do, or to what others are doing. This Audit Exercise usually requires information from outside and inside the firm and assesses the firm's position in product and process technologies. A firm should also examine the fit between its administrative processes and procedures and its strategic goals, looking for areas of competitive advantage to exploit, as well as gaps in the firm's activities that may lead to competitive disadvantage.

ORGANIZATION OF THE TEXT

Figure 1.6 summarizes the flow of the text that was just detailed. Several specifics should be noted. For example, at the beginning of each part, an introductory case will illustrate the ideas that are examined in that part's chapters. In this part, it was the case about GE.

At the end of each chapter, there is a vignette titled "The Real World" to help illustrate the chapter's concepts. In addition, there are Critical Thinking exercises to reinforce the concepts presented in the chapter. These include Relating to Your World exercises, World Wide Web (WWW) Exercises, the Firm Audit Exercise described earlier, plus Discussion Questions. Additionally, there is a short set of questions connecting back to the introductory case at the beginning of each part. For example, in Part One, consisting of Chapters 1 and 2, there are questions connecting the chapter material to the GE case.

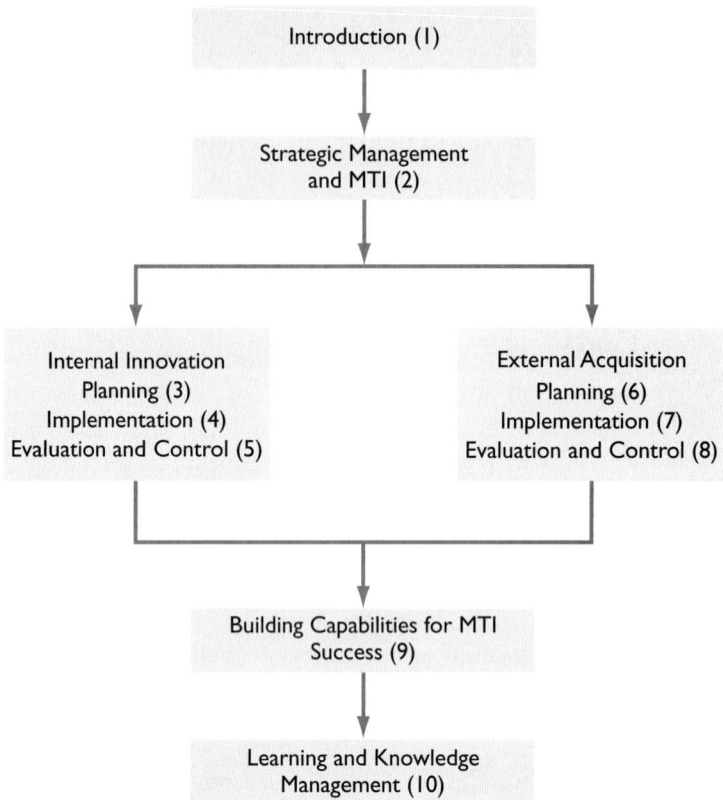

FIGURE **1.6** Organization of This Book

SUMMARY

This chapter has established the foundation for the exploration of management of innovation and technology. The chapter highlighted that the use of technology continues to expand in business in the United States and around the world. This expanding use and impact of technology make the understanding of the management of technology and innovation that much more critical. The chapter has defined both technology and innovation and what is needed to manage them. The focus in these definitions is on multiple dimensions of the concepts, with strategic management playing a particularly critical role.

MANAGERIAL GUIDELINES

To manage its technology and innovation successfully, a firm must be proactive rather than reactive. To promote proactive approaches, a firm should:

1. designate clear technology leaders—individuals who champion change;

2. know how the processes can work to help and to hinder the development of new technology;

3. assess objectively where your firm is on the technology curve;

4. assess the strengths and weaknesses of your personnel and your approach to the management of technology and innovation;
5. set realistic priorities;
6. develop excellent infrastructure to help find and take advantage of potential opportunities;
7. understand what the tasks are and how they are connected and disconnected;

8. be systematic in your search and assessment processes, but review the system thoroughly to be sure it is still applicable;
9. savor every victory and learn from every failure; and
10. be confident that once you have made a decision, it is a decision that will move you in the right direction.

Guiding Questions

As firms begin to lay the foundation for their management of innovation and technology, they should be guided by answering the following questions:

1. Are decisions based on clear goals of where the firm wants to go?
2. Do individuals recognize that being innovative and helping others be innovative are part of their job?

3. Is there an environment of sharing ideas to build and develop ideas?
4. Do development teams bond and truly become teams?
5. How are mistakes handled? Are they seen as learning experiences?
6. Is risk taking prized and supported when expressing new ideas?
7. Is innovation sought and supported?

CASE **1.1** THE REAL WORLD
LEGO

LEGO was founded in 1932. The word LEGO is from the Danish words "LEg GOdt" ("play well"). Later, it was realized that in Latin the word means "I put together." The firm initially made wooden toys as well as other wood products—stepladders, ironing boards, stools, etc. In 1947, the LEGO Group bought a plastic injection-molding machine for toy production and in 1949 the first LEGO Automatic Binding Brick with four and eight studs appeared.

LEGO flourished for many years with its innovative toys. However, in 2004, innovation almost bankrupted the Danish toymaker. LEGO had become concerned about low-cost copies of its plastic building products so it sought to diversify into different activities including: theme parks, Clikits craft sets, action figures (Galidor), and a television show among other efforts. All of these were unprofitable and were eventually discontinued. Today, LEGO is growing in an overall declining toy market with a decidedly old technology toy—plastic parts that connect to build things. How did LEGO manage such a turnaround?

LEGO used an innovative structure to map its turnaround. This new structure strategically coordinates innovation activities and actions through a cross-functional team. The team takes a broad view and splits the firm's innovation efforts into eight distinct types of innovation:

Core processes—sales, operations, financial planning,

Enabling processes—forecasting, market planning,

(continues)

CASE **1.1** *(continued)*

Messaging—advertising, website development,

Offerings—packaging, product presentation,

Platforms—creating new uses and designs for the building blocks

Customer interaction—customer service, customer linkages,

Sales channel—retailers, direct marketing, and

Business model—revenue and pricing.

This new structuring has led LEGO into a new product line—LEGO board games that players build with LEGO bricks. Such a board game can be different every time it is played. The new board game was launched in the United Kingdom and Germany in 2009 and globally in 2010.

1. How do you think the changes that LEGO made will reshape the firm for the long term?
2. What suggestions would you have for LEGO's competitors in the declining toy industry? What should they be watching for?

References

O'Connell, Andrew. 2009. Lego CEO Jørgen Vig Knudstorp on leading through survival and growth. *Harvard Business Review*, 87 (1): 25/ http://www.lego.com/eng/info/default.asp?page=facts

CRITICAL THINKING

Relating to Your World

1. A model is a representation of a complex process or interaction that allows us to use a simplified picture to better understand complex and abstract ideas. This chapter represents the management of technology and innovation as a systems model. After reading this chapter and based on what you know about MTI, how would you model the management of technology and innovation? Explain your model.

2. Consider some new technology or process you have been involved with in the last two or three years. What issues affected you as an individual? What factors enabled you to be successful in adapting to the new process?

3. Make a list of the different ways you have used a paper clip. What were the motivating factors for each of these uses? Is there a pattern?

WWW EXERCISES

1. Use your favorite Internet search engine to find different definitions of technology and innovation. How would you characterize these definitions? How are they the same and how do they differ from the ones in the text? Why are there so many variations?

2. Creativity is important in fostering innovation and in adopting new technologies. How many courses have you taken in creativity? What can you find out about creativity on the web? Find five websites devoted to creativity. What tools did you find that might help you increase your creative thinking?

3. Find an article that provides guidelines for managers on how to manage technology and innovation. What do you think of the advice? Compare the advice that you find with others in your class.

AUDIT EXERCISE

When trying to determine the ability of the organization to manage technology and innovation, it is important for managers to understand the firm's capabilities. **Capabilities** are the set of characteristics an organization possesses to facilitate and support its strategies. In the management of innovation and technology, there are a number of frameworks for determining the innovative capabilities of the organization. The Innovative Capabilities Audit Framework[22] indicates five categories of variables for a business to consider. These categories are:

1. Resource availability and allocation
2. Capacity to understand competitors' strategies and industry evolution with respect to innovation

3. Capacity to understand technological developments relevant to the business
4. Structural and cultural context of the business unit affecting intrepreneurship (internal entrepreneurship)
5. Strategic capacity to deal with innovation initiatives by internal entrepreneurs

What type of information would you need to collect in each of these five areas to determine when, where, how, if, and what innovations should be undertaken in the business? Be specific and justify your answer.

DISCUSSION QUESTIONS

1. Discuss the definition of technology from a strategic point of view.
2. Discuss the role of innovation in the strategic management process.
3. Define management of technology and give an example based on your knowledge.
4. Define management of innovation and give an example of how a firm can manage innovation processes.
5. Give an example of GE's management of technology and how they were able to gain a competitive advantage from those activities.

PART ONE OPENING CASE: GENERAL ELECTRIC

The GE case illustrates the changes a company can go through because of a change in technology and innovation. What changes in technology do you think GE has undertaken? In process? In product? What type of innovation do you think these changes illustrate (see Figure 1.4)?

KEY TERMS

capabilities 29

cognitive institutions 12

innovation 19

just-in-time (JIT) inventory
management 20

management of innovation 21

management of
technology 17

normative institutions 12

platform 18

pulling 9

pushing 9

radio frequency identification
(RFID) 8

regulatory institutions 12

strategic management 24

systems view 15

technology 15

NOTES

1. Greenspan, Alan. 2000. Technology innovation and its economic impact. Speech to the National Technology Forum. St. Louis, MO. http://www.federalreserve.gov/boarddocs/speeches/2000/20000407.htm.

2. Somaya, D. 2004. Firm strategies and trends in patent litigation in the United States. In G. Libecap (ed.), *Advances in the Study of Entrepreneurship, Innovation and Economic Growth*. Greenwich, CT: JAI Press.

3. U.S. Patent and Trademark Office. 2009. All patents, all types report, parts A1, A2, B. http://www.uspto.gov/go/taf/apat.htm.

4. Smith, Brad. 2003. Raising the RFID profile. *Wireless Week*, 9 (Dec. 15): 6–8. And, 2007. Walmart RFID Plans Change. *Supply Chain Management* found in http://www.rfidgazette.org/walmart/.

5. Beyers, William B., and David P. Lindahl. 2001. *The Impact of Technology Based Industries in Washington 2000*. Seattle, WA: Technology Alliance.

6. Baily, Martin and Diana Farrell. 2004. Exploding myths about offshoring. *McKinsey Global Institute Report*, April.

7. Gumpert, David E. 2003. U.S. programmers at overseas prices. *BusinessWeek Online* (Dec. 3).

8. Barsh, Joanna, Marla Capozzi, and Jonathan Davidson. 2008. Leadership and innovation. *The McKinsey Quarterly*, (1): 37–47.

9. McDougall, Patricia and Benjamin Oviatt. 2000. International entrepreneurship: The intersection of two research paths. *Academy of Management Journal*, 43 (5): 902–907.

10. Scott, W. R. 1995. Introduction: Institutional theory and organizations. In W. R. Scott and S. Christensen (eds.), *The Institutional Construction of Organizations*. Thousand Oaks, CA: Sage.

11. Bruton, Garry and David Ahlstrom. 2003. An institutional view of China's venture capital industry: Explaining the differences between China and the West. *Journal of Business Venturing*, 18: 233–259.

12. Hof, Robert D. 2003. Why tech will bloom again. *BusinessWeek* (Aug. 25): 64–70.

13. _____ 2009. Global R&D Funding Forecast *R&D Magazine*. Dec.

14. National Research Council Report, 1987.

15. National Task Force on Technology, 1987.

16. _____ 2003. Stair climbing wheelchair approved. *Journal of Clinical Engineering*, 28 (4: Oct.–Dec.): 198–199.

17. Rubenstein, Albert. 1989. Managing technology in the decentralized firm. New York: Wiley.

18. Tushman, M. and P. Anderson (eds.). 1997. *Managing Strategic Innovation and Change: A Collection of Readings*. New York: Oxford University Press.

19. Delbecq, A. and P. Mills, 1985. Managerial practices that enhance innovation. *Organizational Dynamics*, 14 (1): 24–34.

20. Frohman, Alan L. 1999. Personal initiative sparks innovation. *Research Technology Management*, 42 (3): 32–38.

21. Raber, Linda. 2007. Scotch Tape: An innovation that stuck. *ACS News*, 85 (43): 64.

22. Burgelman, R., T. Kosnik, and M. Van de Poel. 1988. Toward an Innovative Capabilities Audit Framework. In R. Burgelman and M. Maidique (eds.), *Strategic Management of Technology and Innovation*. Homewood, IL: Richard D. Irwin.

CHAPTER 2

Strategic Process

OVERVIEW

This text integrates the strategic process perspective with the management of technology and innovation (MTI) in the firm. The foundation for the understanding of the strategic management portion of MTI is presented in this chapter. The specific issues addressed include:

- The meaning of strategy
- Continuous versus radical technology
- Offensive versus defensive technology
- Key MTI concerns in strategy
- The strategy process
- Understanding an industry and its impact
- Strategic groups within an industry

INTRODUCTION

Chapter 1 provided an overview of MTI. This chapter integrates MTI and strategic process, laying a foundation for the rest of the text. The firm's strategic efforts are the actions that help direct where the firm is going. These actions and activities should fit together[1] to move the firm in a consistent direction. To be successful, the strategy of the firm and its management of technology should be intertwined.

To illustrate the need for individual firms to integrate their strategy with the management of technology and innovation, consider the Polaroid Corporation. Polaroid was mentioned briefly in Chapter 1 as a firm that reinvented its technology. The firm, founded in 1937, was historically one of the leading technology firms in the United States. The products the firm invented range from heat-seeking missiles with microcomputers to a single-step photographic process. In large measure, Polaroid was successful because it had the innovative ability to create cutting-edge technology as well as the strategic ability to build a consistent set of actions around those technologies. This resulted in a strong market position for the firm and the technologies.

However, in the late 1990s, Polaroid's ability to combine technology and strategy fell into disrepair. During this time, the digital camera revolution began. Polaroid knew that digital cameras were increasing in demand, but the firm had not created those products, so it made the strategic decision not to move into that market. Instead, Polaroid focused its strategic efforts on vigorously defending its existing proprietary technology of instant photography.[2] However, during the 1990s, it failed to improve on that basic technology. Therefore, the firm had the wrong product and the wrong strategy for that product. By the end of 2001, the firm filed for bankruptcy and in 2008 Polaroid ceased making film. The firm today is struggling to become competitive in digital electronics; however, they are too late to save what was once their core product—cameras.[3]

Thus, a firm that relies on technology for its competitive success needs to nurture its technology position if it is to succeed. This technological position may rely on technology that is developed internally or is purchased from outside entities, but firm success does not happen by chance. Concerns such as what and where are the markets for the firm's technologies, and what is the right strategy to compete in that market are critical. Thus, it is not a single activity such as making advances in technological capabilities that creates business success; instead, a process of interconnected actions is needed for success.

This chapter looks at two specific issues that managers need to address:

1. Why firms should couple their strategic planning and strategic implementation with MTI. As a result, technology should be approached as a core part of the strategic process, not as a separate concern.
2. Specification of the strategic process that puts into practice the integration of technology and strategy.

WHAT IS STRATEGY?

Strategy is a coordinated set of actions that fulfill a firm's objectives, purposes, and goals. It is a not a single act that occurs in a firm. Frequently, individuals confuse strategy with strategic planning. **Strategic planning** is the process that lays the groundwork and direction of the firm over the next several years. Typically, strategic planning efforts produce a formal written strategic plan. However, strategy is more than the document that results from such planning efforts or the planning effort itself.

Strategic management is an ongoing process through which the organization defines the nature of the businesses in which the firm will be active, the kind of economic and human organization it intends to be, and the nature of the contribution it intends to make to its various constituents. These broad aspects of strategy then serve as an umbrella under which the firm can establish policies and plans to ensure that its efforts are consistent and will lead to achieving its objectives, purposes, and goals.

In establishing a strategy in a technology-focused firm, that firm's technology is not a minor issue. Polaroid Corporation failed because its strategy and technology became separated. The firm did not pursue the cutting-edge technology as it had historically. In fact, it was not even improving on the technology it possessed. Instead, Polaroid's strategy became focused on reducing costs to the point that it missed major changes in the marketplace.

Polaroid is not unique in this behavior; there are many similar examples. Swiss watch manufacturers dominated the global watch industry for over 100 years. In the early 1960s, the Swiss did not believe anyone would be interested in digital watches. They thought a watch needed to have moving parts and a face, not just printed numbers. The Swiss firms were so sure of what watches needed to be like that they actually gave the technology for digital watches to the Japanese who then rapidly became dominant in that industry. In another example, at one time the United States steel industry was the world leader, but the United States industry generally refused to believe that air-fusion technology would impact that dominant position. The United States steel industry today is in a weak competitive position. Thus, technology is not a passive component of a firm. Instead, technology is a critical part of a firm's strategic success that should be planned, actively chosen, and constantly evaluated and adjusted as necessary.

CENTRALITY OF MTI IN STRATEGIC MANAGEMENT

Strategic management's benefit is critical because it helps the entire organization move toward consistent goals. Figure 2.1 shows how strategy, technology, and other organizational factors interact to determine the organization's outcomes.

The resulting interactions in the figure look complex. However, the fundamental point is that technology affects the strategic process in multiple places. Internally, the figure demonstrates that technology affects the organizational

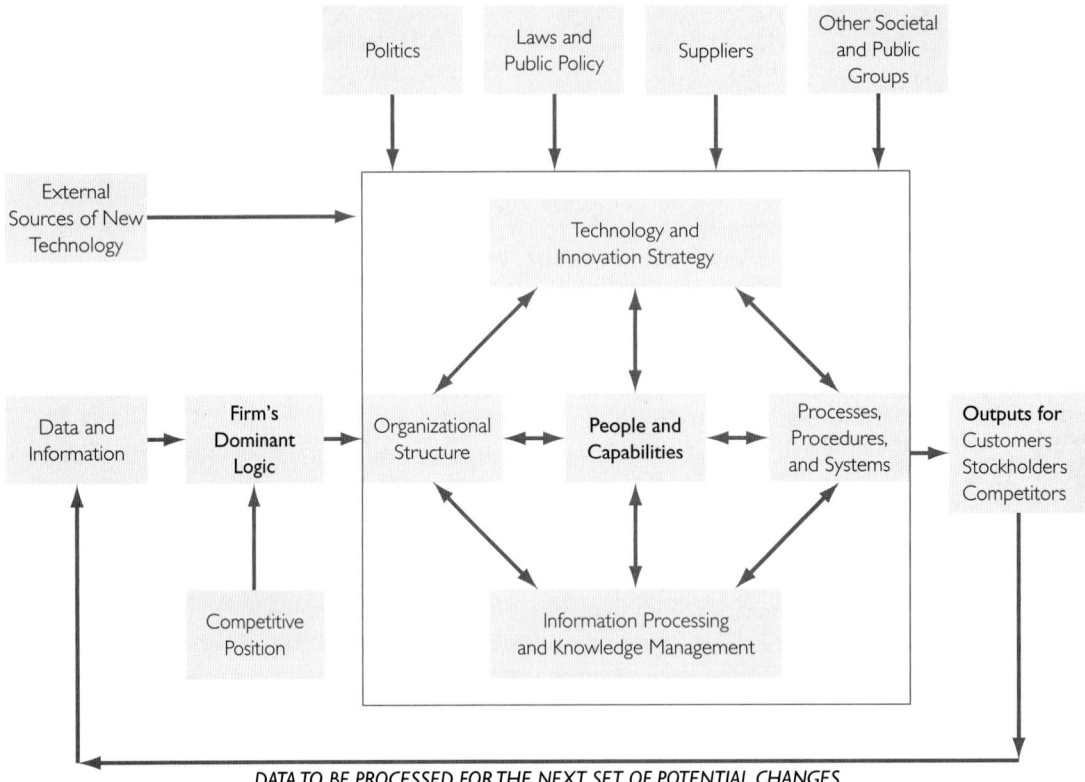

FIGURE **2.1** External and Internal Strategic Interactions

structure, people, processes, procedures, and systems. Additionally, external environmental factors, such as politics, rate of innovation, laws, and public policy, all influence the interaction of people, processes, and structures. These external environmental factors also impact key stakeholders such as customers, competitors, and investors. Thus, a business clearly does not create its strategy in isolation. A business is impacted by, and sometimes can impact, its broader environment. Figure 2.1 appears very full, but in actual practice, there is a much richer interplay among variables than any drawing can provide.

Integrating MTI and Strategy

Capabilities are skills that a firm develops. Firms are similar to their competitors in most areas detailed in Figure 2.1. Therefore, fast-food firms such as McDonald's and Burger King look very similar in many aspects. However, to be successful, there should be five or six capabilities the firm develops and maintains that are superior to its competitors. These **capabilities** are the building blocks for the firm's strategy. It is at the level of capabilities that the firm's integration of technology with strategic concerns should begin because the business ultimately develops its competitive advantage[4] over other firms from its capabilities. The capabilities of a firm can be classified as either technical or market. Each type of capability is examined next.

Technical Capabilities

Technical capabilities address how the firm approaches technology it already has or wishes to have in the future. Therefore, the firm's approach to these capabilities can be classified in one of three ways: destroy, preserve, or develop. The approach to technology is a strategic decision that must be implemented through the firm's choices, including its people, structure, and processes.

Destroying is concerned with eliminating certain technological capabilities in the organization and replacing them with others. Although destroying capabilities seems counterintuitive for a firm's strategy, perhaps the technology that has been employed is flawed, and improvement must take place. After the *Exxon Valdez* accident, many tanker companies viewed the old technology of single-hull design as too risky to continue using. Therefore, many old, still usable tankers were taken out of commission and replaced with ones with double-hull technology. The development and management of these double-hulled ships required different capabilities than did the old single hull ships.

Developing new technology capabilities can give a firm a competitive leap over others in the industry by changing the playing field. These capabilities can be purchased externally or developed internally. Many firms pursue new technology capabilities to maintain or enhance their competitive position. An example of this includes retailers who pursue new Internet capabilities to complement their existing store locations, such as Sears, Walmart, and Target.

Alternatively, a firm may seek to preserve its technology. In these situations, the technology may be old, but the firm believes it still has utility. Such firms may practice continuous improvement, but they preserve some aspects of the technology. Crayolas are still a viable product even though new technologies have emerged. Binney and Smith, the makers of Crayola, have improved the product with new colors, washability, and so on, but the fundamental technology has been preserved. This has allowed the product and the firm to prosper. However, Binney and Smith also continues to seek related products and to expand their market to older children—tweens—with new products such as Girlfitti and Gadget Hedz. In addition, they have added Color Wonder finger paints. The products are all tactile and experiential—just like Crayolas.[5] This continuous improvement process is part of the firm's technology strategy.

Market Capabilities

The firm must not only have direct technical capabilities; it must also have market-relevant skills that indirectly impact the technology of the firm. Engineers may develop tremendous new products but may have ignored issues such as how to distribute those products. To illustrate, a start-up medical device firm developed a product associated with hip replacement. The firm had good technology, but it could not get orthopedic surgeons to use its products. The firm could not understand why they had this problem since the company representing it and doing the marketing of the product was one of the leading distributors of orthopedic products in the country.

The start-up firm realized only later that the sales representatives of the firm with which it had partnered to distribute the product focused on customers who carried orthopedic products used to treat sports injuries. Orthopedic doctors who treat sports injuries often do not perform surgery, and when they do, the standard is that the sales representative is not in the operating room. In contrast, orthopedic doctors who do hip replacement are all surgeons, and commonly have the sales representative come into the operating room and coach them through the use of a new product. The firm's failure to have sufficient market knowledge led to its decline. A fresh management team was hired to rescue the firm, and they addressed this critical difference by obtaining new distributors. Today, the start-up firm is doing well. Technological capability without market capability typically will not succeed.

In summary, technology is viewed in some texts as an input to strategy but not as a central factor. The argument here is that technology should be considered a central component of the firm's strategy. In fact, technology should be considered even at the most basic level of the firm. The firm's various proficiencies must be consistent and intertwined with its technological capabilities. The firm's capabilities, including technology, provide the firm with its competitive advantage. The goal is that the competitive advantage be sustainable by the business over a significant period of time. Thus, the goal is a **sustainable competitive advantage**.

Technology and Competitive Advantage

A **competitive advantage** is what the firm does better than any of its competitors. However, the ability to perform an activity better than competitors will lead to a sustainable competitive advantage only if the activity is something that the customers value and other firms cannot easily duplicate. To illustrate, the ability to have faster processing by computer chips can be a competitive advantage for a chip manufacturer only if there is a demand for such chips. Thus, a competitive advantage must not only be something a firm does better than its competitors, but it must be something that impacts the customers' purchasing decisions so that they buy the firm's product over its competitors' products. From such a competitive advantage, the firm can build value for its shareholders or owners.

As we think of technology and competitive advantage, there are several ways to analyze technology. Specifically, technology development can be viewed as either continuous or radical; plus the technology can be used in an offensive or defensive manner. These different aspects of technology are not mutually exclusive.

Continuous versus Radical Technology

Technology development can be classified as either a continuous or radical.[6] An example of **continuous technology** development is the personal computer.

It seems personal computers become lighter and more mobile every year. These changes in technology are not a constant progression; instead, they happen over relatively short periods of time. Therefore, they are viewed as continuous improvements in the technology by consumers because there are no major changes that occur at one time. This progression is designed to change an existing technology but not to change its functionality. The innovation is aimed at improving performance, function, and/or quality at a lower cost.

On the other hand, **radical technology** development causes a dramatic change in the way things are done. The initial introduction of computers altered the way information was processed and stored in organizations and by individuals. The automobile was a radical technology when introduced. It provided an extreme change in modes of transportation. No longer were individuals dependent on horses, nor were they limited to where the railroads went. In the same way, when Henry Ford took the theory of assembly lines and began using it to make automobiles, he radically changed how products were made. More recently, the smart phone has changed the way we communicate and work. For example, iPhones and BlackBerries are widely changing many industries by speeding complex information to other locations such as heart monitor information instantaneously to multiple doctors and medical centers.[7] These radical technologies established a new functionality and a new way of doing things in business and society.

Between continuous and radical technologies, a third type of technology development exists that is not often recognized. Continuous and radical technologies can be viewed as the ends of a continuum. In between on this continuum are **next-generation technologies**. These changes in technology and their impact on society are more than the small step experienced in continuous change, but they are not revolutionary either. For example, the personal computer is a next-generation technology from the mainframe computer, made possible by the radical innovation known as the silicon chip. Before the silicon chip, computers used tubes for connectivity and then wires and contacts. These were awkward and much less dependable than the silicon chip.

As the discussion of computers illustrates, technology can be radical, next generation, and continuous at different points in time. The type of technology and innovation can also be different for various industries. Radical technology for one firm or industry may be continuous technology for another. Finally, an improvement in one industry may cause another industry to fail. The use of LED displays and silicon chips in calculators was an application of an existing technology; however, these technologies caused the discontinuance of slide rules. Continuous technological change reinforces the existing industry structure, and competitive advantage can be gained by leading the way. However, radical technological change creates new industries and alters or destroys old industries. Leading the continuous change process in the old technology when a radical technology appears is a recipe for disaster as illustrated by our discussion of Polaroid and its missing of the digital camera revolution.

A concept closely related to radical technology is **disruptive technology**. There are similarities and differences between the two that are recognizable.

The concept of disruptive technology was popularized by Clayton Christensen of Harvard in his book *The Innovator's Dilemma*.[8] This concept is similar in many respects to a radical technology because they are both technologies that change how an industry competes. However, Christensen differentiates his concept by arguing that a technology does not always have to be radical to be disruptive. Open source software is an example of a potentially disruptive technology. A **low-end disruption** is technology that enters the market with lower performance than the incumbent but exceeds the requirements of certain segments of that market often at a lower cost than existing products that are used by the segment. Thus, the technology disrupts that market although not a radical technology.

Maturing Process of Technology

A tool often used to examine where technological change is going is the S-curve.[9] Initially, innovation in a domain occurs and new products and processes are introduced as firms seek to translate that innovation to the marketplace. However, typically, no single product that uses the technology in a particular way is dominant. Instead, there are competing designs that may use the same technology in different ways. It will take time for a dominant design that uses the technology in a product or process to emerge. Over time, the amount of product innovation in this domain decreases as the process innovations (structure, etc.) associated with that product improves. However, over time there are fewer product or process innovations taking place in this technology domain. The use of the technology still continues though so the top of the S declines slowly in the S-curve. The S-curve is summarized in Figure 2.2.

The technology life cycle in the **S-curve** has four phases: embryonic, growth, maturity, and aging. The embryonic phase includes the invention and application of the invention through innovation. Improvement in the

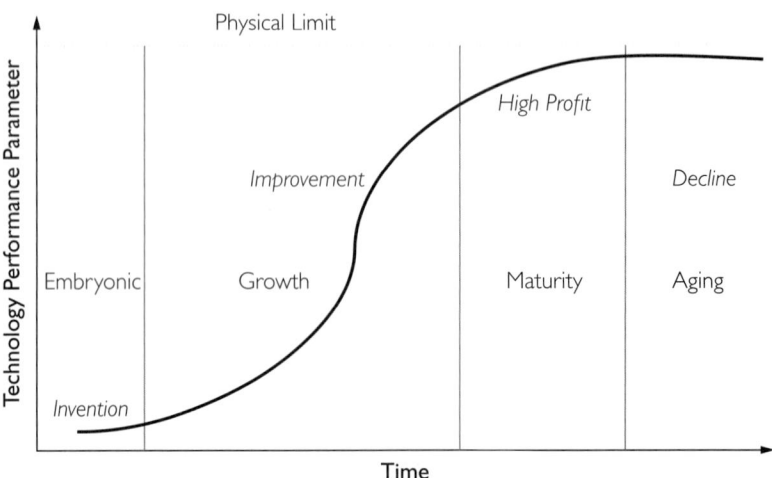

FIGURE **2.2** The S-Curve of Technological Progress

2.1
REAL WORLD LENS

Linux

Linux is an operating system created by Linus Torvalds, who was a Finnish university student at the time of its development. The operating system is offered free and has become the leading competitor to proprietary operating systems like UNIX and Microsoft. The heart of the system is referred to as the Linux kernel, which is the code that forms the basis of any firm's operating system. The firm is then able to take that code and build on it. This adaptability has led firms such as IBM and Hewlett-Packard to use Linux as a base operating system. The development of a system like Linux demonstrates a consistent theme in many technological areas: the pattern of development is difficult to predict. Firms must constantly scan the business horizon for changes that are occurring and look for the unexpected. A free operating system that becomes the backbone of many firms' efforts would have been difficult to predict five years ago. Today, however, the resource is widely used. Other firms such as Red Hat, VMware and Novellus Systems have specialized in developing Linux applications. These smaller firms are subject to acquisition. This raises many technology management issues for the users of the applications.

1. What type of technology does Linux represent—continuous, disruptive, or next generation? Explain what such a classification of type of technology would mean for competitors and consumers.
2. If your competitor acquires a firm that owns an application that is key to your business, what issues would you face? What would happen to the value of the acquired firm?

References

Shipley, Greg. 1999. Is it time for Linux? *Network Computing*, 10 (11): 54–55.

Shipley, Greg, and Kevin Novak. 2000. The Linux challenge. *Network Computing*, 11 (12): 48–58.

Veverka, Mark. 2009. Red Hat: Tempting target. *Baron's*, 89 (19): 42.

uses and the processes directly related to the technology mark the growth phase. During the maturity phase, firms that have done a good job of managing the first two phases can enjoy high profitability. In the aging phase, there is a decline in the utility of the technology. The technology may be rejuvenated and a modification of the S-curve will take place, or the product or process may become obsolete. To illustrate, digital music format has evolved rapidly in recent years. In two generations, silicon has replaced nylon, and memory cards are replacing CDs. Ultimately, the quest is to provide music with no moving mechanical parts and huge storage volume. As a result, the S-curve for digital music has been rejuvenated constantly, but individual products in the domain have become obsolete.

Offensive versus Defensive Technology

A firm can employ technology in either an offensive or defensive manner. The firm uses an **offensive technology** in a way that is not being used by competitors so that it gains a competitive advantage. This advantage may come from lower costs for the firm or from providing value more effectively or efficiently to customers. For example, Sotheby's and Christie's are two of the leading auction houses in the world. These two firms have dominated the high value auctioning of art and historic artifacts. The competition between them has been fierce for decades. In the late 1990s, both firms realized that the emergence of the Internet had the potential to completely reshape their competitive landscape. The fear of the two firms was that the Internet would ultimately make their businesses irrelevant because individuals could sit in their own homes and bid worldwide over the Internet for any item. Both firms came to this realization late and tried quickly to start up Internet auction capability in the late 1990s. However, these efforts were not successful.[10] In late 2002, Sotheby's was able to gain competitive advantage by signing a partnership with eBay. The result today is that while Christie's has no effective effort in this area, Sotheby's is able to offer simultaneous bidding of its goods being sold at auction over Internet.[11] A new technology that has the potential to place Christie's at a significant disadvantage has now been effectively controlled by its competitor, Sotheby's.

Alternatively, a firm can have a **defensive technology** and obtain technology that others already employ. The firm making the purchase in this situation feels it must employ that technology to be competitive. This use of technology will not give the firm an advantage, but it allows the business to match its competitors. Another defensive use of technology can occur when a firm acquires or employs a particular technology to block its use by others.

An example of the defensive use of technology to block its use by others occurred in the wound-closure industry. Johnson & Johnson, through its Ethicon subsidiary, in 1999 was the world's largest provider of surgical staples, stitches, and special-purpose wound bandages for major surgery. However, Closure Medical Corporation patented a completely new technology for the closure of wounds in 1999. The firm's DermaBond product was developed to glue the skin back together. The technology had the potential to make most of Johnson & Johnson's wound-closure products irrelevant. As a result, Johnson & Johnson sought out and obtained a marketing agreement with Closure Medical Corporation. Johnson & Johnson realized that its Ethicon unit was at a competitive disadvantage after Closure Medical developed its new technology. However, rather than ignore the change, the firm sought out the means to gain some benefit from the presence of the technology in the market. Johnson & Johnson would profit more if it owned the new technology, but Johnson & Johnson did not and could not ignore the new technology. Recall that Polaroid did ignore a new technology in its industry and failed. Therefore, Johnson & Johnson sought to be a part of this new technology through obtaining a license for it.

THE STRATEGIC PROCESS IN MTI

From the discussion in Chapter 2 to this point, it is clear that technology should permeate the strategic process of a firm. But what exactly is that process?

The strategic process of a firm can be broken down into three principal activities. In practice, a well-managed firm performs these activities simultaneously and continuously. Thus, while the components are presented here in discrete units, it should be recognized that in a firm these are part of an ongoing internal process. The three components are:

1. Planning
2. Implementation
3. Evaluation and control

Each of the three components (shown in Figure 2.3) will now be discussed in further detail.

FIGURE **2.3** Key Activities in the Strategic Management Process

Source: Adapted from UC Santa Cruz Leadership Convocation, Kristine Hafner, Director Business Initiatives, UCOP, February 4, 1999.

Planning

Planning is defined as the systematic gathering of information that leads to the generation of feasible alternatives for the firm, selection of the most appropriate action among the alternatives, and ultimately to the setting of direction for the firm. Activities in the planning process include

1. Data gathering
2. Mission generation
3. Objective setting
4. Strategy establishment

During strategic planning, the firm gathers extensive information on the external environment and about its internal capabilities. This information-gathering process is critical because it helps establish the foundations on which the firm bases its plans. The information gathered is critical to the firm so we shall spend more time on the types of information to gather after briefly discussing mission, objectives, and strategy.

From the information generated, the firm sets its mission. The mission of a firm is a simple statement of the basic purpose or reason for existing. The mission should identify, in relatively few words, what is unique about the firm and the scope of activities it wants to pursue. Limiting the firm's activities helps ensure that the firm stays focused on what it principally sees as its reason for existing. For example, 3M's mission statement is "To solve unsolved problems innovatively." 3M, in part as a result of this mission, has been one of the consistent innovators of new products from the reflective paints that now mark the roads to Post-it Notes. Compaq Computers and Intel formed a joint venture (an organizational form that will be discussed further in Chapter 6) in hand-held devices, and the mission statement for their joint venture was "To collaborate in bringing to market innovative industry standard solutions for small hand-held access devices to large enterprises servers." The mission statement here clearly defines the joint venture's reason for existence—the new entity was created to cooperate. Thus, the new entity could focus its efforts, and managers understood what they were to achieve.

Once the broad mission is set, the firm builds on that mission to establish measurable objectives and performance targets that will help it fulfill its mission. These objectives and targets state in specific terms what is to be accomplished in a given time period. The time period over which objectives and targets may extend can be as long as five years. However, even if the objectives and targets extend that far in the future, the firm also establishes short-range and mid-range objectives and targets. These short- and mid-range objectives and targets can be visualized as small steps that ensure the longer range objectives and targets will be met.

Finally, the firm establishes a strategy that helps ensure that the actions it takes will accomplish the objectives and targets that are set. There are three different levels of strategy. The corporate strategy is the overall firm direction and the actions pursued to move the firm in the direction chosen. The corporate strategy determines the positioning of the firm in different types of business. Thus, the corporation will establish how diversified it is to become and in what domains that diversification will occur as part of the corporate strategy. The business strategy is how the individual businesses in the corporation will be operated. If a firm has multiple divisions or strategic business units (SBUs), each will have its own business-level strategy. Wide ranges of ways to describe different business strategies have been proposed. However, in the simplest terms, the business strategy can be perceived as either low cost, typically when the business can sell a commodity product at the cheapest price possible, or the business can sell a differentiated product with special features that allow it to charge a premium price.

Functional strategies are those of the different departments, such as accounting, engineering, and marketing, that act in support of the given business strategy. For example, if the business strategy of the firm is low cost, then the engineering department needs to design products that are consistent with that goal. Cutting-edge products that have a narrow market are not appropriate. Figure 2.4 summarizes the different levels of strategy.

Information Gathering as Part of Planning

As noted, information gathering is a critical part of the planning process. Due to its critical role, more time is spent on this aspect of strategic planning than on other aspects. The planning effort requires that the firm understand its internal capabilities and the opportunities that exist in the external environment.

Internal Analysis The internal capabilities are the easier of the two to understand. The internal environment focuses on the internal operations and resources of the firm. There are many internal resources that can create a competitive advantage such as a firm's creativity, culture, and ability to integrate business units that are purchased. However, a key part of the internal analysis is a firm examining its current technologies and determining whether

FIGURE **2.4** Levels of Strategy

there is a better way to add value to the organization. Some of the broad technological questions a firm should ask include:

- Does the firm focus on dominating a market by increasing buyer value through its management of technology and innovation?
- Does the firm search for key value creation opportunities that can increase the size of the market (even if some customers might be lost)?
- Does the firm search for complete solutions and complementors as it pushes beyond accepted industry practice?

By examining such questions, the firm will develop an understanding about where it wants to add value, what it can do to add value, and how it will add value. The understanding of the firm's innovation policies, structures, and processes can then be integrated with the understanding of the cycles of innovation and technology development revealed in the external analysis. Figure 2.5 shows the elements of technological analysis at the strategic level.

In MTI, all the elements of technological analysis are important. However, as stated earlier, the financial aspects of the firm are a critical part of that technological analysis. Without the money to support its choices the firm will be unable to exploit its innovations and gain competitive advantage. Financial resources are also necessary to acquire technology in the marketplace. The firm will typically have many potential options but not enough funds for them all. The firm will have to make choices—based not only what the cost may be but also on the potential financial benefit of the choice. Therefore, we will focus on financial analysis as part of internal information gathering next.

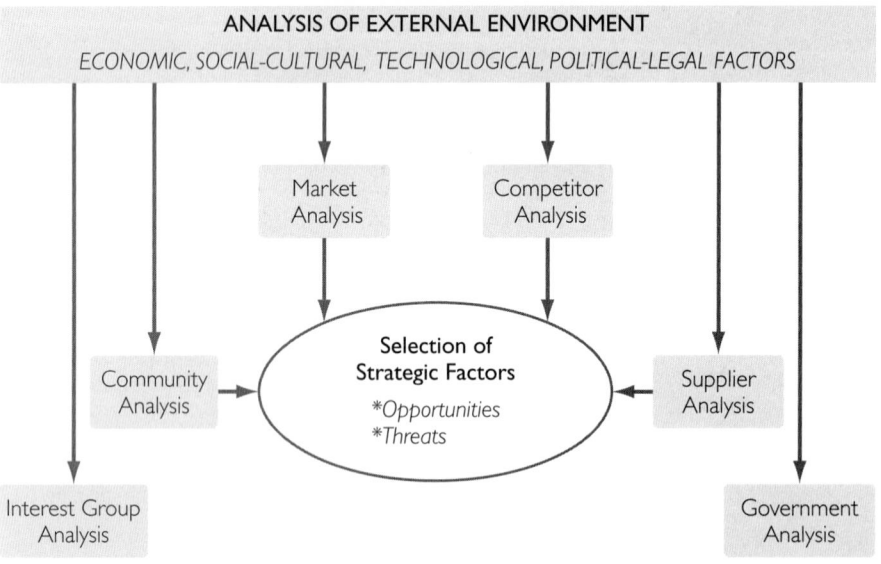

FIGURE **2.5** External Environment

Financial Analysis A financial analysis examines the income statement and balance sheet of the firm to understand how it is performing. In examining the firm's performance in its environment, the firm should compare its productivity to other firms in its strategic group and to industry norms. The most common financial analysis is based on ratio comparisons. These ratios can be classified in a number of ways, but the most common areas of interest in determining the relative performance of the organization are profitability, liquidity, efficiency, and other ratios. There are some basics of financial analysis that should be recognized prior to examining the specifics of financial analysis.

The basic accounting formula for a firm is:

$$\text{Assets} = \text{Liabilities} + \text{Paid in Capital} + \text{Retained Earnings}$$

This formula provides the base for beginning all financial analysis and decision making. Accountants generate financial documents such as 10-K reports to provide the information for decision makers who must then decide what actions the firm needs to take. In the realm of management of technology and innovation, the central decisions this information will impact are:

- Financing technology internally or externally
- Determining which technologies show the most promise for the firm
- Balancing the allocation of resources among technologies of the firm

A central part of the management of technology and innovation is determining what projects to fund, when to fund them, and how to fund them. The firm has to manage its current resources and its ability to attract future resources. Therefore, while the investment in a given technology may appear critical to those who will use the technology, the firm needs to take a broader view before it approves such expenditures. The expenditures on technology must provide value to the firm, or the expenditures are not warranted. Therefore, the analysis must include the potential payback from investments.

The financial resources for the firm's technology and innovation efforts principally come from retained earnings or existing assets. The **income statement** and **balance sheet** are two internal accounting documents that all firms generate and that provide information on these two items. The **retained earnings** (net profits retained in a business after dividends are paid) are delineated in the income statement. The balance sheet indicates the existing assets and liabilities. The decision-making process requires information from both. However, the income statement will be reviewed before the balance sheet. Then the two internal ways of financing MTI (retained earnings or assets) will be discussed in turn.

INCOME STATEMENT AND RETAINED EARNINGS To obtain information on the retained earnings, the analyst needs to focus on the income statement of the firm. This statement reflects the revenues and expenses of the firm over some specific period of time. Information from the income statement can be used to determine more than retained earnings, including:

- How profitable the firm is
- The nature of the expenses of the business
- If the firm has sufficient resources to conduct other strategic activities that may arise

Specific parts of each statement are central to the decision-making process. Listed below are the key elements on the income statement.

The information on the income statement is combined to develop key information about the direction and health of the company. If the company does not have a healthy income flow, then it will not be able to undertake MTI projects that are funded internally. It will have to find other means of funding the projects (partnerships, etc.) or risk falling behind in technology (a competitive disadvantage).

- *Gross Margin = Net Sales − Cost of Goods Sold*: This is a very useful calculation because it shows the profitability of the company in its core business. A firm that has a high gross margin has resources to put into items like research and development. Reductions in gross margin are of concern because that may reflect increased competition in an industry or a price war.
- *Operating Income = Gross Margin − Operating Expenses*
- *EBIT (Earnings before Interest and Taxes) = Operating Income + Other Income (loss) + Extraordinary Income (loss)*: This variable is useful because it indicates how well a firm can cover its debt.
- *Profit (loss) before Taxes = Gross Margin − Total Expenses*
- *Net Profit = Profit before Taxes − Taxes*

As can be seen, the process of getting to net profit is a series of calculations for a firm. The firm needs to know its net profit because the available retained earnings for investment in new opportunities are determined by the following formula:

Beginning Balance of Retained Earnings + Net Profit − Dividends
= Ending Balance of Retained Earnings

The retained earnings of a firm are defined as the net profit minus the dividends paid to stockholders. The funds the firm keeps in retained earnings are then available for investment in new capital items (machines, buildings, etc.) or new projects (R&D, mergers and acquisitions, etc.). In other words, the retained earnings are used by the firm to fund future investments.

BALANCE SHEET AND REALLOCATING RESOURCES The balance sheet provides information on what the company owns (its assets), what it owes (its liabilities), and the value of the business to its stockholders (the shareholders' equity). The information on the balance sheet helps determine:

- The types of assets the company has—how money has been used in the past
- Whether the firm can meet its financial obligations
- The amount of money that has been invested in the company
- The amount of debt the company has

The firm can use the resources it has currently allocated into existing assets and directly invest those assets into the venture if the assets are relevant.

Interpreting Financials   Managers need to do more than simply determine the retained earnings and the assets of the firm. Managers need to have a rich understanding of the firm's financials before choosing if and how to finance a given technology project. Thus, a financial review should include more than looking at the numbers and saying the company is making or losing money. A firm may make money, but if similar firms are making more money, then the firm is actually not performing very well. Similarly, if a company loses money but loses less than similar firms in the same industry during a decline period, the company may be performing well. Therefore, two types of comparisons are normally conducted: one for the firm and one for a strategic group.

First, what is happening to a firm's profitability and asset structure over time is critical. The pattern for at least three years needs to be examined. How a firm is doing relative to other firms in the same industry is the second important comparison. The key is to obtain a good comparison group. One frequently used printed source for comparison group information is the Risk Management Association. However, anyone doing this type of analysis needs to be sure they are making comparisons to a relevant group of firms. For example, if the manager simply asks for retail performance numbers, he or she will find that the numbers combine firms in retail groceries, retail electronics, retail furniture, and so forth. Clearly, the performance numbers for each of these groups is very different. The analyst needs to be sure that the group is well defined and comparable.

To illustrate, Actel is a technology firm that provides programmable logic solutions, including field programmable gate arrays. In analyzing their financial performance, not all firms that do the same thing are relevant. Rather than use firms that are too large or too small, the analyst should generate a comparison group that is composed of firms similar to Actel on many dimensions. For example, firms like Lattice Semiconductor, Altera Corporation, Atmel Corporation, and Xilinx Inc. could provide relevant insights.

When the strategic group is developed, there are several ratios that should be calculated for the firm and for that group. A **ratio** where the number of interest is divided by some relevant measure, such as total assets, sales, or equity, should be employed because it controls for issues such as size. Comparing the absolute profit of a midsized firm and a large firm tells you little. The larger firm should make more money. Comparing what a firm makes relative to the size of its assets, sales, or equity allows more direct comparisons to be made. There are a few general groups of ratios that are important. The ratios listed below are the most useful in the management of technology and innovation. This list is not all inclusive, but it provides a good basis for understanding how the firm's financial resources can aid or hinder investment in technology and innovation.

Profit Ratios   **Operating profit margins** show the profits from operations, and firms want higher ones because they show greater efficiency. The impact of taxes can be significant on a firm. Therefore, examining the *net profit margin* allows taxes to be taken out of the analysis of profits. Again, a higher ratio is desired. It is useful for firms to understand not only their profit but to compare that profitability relative to others based on how many assets (*return on assets ratio*) are and how much equity (*return on equity ratio*) is invested in the firm. Again, higher numbers for these ratios are desirable.

LIQUIDITY RATIOS Liquidity ratios are used to judge how well the firm can repay its debt. These ratios examine both short- and long-term debt. The *current ratio* tells the firm's ability to cover current debts and should be between 1.0 and 1.5. If a firm has to sell all of its inventory to cover current liabilities, the firm might have difficulty finding a buyer. Therefore, the *acid ratio* (this ratio is also referred to as the *quick ratio* by some analysts) removes inventory from the calculation of a firm's ability to cover its debt. Ideally, this number should be less than 1.0 (between 0.8 and 1.0). The *debt to equity ratio* examines the firm's ability to repay its total debt, including long-term debt.

EFFICIENCY RATIOS A firm also needs to determine if it is acting in an efficient manner in areas such as inventory and managing payables. (In the calculations here, we use sales, but cost of goods sold can also be used.) The *inventory turnover ratio* provides information on how fast the firm's inventory turns or is sold. A firm does not want to have a lot of inventory that is not selling, because it costs money to retain. If the *inventory turnover ratio* is much slower than other firms in the industry, it can indicate a potential problem in product quality or market efficiencies. Likewise, if a firm has higher levels of fixed assets per unit of sales, it may need to determine the value added by the assets that are not contributing to sales. If the *fixed assets turnover ratio* is lower than competitors, then the firm is probably not in a good competitive position. Finally, a firm wants to be a good customer and pay its debts (*payables turnover ratio*). However, a firm can also make use of the cash that goes to pay suppliers. The interest income on such money for even a few days for a large company can be substantial. Therefore, a firm wants to pay but not much faster than others in the industry.

OTHER RATIOS A *price-earnings ratio* tells whether the market expects growth from this firm, and thus, it is a forward-looking ratio. A higher *price-earnings ratio* indicates that the stock market has a more positive view of the firm. This is desirable. Companies have to generate cash to pay their bills. It is possible for a firm to be profitable but have poor cash flow and go out of business. When bills need to be paid, the firm must be able to pay cash, not give payees a promise of future income. Cash flow is calculated using only cash that has actually been received less the expenditures actually paid out. The *cash flow to assets ratio* indicates if the assets in which the firm has invested are generating enough cash for the firm to continue to be viable.

Understanding the Financial Analysis Outcome The condition of the firm and the actions of its competitors can be better understood after the analysis of the financial documents. However, the analysis is just the beginning of the decision-making process. The decision makers must determine what the results indicate and whether that information supports the qualitative analysis.

Some criteria for deciding how to use resources after the financial analysis is finished are:

- Is the firm's position improving or declining? An improvement indicates that the technology and processes in place are better than other firms. A declining position means the firm needs to examine its technologies for change. The retained earnings, if any, should be used for improvements in technology or in building more efficient processes.
- If the firm is low on retained earnings, does it have areas that are using resources that are not providing appropriate payback (and have little or no hope of providing payback)? Firms can get "stuck" in a given industry or product category for a variety of reasons, including firm history. 3M started out as a firm that mined sand grit and made sandpaper. It still makes sandpaper over a century later but is it the best use of resources? The making of sandpaper is part of the history, culture, and mindset of the firm. Although 3M can afford to continue to make sandpaper, is it the best investment for the firm given that other domains, even in its abrasive unit, are more profitable?
- Is the technology currently used being threatened by substitutes? If so, does the firm have retained earnings to implement new technologies, or has the firm identified resources that can be reallocated? The ideal occurs when the firm needs to make a change in technology, and the funds are available because of wise management of earnings and equity. This is not always the case, so personnel and other resources need to be shifted.

Using Financial Assessments Although financial analysis is a key ingredient for determining the strategic positioning of the firm in the technology and innovation arena, some cautions need to be taken into consideration. There is general agreement that financial analysis provides needed information for good decision making, but it is not all that is needed. Some of the cautions that need to be considered when examining financial assessments include:

- One set of metrics (financial measurements) does not fit all situations in the firm. What may be an acceptable profitability ratio for one product line may not be for another.
- Most measures can be interpreted multiple ways, especially when linked with other outcomes. There needs to be a balance of viewpoints considered.
- At some point, a decision needs to be made. One can spend too much time on analysis (**analysis paralysis,** or not getting anything done by focusing solely on analysis) and not enough time on decision making.
- Any financial analysis is based on the accounting practices of the firm. Although the data may conform to accepted accounting practices, the results may be distorted by those practices. Thus, it is valuable to read the notes to the accounting statements because these notes report unique positions taken by the firm which effect the financials reported.

External Analysis Once the financial analysis of the firm is complete, there is a need to also gather information on the firm's external competitive environment.

The various elements of the external environment that must be examined are summarized in Figure 2.5. These include the economic, social-cultural, technology, and political–legal factors. The goal in gathering data is to understand the industry's evolution to date, resources available in the industry, competitors, and the general environment that impacts firm success. In addition, the firm is looking for possible future trends and opportunities.

The first step in the external analysis is defining the industry in which a firm operates. Abell and Hammond suggest that industries should be characterized along four dimensions:[12] products, types of customers, geography, and stages in manufacturing to retail chain. An analyst should consider all of these dimensions. The analysis of "what industry" a company is in can be complicated and changing. However, it is the foundation for much of the analysis that follows; it is critical that this is done carefully. GE has changed its product mix a great deal throughout its history. As the changes occurred, especially before it divested itself of all product lines not related to telecommunications, it would have been easy for GE not to change the firms to which it compared its performance. If GE kept its competitors in the light bulb industry as its only key strategic group while expanding its medical imaging business, it would have been making decisions based on the wrong context.

Once a firm's industry is defined, then broad information on that industry should be gathered. This information includes general data on trends and the nature of the competition in the industry. To illustrate, demand growth, foreign competition, and concentration levels are key factors that should be determined.

Industry concentration is determined by finding the market share (firm sales in the industry relative to total industry sales) for different firms for a product. For example, if one knows three firms make up 50 percent of the market share but it will take twenty-eight firms to reach 75 percent of the market, insights about the nature of competition in the industry are gained. Such concentration figures tell management there are three large competitors (50 percent) and a number of small competitors. The managers in the three largest companies have a different context for making decisions.

One means to analyze the overall industry is referred to as Porter's Industry Model. This model will be discussed in detail below.

Porter's Industry Model A useful tool for gathering and organizing much of this information in an industry is Michael Porter's five-forces analysis.[13] This model builds on industrial organization economics to analyze how various parties influence an industry. It is important to note that this is an industry model, not a firm model. The model seeks to understand how five forces in an industry (buyers, suppliers, new entrants, substitutes, and rivalry) impact each other, not how they impact an individual firm. A sixth force, complementors, is now widely used when examining technology-focused concerns (Figure 2.6).

The forces in this model are analyzed in terms of which are powerful. In other words, which can lead to an economic benefit for the firms in this industry? For example, where there is higher than average profitability in an industry, there must be some economic inefficiency. Thus, for there to be

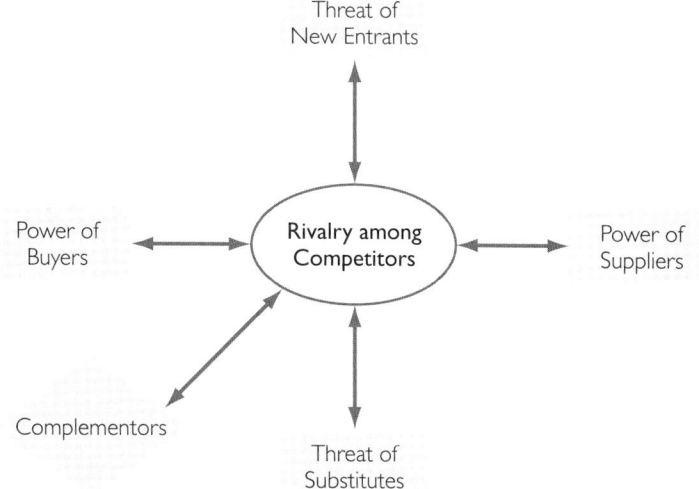

FIGURE **2.6** Porter's Five-Forces Model Plus

Sources: "Co-opetition: Competitive and Cooperative Business Strategies for the Digital Economy," Nalebuff B., Brandenburger A., *Strategy and Leadership* (1997, Vol. 2, No. 6) © Emerald Group Publishing Limited. http://www.emeraldinsight.com/sl.htm. Republished with permission, Emerald Group Publishing Limited; adapted with the permission of The Free Press, a Division of Simon & Schuster Adult Publishing Group, from COMPETITIVE STRATEGY: Techniques for Analyzing Industries and Competitors, by Michael E. Porter. Copyright © 1980, 1998 by The Free Press. All rights reserved.

above-expected profits, at least some of the forces must be weak so that the industry can take advantage of them and generate excess profits. If there are no weak forces, there will be nowhere for relevant firms to obtain the financial benefit. If all of the forces are strong, it is likely that the industry has low profits. Each force in the model is described in turn.

BARGAINING POWER OF BUYERS **Buyers** are individuals who actually buy the output of the industry being analyzed. Identifying the buyers in the model can sometimes be confusing. For example, for pharmaceuticals, an individual may purchase the drug at a pharmacy, but an insurance company pays for most of that drug. However, beyond that, it is typically the physician who decides which drug will be purchased. The fact that the drug purchased is decided by the doctor makes physicians the buyers in this model. The characteristics that determine if a buyer of an industry's output is strong include:

- Percentage of the industry's output the buyer purchases
- Costs of switching to competing brands or substitutes
- Number of sellers available

If the technology of an industry is relatively mature and there are few buyers, then the buyers' power will tend to be high. This is because the technology will be relatively the same across the industry, and the number of buyers will be less than the number of sellers.

BARGAINING POWER OF SUPPLIERS Suppliers are firms or individuals that provide input into the ultimate output of the industry. It is important in visualizing this model that the manager identifies the principal industry under investigation. It can be argued that every industry is someone else's supplier. The concern in this model is to examine the direct inputs into the product manufactured by the industry under investigation. The inputs will include widely recognized items such as raw materials, fixed assets, and financial support. One frequently forgotten input is labor. The concern with labor is not whether the employees are good workers or important to the process but, instead, whether they are powerful enough to demand a premium for their services. To illustrate with our example of the pharmaceutical industry, suppliers are firms that supply the necessary equipment, raw materials such as chemicals, and highly skilled personnel. The factors that make suppliers powerful include:

- High demand for supplier's products
- Quality and performance of product supplied are unique
- Inability of customer to vertically integrate

NEW ENTRANTS If an industry is experiencing high returns, then others will wish to enter that industry. These potential new rivals believe they may be able to make similar or better returns. These new entrants typically have improved products or processes incrementally. In addition, they may not have the sunk costs of those that have been in the industry for a period of time. Often, the ease of entry into an industry comes from its structural characteristics. For example, in the pharmaceutical industry, the equipment necessary to conduct the research and development is quite costly. In addition, it takes a long time to receive government approval to enter the pharmaceutical industry. Thus, the structural characteristics frequently impact access. The structural characteristics that lower the ability to enter an industry and the power of new entrants include:

- Brand loyalty by consumers
- Economies of scale increase the size at which a firm must enter the industry
- Capital requirements make it more expensive to enter the market
- Inability to access distribution channels
- Proprietary process technology from patents

Furthermore, the reaction of the existing firms in the industry can impact the ability of others to enter. Firms in the industry may invest in structural **barriers to entry** that discourage potential new rivals. For example, firms may expand into new regions of the country or market niches to discourage other firms from entering those markets because that would give their competitors a foothold in the industry. In addition, the incumbent firms may react aggressively to a new entrant, and this can discourage other firms from entering.

SUBSTITUTES Substitutes are products that perform a similar function but not in the exact same way. The effect of substitutes is to develop a ceiling for the pricing and profitability of the industry under investigation. At some stage, the gap in the price-quality trade-off becomes so great that buyers of one product begin to explore other options to satisfy their need. Thus, substitutes form a ceiling on the price that can be charged for a given product. In the pharmaceutical industry, the substitutes would be herbal medicines and other natural cures. The factors that impact the power of substitutes include:

- Ability of customers to compare quality, performance, and price
- Switching costs, or the cost of switching from the industry's product to a substitute

RIVALRY The rivalry among firms in the industry under examination is the last of the original five forces. The higher the rivalry, the more likely firms are to cut prices. This negatively affects profitability unless other factors such as process technology change. If rivalry among the competitors is reduced, it is more likely the industry will have higher profits. Continuing our example of the pharmaceutical industry, rivals in the industry are firms that also sell pharmaceuticals. In this example, over-the-counter drugs are substitutes, not competitors in pharmaceuticals. A new bioengineered cancer drug does not compete against over-the-counter drugs because they do not perform the same activity. The factors that increase rivalry include:

- An increasing number of competitors
- A growing demand for the product
- Producing an increased volume to obtain economies of scale
- Switching among producers results in low costs to the customer
- Increasing payoff from successful strategic moves
- Exiting the industry costs more than staying in it (exit barriers)

SUMMARY OF PORTER'S MODEL In the strategic planning process, the firm should gather information on these various elements and understand which forces are strong and why the industry profitability is where it is. This external analysis of industry can aid the firm in understanding where it, as an individual firm, needs to act in the future to gain a competitive advantage.

In the example of the pharmaceutical industry, a review of the five forces demonstrates that each of the forces is weak. The suppliers are typically plentiful and not organized. The buyers are the doctors, and they are not price sensitive. New entrants into the industry are not a threat due to the cost and time required to have a drug approved for use and the lack of adequate substitutes. Substitutes are natural remedies that are not much of a threat. Finally, the rivalry in the industry remains limited. The time frames for most new drug development are so long that firms know clearly what their competitors are doing, and they actively avoid challenging each other directly. Additionally, most pharmaceutical firms focus their drug development in specialized areas, such as heart disease, and do not offer a full range of drugs. As a result, the drug industry has been one of the most profitable industries for many years.

But recently, the buyer of pharmaceuticals has been changing. Doctors no longer have a unilateral ability to choose drugs. Insurance companies use formularies to dictate which drugs may be prescribed. In addition, new entrants to the industry have included generic drugs that have increased the competition. The result has been growing pressure on the profitability of firms in the pharmaceutical industry.

COMPLEMENTORS   Complementors, a sixth force that has been added to Porter's original model, are products that sell well with another product.[14] To illustrate, if a firm makes crackers, the product will do better if somehow it can be connected with a soup company. The two products complement each other. For technology firms, such complementary activities are increasingly important. For example, personal computer manufacturers typically do not generate the microchips used in their machines. However, the ability of firms to sell more computers is directly connected to the ability of chip manufacturers to create the next generation of microchips that allow personal computers to accomplish more and operate faster. But greater sales of computers will also increase the demand for chips. Thus, the microchip manufacturer's and the computer manufacturer's products complement each other and impact each other's success.

The complementor can come from the same parent company. For example, the crackers and soup illustration is demonstrated by Pepperidge Farms and Campbell Soup, which are owned by the same parent company. Alternatively, complementors can be different companies whose products support each other. For example, Microsoft software runs on Intel's chips. The two firms help each other. Success, or failure, in one will impact the other.[15] The characteristics that influence the power of complementors are:

- Ability to integrate backward or forward to replace the complement
- Availability of substitute complements
- Buyer or supplier switching costs
- Relative concentration

Strategic Groups   Another useful tool for gathering information in the planning process is to analyze firms within their strategic groups. A firm does not compete directly against all firms in its industry. A manufacturer of generic low-cost drugs such as penicillin does not compete against all other drug companies. Instead, it is producing a commodity product that is competing against other commodity products like penicillin. In contrast, a firm that makes part of the drug cocktail that fights the effects of AIDS is competing against only those firms that make similar products. A **strategic group** is a group of firms that compete in a similar manner (i.e., customer, product, geography).[16] There can be a number of ways to divide a given industry into strategic groups. However, the firm should do so in a way that provides information on the firms with which it competes most directly. This information can then help the firm identify a gap in the marketplace that it fills or wants to fill.

Once the firm identifies the firms that are in its strategic group, the next step is to determine the strategic posture of rivals that provide the most direct competition. This involves qualitative analysis as well as review of the financial information of the relevant firms. The concern is to determine how competitors compete today and how they wish to change the industry environment. Comparisons between the firm and others will give an indication of the firm's ability to remain competitive.

By knowing the characteristics of the industry and the relative strengths and weaknesses of the competitors, an analyst can predict the potential strategic moves of rivals. Once the competitive environment has been analyzed, then the managers of the firm must combine the knowledge garnered about the external and internal environments to determine the optimum course of action.

Information Gathering Process Overview   We have seen that in gathering information for the planning process that financial information and industry information is critical. The planning process should gather information about the environment as it is today and as it might be tomorrow. Too often, firms look to the past and assume the patterns of the past will continue. This occurs for several reasons:

1. It is easier to know what happened in the past than what will happen in the future.
2. If the firm has been successful, then it will want to "keep being successful" even if signs from the environment indicate that change is coming.
3. The leaders of the firms became top officers by doing certain things; they are good at those activities and continue them.

Instead, the firm should focus on challenging itself to look beyond what it now does and consider what the future could look like. Appendix 4 provides specific means to accomplish this forward-looking analysis. It is sufficient for now to know that the process should fully involve the firm's employees, major customers, major suppliers, and other interested parties. The generation of a wide range of views and potential options makes it more likely that the firm will generate a better informed plan.

Implementation

After the strategic planning (information gathering, mission generation, objective setting, and strategy selection), the firm must implement the plans. Figure 2.7 summarizes the strategic implementation process for the firm. As shown in the figure, once the firm has gathered information; identified a gap in the market; and developed a mission, goals, and strategy to be successful in that market, it will ultimately need to implement its strategy. Activities in a firm are not isolated from each other. The actions in one area have implications for employees in other sections of the business.

The result is that the implementation of the strategy requires the firm to conduct activities that are consistent with the given strategy.[17] The following chapters will develop in greater detail much of the information about how to implement strategies successfully. However, it is important at this point to be clear about the need for a fit between all of the various actions that the firm takes to implement the strategy. The true impact of a strategy comes from the firm setting a clear direction and taking actions that are consistent with that strategy. The firm's common implementation concerns include:

- Structure
- Employee hiring and relations

- Decision making
- Communication
- Culture
- Employee incentives

To illustrate, if a firm develops a strategy that employs a given technology, it needs to have people who understand that technology. To attract those individuals, a different type of compensation system may be required. If the firm develops software, it will need to hire individuals with the skills to write that software in the most current language. However, such a technology firm may not have the resources to hire the necessary individuals. Therefore, the firm will include in the compensation not only money but also stock options that may become quite valuable if the firm grows and is successful. Such stock options are often employed by new firms trying to implement next-generation technology. The stock options have the benefit of tying the employees to the long-term success of the firm. If an individual receives stock options, he or she will need to stay longer to ensure the firm is a success. Companies such as Sun Microsystems and Microsoft used this strategy when they were start-up firms. The successful firm actively manages implementation issues to ensure that it maximizes its value added.[18]

A tool to conceptualize how the elements concerned with the implementation of the firm's strategy fit together is a value chain analysis.[19] A value chain analysis breaks the firm's activities into two categories: primary and support. The primary activities are the major categories of activities that must take place to produce a firm's given product. These commonly include the inbound logistics of inputs, the operations or actual transformation of the inputs into a product, shipping the product or outbound logistics, marketing

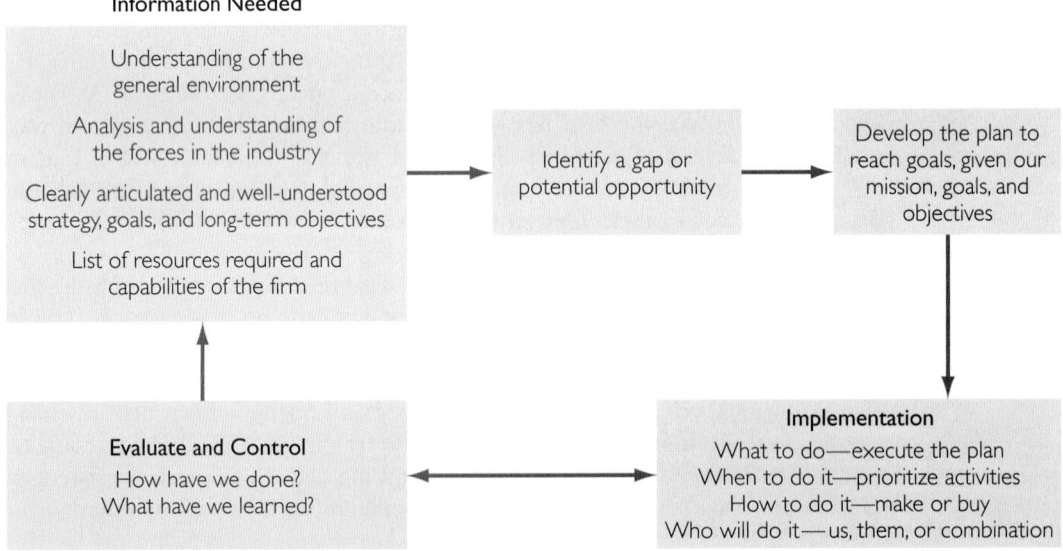

FIGURE **2.7** Strategic Implementation Process

the product, and servicing the product. The support activities sustain all of the primary activities. For example, human resource management hires individuals concerned with the product from inbound logistics to service. The support activities commonly considered in a value chain analysis include the firm's infrastructure, human resource management, technology development, and procurement.

The component parts of the value chain are illustrated in Figure 2.8. The diagram shows where technology impacts a firm's value chain. Technology and technology development could add value in both product and process. Where they are most critical, conducting an analysis of the quality of fit with the general goals and strategy of the firm should be conducted.

Applying the value chain analysis to electronics manufacturing illustrates the primary activities of the firm are widely impacted by technology. The inbound logistics to the manufacturing process are monitored to ensure timely delivery of materials and the quality of those materials through the firm's technology. The processing of the inputs into the firm's product requires technology not only in the manufacturing process but also in monitoring the quality of output. The outbound logistics rely extensively on technology. Today, just-in-time inventory is the norm. Thus, the manufacturer must work intimately with the customer to ensure that the "right amount" of product is available—not too much and not too little. This helps the firm by limiting investment in inventory and also helps the quality because the inputs are less

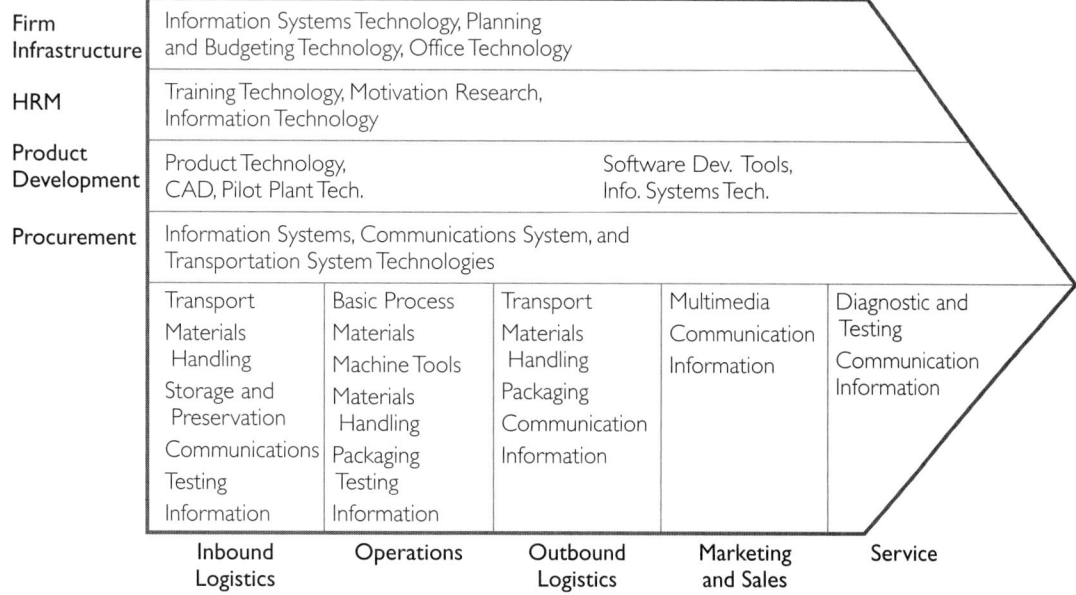

FIGURE **2.8** Technologies in the Value Chain

2.2 REAL WORLD LENS

IBM

IBM's strategy has become to deliver products on demand. In other words, they want to create technologies, products, and services that meet their customers' needs better and faster than any of their competitors. To accomplish this, IBM has increased its information gathering with customers and implemented a set of software application acquisitions. The firm has always had good creative ability. However, that ability has not always been focused on consumers. As a result, today, researchers spend as much as 25 percent of their time with customers. One means of R&D–customer collaborations is through IBM's two Industry Solutions Laboratories in Hawthorne, New York, and in Zurich, Switzerland. A customer will come to one of these facilities for a day, hear presentations by IBM scientists, and have a collaborative dialogue on specific business issues and demonstrations of key strategic technologies. The second thrust for IBM means it is now the second largest software company (behind Microsoft). IBM dominates in middleware brands with Lotus, Tivoli, Websphere, Rational, and DB2 database businesses.

1. Do you think these approaches will help IBM create a competitive advantage? Why or why not?
2. If you were with a competitor, how would you respond to IBM's two-pronged strategy?

References

Bradshaw, R., and C. Schroeder. 2003. Fifty years of IBM innovation with information storage on magnetic tape. *IBM Journal of Research & Development*, 47: 373–383.

Teresko, John. 2003. Interconnecting innovation. *Industry Week/IW*, 252 (12): 27–33.

Waters, Richard. 2007. Bib Blue looks to be more in the pink after changing tack: A focus on "middleware" and acquisitions has helped IBM into lucrative teritory. *Financial Times, London*, Feb 28: 30.

likely to be damaged before they are used in production. Just-in-time inventory uses technology to closely monitor the status of the delivery of the product. Similarly, the servicing of the firm's electronic manufacturing output also requires extensive technology to diagnose and solve problems with the firm's product.

The support activities of the electronics manufacturer also rely on technology. The two most obvious examples of technology's impact in electronics manufacturing are the development of new products and the procurement of the firm's inputs. Thus, to be successful, electronics manufacturers must ensure that their technology is employed intelligently and integrated effectively throughout their value chain.[20]

Evaluation and Control

As noted earlier (Figures 2.3 and 2.7), the strategic process is circular. Once the strategy is implemented, the firm must make sure that its strategy is working. The firm, through planning, establishes goals and objectives. After the strategy is implemented, the firm must ensure that the goals and objectives are met. If they are not met, then adjustments are required. This process is referred to as evaluation (comparison of actual outcomes with expected outcomes) and control (adjustments, as needed). The firm must determine why it is not meeting its goals and objectives and either change what it is doing or change what it wants to accomplish. Determining if the goals are not met is a straightforward evaluation process. The control process is more difficult and frequently requires revisiting the planning process. The feedback then must be given to the appropriate areas in the firm and changes pursued.

A key part of evaluation and control is establishing means to determine whether the firm is successful or not. There are a number of key issues the firm needs to address about the relationship between strategy and technology. First and foremost, the firm must ensure that its strategic efforts will be successful. These efforts include having processes that work, having the right resources in the right place at the right time, and having a well-thought-out strategic direction. There are two keys aspects in any such measures:

- Defined quantitative objectives in the production process and in the marketplace
- Defined qualitative measures focused on strategic concerns

The strategic process of the firm is active; it is not a process that can be performed automatically with little analysis. The measures of success need to be richer than simply sales or profits. Although these are critical, they tell little about the future. Instead, a rich variety of measures of the firm's performance should be used. In addition, the various aspects of the strategic process are interconnected. If the organization does not follow a process of evaluating, then the firm's actions, including its technology management processes, will be undefined with little cohesion or reasoning in action plans. Without these elements, the chances of success are reduced greatly.

A number of evaluation tools have been identified in this chapter. The five-forces analysis, the strategic group identification, and the value chain all provide guidance for issues that may need to be monitored by a given firm. For example, the number of substitutes for the products of an industry may change over time. Such a change can affect the attractiveness of that industry. Periodically, a company should evaluate how the forces in the industry are evolving and how its positioning is affected.

Another tool that has recently grown in popularity, which may be helpful to students in such monitoring, is the Balanced Scorecard (BSC) approach.[21] The basis for this evaluation technique is that financial returns give an incomplete picture of the performance and prospects of a firm. Kaplan and Norton, the developers of BSC, identified four key perspectives to be analyzed: financial, customer, (internal business) process, and learning and (innovation)

Perspective	Issues
Financial	Shareholder value; Revenue growth strategy; Cost structure strategy
Customer	Value basis—quality, price, availability, variety, function, service, brand, image
Internal	Processes—Production/Operations; Customer relationship management; Innovation; Industry based
Learning and Growth	Skills training and development; Information capital investment; Organizational culture and alignment

FIGURE **2.9** Balanced Scorecard Issues

growth. Figure 2.9 delineates some of the issues that need to be addressed in this type of evaluation of the MTI strategic considerations. The BSC will be discussed later in this text.

THE NEXT STEPS IN INTEGRATING MTI AND STRATEGY

This chapter concludes Part One of this book. The next six chapters address planning, implementation, and evaluation and control issues for the two major strategic approaches to bringing about major change in an organization: internal innovation and the acquisition of technology. The nature of innovation requires a more internal approach for the organization. Although the firm must examine opportunities in the environment, it must also emphasize development of an internal environment that encourages risk taking and accepts failure. On the other hand, the acquisition of technology through various methods of strategic alliance requires more analysis of external factors and the balancing of costs and benefits in an entirely different way.

The major questions, then, for the organization trying to strategically manage its technology become:

1. Should we create our own new technology and innovations internal to the firm?
2. Or should we acquire technology from others through acquisitions or strategic alliances?

The answers are determined by the costs and the likely outcomes as well as how the approach fits into the goals and future direction of the firm.

As noted financial analysis can be critical to this evaluation. There are ways to take and build on the financial analysis discussed as part of internal data gathering to judge whether an individual project or investment in a technology should be made. Only a few of the key methods to make this evaluation are discussed here. There are others that specific firms may wish to employ.

This review provides a basic understanding of how financial information can be used to make such evaluations. Specifically, two key methods are examined: net present value and payback period.

Net Present Value

In making an investment in technology, a firm expects to receive some financial benefit over time. However, money received in the future is not as valuable as money received today because the future money needs to be discounted for inflation. Thus, the technology may provide some monetary benefit to the firm, but that amount of money comes over time so it must be discounted.

There also is a need to discount that potential money for the risk factors associated with the technology. The firm may expect to receive that monetary benefit from the investment in technology, but a better technology that makes the current method obsolete may come to market without any warning. The result is that the firm's investment is totally worthless. Thus, this risk must also be included in the discount of the expected future benefit. The discounting of the expected benefit in this manner is referred to as a net present value justification.

To calculate the net present value of the firm's investment, the manager must initially predict the expected cash flow from the investment. The relevant formula is:

$$\text{Net Present Value (NPV)} = -I_0 + \sum_{t=1}^{n} F_t / (1 + k)^t$$

where I_0 = the initial investment; it is negative because it is an outflow
F_t = the net cash flow in period t
k = the required rate of return (hurdle or cutoff rate)

Recall that cash flow is the actual cash generated by an investment (not simply the revenues booked). The net present value predicts cash flow for the firm typically for four to five years in the future. Analysts typically do not predict further into the future because data become less reliable as the time frame lengthens. The cash flow needs to be discounted. The predicted cash flow is divided by the discount rate, but the discount rate $(1 + k)$ is raised to the power of the year where the prediction is made. The discount rate is chosen based on factors such as size of the market and its growth, risk of obsolescence and deterioration in the equipment, risk of technological breakthroughs, operating and fixed costs of the firm, and so on. Figure 2.10 illustrates NPV calculations. As can be seen, depending on the discount rate chosen, there can be considerable differences in the expected return and whether or not a technology purchase is viewed positively.

In looking at investment in technology and innovation, net present value has one major drawback. Typically, the discount rate is based on risk factors. For example, newness in the market is highly correlated to risk. Therefore, innovation, especially radical innovation, may be considered too speculative when net present value is calculated. In addition, NPV does not take into account the judgment of management over time. It is quite possible, and should

Year	Inflow	Outflow	Hurdle Rate 5%	Annual Discounted Cash Flow	Net Present Value	Hurdle Rate 10%	Annual Discounted Cash Flow	Net Present Value
0	$0	$75,000	1.0	$(75,000)	$(75,000)	1.0	$(75,000)	$(75,000)
1	$45,000	$0	1.05	42,857	$(32,143)	1.1	40,909	$(34,091)
2	$35,000	$0	2.071	16,900	$(15,243)	2.144	16,325	$(17,766)
3	$35,000	$0	3.169	11,044	$(4,199)	3.348	10,454	$(7,312)
4	$30,000	$0	4.287	6,998	$2,799	4.595	6,529	$(783)
Total				NPV is $2,799 with 5% hurdle rate			NPV is $(783) with 10% hurdle rate	

FIGURE **2.10** Old A1.5—Calculation

be expected, that managers make adjustments that impact the value of the investment.

Payback Period

A payback period calculation is another method to evaluate whether or not to buy or invest in a given technology. This technique compares the payback period of a new technology with the expected lifetime of the equipment or investments that need to be made. To calculate this figure, the following definitions are relevant:

PBP_i = payback period for typical technology or equipment i

IC_i = initial cost of technology or equipment i

B_i = annual benefits involved in using technology or equipment i

C_i = annual cost involved in using technology or equipment i

$CF_i = B_i - C_i$ = annual cash flow involved in using technology or equipment i

$$PBP_i = \frac{IC_i}{CF_i}$$

Typically, a payback period of less than one-half the lifetime of technology or equipment is considered a viable investment when the lifetime is ten years or less.

SUMMARY

This chapter has laid the foundation for the study of MTI by connecting technology to the concept of strategy. Strategy is the direction the firm sets for itself. Technology and innovation are critical parts of the process and the capabilities of the firm. The chapter has established several concepts key to MTI such as continuous and radical technology development and offensive and defensive technologies. The chapter has also established that strategy is not a single event but instead is a process. This process is connected to both the external environment and industry of the firm and to the internal capabilities of the business. The next chapters will examine in detail specific ways that firms can obtain technology either internally or externally.

MANAGERIAL GUIDELINES

For a firm to navigate successfully the strategic processes involved in the management of technology and innovation, it must keep certain actions in mind. These include:

1. Forget traditional organizational functions—judge ideas, not positions.
2. Know where the firm is in the life cycle of the technology and where its competitors are.
3. Be willing to assume risk if the potential long-term reward is great.

4. Utilize all resources in the environment. Do not get caught by the "not invented here" syndrome.
5. Break down communication barriers. Many firms lose opportunities because of a "not shared here" approach to lessons learned.
6. Keep expectations realistic. Too often, firms abandon technologies too soon because unrealistic expectations cannot be met.
7. Establish processes for new initiative approaches to management.

Guiding Questions

There are questions managers should ask themselves and the organization as they begin to evaluate the strategic technology position of the firm. Some of the key concerns these questions should include are:

1. What approaches to technology development have worked in this organization before? What value was created? Did new products and/or processes emerge?
2. What approaches have not worked well? Why? What lessons were learned?

3. What are the ideal structures and processes for creating an innovative context in the organization? How do these affect the view of the environment and the organization's potential?
4. Does the firm have a formal initiative process for innovation?
5. Who are the champions of innovation in the firm?
6. How are decisions made about what to work on and what to stop working on?
7. How are innovation and technology "driven" in the firm?

CASE **2.1** THE REAL WORLD
UPS Store

The UPS Store grew out of United Parcel's acquisition of Mail Boxes Etc. in 2001. Mail Boxes Etc. provided copying services, a place for individuals to pick up their mail, and mailing services with firms such as UPS. The purchase by UPS was intended to give the package delivery firm a retail connection. Shortly, thereafter, UPS allied itself with Office Depot and Staples to provide shipping services in their stores. To enhance its services, UPS has been consistently adding on-line tracking and other services.

In developing these connections to the customer, the goal has been to create a seamless integration of technology that makes it easier for consumers to choose UPS. Thus, there was not a separate technology strategy, but instead, technology was a key part of what enabled the strategy.

The way UPS achieved this is through the development of a consistent set of business activities. For example, information technology is not treated as a separate functional area. Instead, it is integrated into all of the working teams in the firm. Thus, rather than asking technology support individuals if there is technology to support the goal after the goal was already generated, there are technology professionals involved in all aspects of the firm. Goals are generated with technology in mind. This role for technology is enhanced by the fact that the chief information officer is on the business strategy steering committee. All strategies that come out of the firm have input from a technology perspective.

UPS leads in delivery-tracking technology, package-flow technology, and data analysis. UPS has one of the largest DB2 databases in the world (an IBM product that is a relational database management system) to help track all of its packages from the different customer contact points. In addition, UPS uses Oracle databases to provide package information to drivers and customers as well as to store and analyze sales patterns, financial information, and marketing data. Because the trend in this industry is toward smaller, more frequent shipments by and for individuals, the detail of the UPS system should afford it a competitive advantage over its competitors.

1. What other areas do you think UPS would align to ensure that technology was fully integrated into the firm's strategy?
2. UPS has two major competitors—the United States Postal Service and Federal Express. What does UPS' lead in technology mean for them?

References

Alghalith, Nabil. 2005. Competing with IT: The UPS Case. *Journal of American Academy of Business, Cambridge.* 7 (5): 7–15.

Costides, Nick. 2004. There are no longer independent technology investments … just investments. *Franchising World*, 36 (1): 69–71.

Podmolik, M. E. 2004. UPS promotes commerce, supply chain capabilities. *B to B*, 89 (12): 18–20.

CRITICAL THINKING

Relating to Your World

1. Figure 2.11 illustrates the concept of a product life cycle from the perspective of technology: start-up, growth, maturity, and decline. The issues the firm needs to address during each of these stages are different. Identify potential strategic issues your firm would need to address in each of these stages.

Technology-Driven Stage	Potential Strategic Issues
Start-up	
Growth	
Maturity	
Decline	

How would you address these issues?

2. "Buy-in" of the strategic direction of the organization is important. How should managers develop buy-in from the various groups and areas within the organization? Does getting the support of employees and other stakeholders require strategic planning? Explain your answer.

3. Develop a strategic plan for your life. Using the definition of technology and the management of technology we have learned, how would you expect the technologies in your life to change as you implement your strategic plan?

	Start-Up	Growth	Maturity	Decline
Type of Innovation	Major product changes	Incremental product change; Process change	Incremental process change; Cosmetic design	None
Stimulus of Innovation	User's needs; Feedback	Internal technical capability: cost and quality	Cost and quality improvement	None
Competitive Emphasis	New product performance	Some product change; Market share and quality	Cost reduction; Standard products	Defensive; Protectionism
Market	Market niches	Growth in the mass market	Market becoming saturated	Replacement markets
Organization and Management	Small; Flexible	Expanding; Highly organized	Rigid	Rigid

FIGURE **2.11** Characteristics of a Technology-Driven Corporate Cycle

Source: Girifalco, L. Dynamics of Technological Change. © 1991, Van Nostrand Reinhold, p. III. Reprinted with permission of Springer Science and Business Media.

WWW EXERCISES

1. Identify a company that is well known for its excellence in the management of technology and innovation. Go to that company's website and track how many times they mention technology and innovation. How many articles or comments about the company can

you find that relate MTI to the company's strategy?

2. Find a website that addresses the MTI issues for a specific industry. Then select two or three companies in the industry and visit their websites. How does the industry website differ from those devoted to company issues? Are the issues different? Why or why not?

3. Find an article or website that provides guidelines for the strategic management of MTI. What do you think of the advice? Compare the advice you find to the advice your classmates find.

AUDIT EXERCISE

In the chapter, we discussed S-curves. This tool is used to examine where a technology is in its development and to compare how the competitive positioning of a firm is related to its technology. Following are two different types of scenarios for S-curves and a description of what each means. Read each scenario, review the relevant chart, and then answer the questions that follow.

Scenario I
A new technology develops because the physical capacity of the old technology is inadequate. In this case, a copper cable with 1,000 copper lines could be replaced with a fiber-optic cable about the size of one copper line. This led to much greater capacity for data and information transmission.

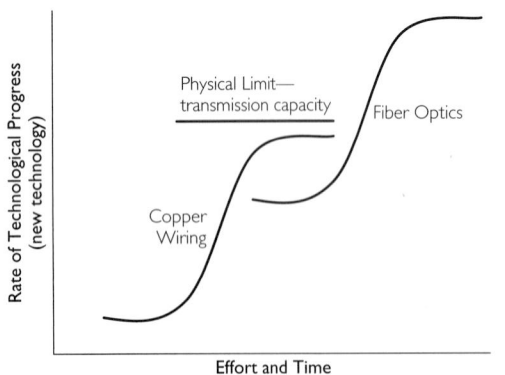

Scenario II
With sibling curves, the change in technology affects a range of products. If you are the developer of the new technology (in this case, silicon chips), then you want to identify a range of products that your new technology could improve.

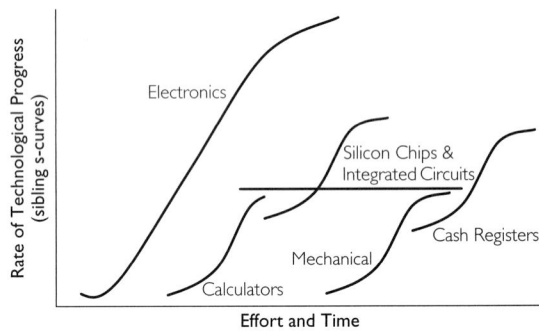

1. What factors does a firm need to monitor for each type of scenario?
   a. New technology
   b. Sibling curves
2. How would you develop an audit tool to accomplish this?
3. What environmental factors would you analyze to transfer a product from a business environment to the home and to the car? What technology breakthroughs moved the telephone from a business tool to a personal, portable necessity? How would you model those S-curves?

DISCUSSION QUESTIONS

1. Why are strategy and the management of technology and innovation so interconnected?
2. Describe the three stages of the strategic planning process and how they impact the management of technology and innovation.
3. What is the role of integration of the different activities in a firm in the strategic management process of technology and innovation?
4. What are the major decisions that impact the strategic management of technology and innovation?
5. What would be some of the strategic issues that a firm like GE would need to focus on as it seeks to improve its strategic management of technology and innovation?

PART ONE OPENING CASE: GENERAL ELECTRIC

1. How has GE combined its technology and strategic management to be successful?
2. What strategic concerns would you have for GE in the future? What technology and product changes should it monitor? How well has GE balanced the marketing and technology capabilities within the firm?

KEY TERMS

analysis paralysis 49

balance sheet 45

barriers to entry 52

buyers 51

capabilities 34

competitive advantage 36

complementors 54

continuous technology 36

defensive technology 40

disruptive technology 37

income statement 45

liquidity ratios 48

low-end disruption 38

next-generation technologies 37

offensive technology 40

operating profit margins 47

radical technology 37

ratio 47

retained earnings 45

S-curve 38

strategic group 54

strategic planning 33

strategy 33

strategic management 33

sustainable competitive advantage 36

NOTES

1. Henderson, J., and N. Venkatraman. 1993. Strategic alignment: Leveraging information technology for transforming organizations. *IBM Systems Journal*, 32 (1): 4–16.
2. MacRae, D. 2002. Miss the picture, fade like Polaroid. *BusinessWeek Online* (Oct. 29).
3. Baer, J. 2008. Polaroid film faces the final shutter. *Financial Times* (Feb. 9): 13.
4. Barney, J. 1991. Firm resources and sustained competitive advantage. *Journal of Management*, 17 (1): 99–120.
5. Lees, N. 2005. Crayola draws new brand parameters. *Kidscreen* (Mar/Apr): 76.
6. Cox, M. 2002. Rates of penetration of new technology. *Federal Reserve Bank of Dallas*.
7. Desmond, P. 2008. Seizing the opportunity. *Network World*, 25 (30): 24–25; Page, D. 2008. Cell phones are quickly becoming cutting-edge medical devices. *Hospitals & Health Networks*, 82 (8): 13.
8. Christensen, C. 1997. *The Innovator's Dilemma*. Cambridge, MA: Harvard Business School Press.
9. Foster, R. 1986. *Innovation: The Attacker's Advantage*. New York: Summit Books.
10. Bensinger, K., and D. Costello. 2000. Sotheby's push onto the web faces problems. *Wall Street Journal—Eastern Edition*, 235 (48): B1.
11. Thomas, D. 2002. Sotheby's joins eBay in online auction venture. *Computer Weekly*, 2 (7): 8.

12. Abell, D., and J. Hammond. 1979. *Strategic Market Planning: Problems and Analytical Approaches*. Englewood Cliffs, London: Prentice-Hall.

13. Porter, Michael. 1985. *Competitive Advantage*. New York: Free Press.

14. Nalebuff, B., and A. Brandenburger. 1997. Co-opetition: Competitive and cooperative business strategies for the digital economy. *Strategy & Leadership*, 25 (6): 28–33.

15. Brandenburger, Adam, and Barry Nalebuff. 1996. Inside Intel. *Harvard Business Review*, 74 (6): 168–173. In this article, the authors refer to complementors as value net.

16. Daems, H., and H. Thomas (eds.). 1994. *Strategic Groups, Strategic Moves and Performance*. Oxford: Elsevier Science.

17. Hong, K., and Y. Kim. 2002. The critical success factors of ERP implementation: An organizational fit perspective. *Information & Management*, 40: 25–40.

18. Montalbano, E. 2002. Sun ups the ante. *Computer Reseller News* (Aug. 26) (1009): 3; Montalbano, E. 2003. Sun professional services to forge ties with Channel. *Computer Reseller News* (Jan. 13) (1028): 6.

19. Porter, M. 1998. *Competitive Advantage: Creating and Sustaining Superior Performance*. New York: Free Press.

20. Fargo, M. 2002. Managing the EMS value chain. *Circuits Assembly*, 13 (11): 36–37.

21. Kaplan, R., and D. Norton. 1996. *The Balanced Scorecard*. Cambridge, MA: Harvard Business School Press.

PART 2

Internal Strategy

- **GlaxoSmithKline**
  Successful Internal Innovation

- **Chapter 3**
  Planning for Innovation

- **Chapter 4**
  Implementation in Innovation

- **Chapter 5**
  Evaluation and Control in Innovation

GLAXOSMITHKLINE: SUCCESSFUL INTERNAL INNOVATION

A firm may pursue new technology in two principal ways: internally, through research and development, or externally, by purchasing the technology. Each activity has benefits and drawbacks, and each takes different skills. Therefore, although a firm like Cisco or GlaxoSmithKline may have both internal and external development efforts, the firm typically will develop greater skill in one area than the other. For example, Cisco's competitive advantage has come from obtaining technology from external activities, in particular, through a large number of acquisitions each year.

This part of the book focuses on obtaining technology through internal innovation. GlaxoSmithKline is an example of a firm that has excelled through the use of an internal innovation strategy. The firm was formed through multiple mergers and acquisitions during the merger boom in the pharmaceutical industry at the beginning of the 21st century. The firm seeks to excel in internal innovation through research in the three priority diseases that were identified by the World Health Organization—HIV/AIDS, tuberculosis, and malaria. Here is a brief overview of the company and its internal innovation effort.

GlaxoSmithKline: The Firm's History

GlaxoSmithKline is an interesting global corporation—it is headquartered in the United Kingdom but operations are based principally in the United States. With 7 percent of the world market in pharmaceuticals, it is a leader in the industry. GlaxoSmithKline's mission is "to improve the quality of human life by enabling people to do more, feel better and live longer." They do this by planning and implementing programs of research and development. Founded in 1715 as Plough Court Pharmacy in London, GlaxoSmithKline has transformed itself into a global presence through a number of strategic actions over the years. In the 1830s, John K. Smith and his brother George formed what became the leading drug wholesaling company in the United States in Philadelphia. In 1842, Thomas Beecham launched the Beecham's Pills laxative business in England. The foundation of Glaxo emerged in New Zealand under the direction of Joseph Nathan. From these three very diverse beginnings, GlaxoSmithKline was formed when Glaxo Wellcome and SmithKline Beecham merged in 2000. In part, the purpose of the merger was to improve R&D. To help ensure that the new firm accomplished this in 2001, GlaxoSmithKline (GSK) reorganized its research and development efforts into Centers of Excellence for Drug Development (CEDD), small business units that emphasize flexibility, innovation, and therapeutic focus. Since the merger, GlaxoSmithKline has been a leader in innovation in pharmaceutical development and distribution. For example, GlaxoSmithKline has made a ground-breaking effort to provide HIV/AIDS drugs in developing countries at significant price reductions. Since the initial merger, GlaxoSmithKline has developed leadership in pandemic flu readiness, and made other treatments available at reduced prices to people in the poorest countries of the world. Besides seeking to help with the treatment of the key diseases identified by the World Health Organization, GlaxoSmithKline concentrates on medicines that treat asthma, viruses, infections, mental health, diabetes, and digestive conditions. In addition, they explore treatments for various types of cancer. In 2005, GSK was recognized by Bill Gates of the Bill & Melinda Gates Foundation for the company's commitment to R&D on malaria and other neglected diseases.

GlaxoSmithKline Today

GlaxoSmithKline implements its strategies by employing 99,000 people in over 100 countries. Over 15 percent of GlaxoSmithKline's employees work directly on research to discover new medicines. They screen approximately 65 million compounds annually in their search for new pharmaceuticals to cure the diseases focused on. GlaxoSmithKline's commitment to prevention is illustrated by the fact that they supply 25 percent of the world's vaccines.

The strategies they pursue or use to frame their business planning and implementation are summarized by three words—grow, deliver, simplify—and are articulated as follows:

- Grow a diversified global business
- Deliver more products of value
- Simplify the operating model

Today, GSK performs the following tasks:

- Every second of every minute, they distribute more than 30 doses of vaccine throughout the world.
- Every minute of every hour, doctors write more that 1,000 prescriptions for a product of GSK.

- Every hour of every day, GSK spends over $500,000 in research for new medications.

In addition to its prescriptions drugs, GSK produces consumer brands such as Gaviscon, Panadol, Nicorette, Ribena, Horlicks, Tums, Aquafresh, and Sensodyne.

Building an Internal Innovation Foundation

The GlaxoSmithKline (GSK) internal research and development efforts formed the basis for this extensive set of products and this innovative process for distributing pharmaceuticals to the poorest countries. Consistent with this goal the firm spent over 10 percent of its revenues on research and development in 2008. In addition, GSK is changing its R&D structure to ensure that it can deliver new pharmaceuticals, vaccines, and healthcare products in the future. One of the major problems facing GSK and other companies in this industry is patent expiration. Because GSK has 30 patented drugs that are in the late stages of patent protection, they are redefining the portfolio of drugs that they want to pursue—they are concentrating their R&D on developing a higher volume of mid-size products in particular patient populations. This will lower the risk of the portfolio of drugs because the revenues of the firm will not be dependent on one or two major successes.

GSK also wants to ensure that the firm focuses on the best science. In 2008, approximately 75 percent of new products in the pipeline were entirely new compounds/vaccines. Thus, to be successful in the future there is today a strong drive to be more innovative. To accomplish this, the R&D area of GSK has been reorganized to improve efficiency and focus after the merger. GSK focuses on eight therapy areas—biopharmaceuticals, immuno-inflammation, infectious diseases, metabolic pathways, neuroscience, oncology, ophthalmology, and respiratory. To address these areas the firm in 2008 created 70 Discovery Performance Units (DPUs). Most of these units are inside the firm, but some are also external to the firm. The sign of the success of this organization is that GSK completed or expanded 21 new drug discoveries in the 2008 fiscal year.

Building Return on Investment in Innovation

In addition to hiring the individuals to conduct the research and to develop new products, the firm has a culture and the processes that support innovation. The first part of these processes is building a structure for innovation. As noted above, GlaxoSmithKline has identified eight broad areas to focus its future growth on and in 2008 created DPUs within The Centers of Excellence for Drug Discovery (CEDD). More detail on the DPUs is that they are to be compact, fully empowered, focused and integrated teams, which have responsibilities for a small part of the pipeline associated with the production of a drug. These teams are to be cross-functional teams that include scientists, marketing specialists, and others from key domains in the business to work on innovations.

The firm encourages cross-fertilization of ideas through building alliances and meetings among its own scientists where different compounds and potential products are showcased. In these settings, R&D groups of individuals gather, and learn about new products and processes. The firm also has policies that support innovation.

GSK generates far more ideas and products than it can pursue during any given time. The result is a need for a process to evaluate the different ideas and products so that judgments can be made on which innovations to initiate support for or to continue support of. The ability to justify the product at each step of the process is critical for the team promoting it. The evaluation system is designed to be flexible as well as to avoid the continuation of projects that are not meeting expectations.

GSK has a disciplined approach to how and where resources are allocated within R&D. More than 35 percent of discovery projects have been terminated following the reorganization in 2008. After the elimination of these projects, the DPUs were given three year funding guarantees. The certainty of funding helps the R&D group focus on providing the best science and the best product for consumers but also gives them hard timelines to generate a marketable product.

The result of these innovation efforts is that in 2008 GSK received over 30 percent of its revenue from products that had been in existence fewer than three years. It is worth noting that GSK's extensive internal development efforts often lead to innovations that do not fit with the company's primary focus. However, the firm does not abandon those ideas; instead, it develops external discovery teams with other firms or universities or research labs. Thus, GlaxoSmithKline may yet gain a benefit from the innovation but is still able to maintain its focus on its own strategic goals and its eight primary areas of research for innovations.

Overview of Part Two

Part Two of the text will employ the strategic process model covered in Chapter 2 to discuss internal innovation. Thus, the text initially will discuss the planning process that must accompany internal innovation in Chapter 3. This examination will look at topics such as understanding the types of decisions that need to be made in internal innovation, the types of internal innovation that can occur, and creating an environment that encourages creativity. Chapter 4 will focus on the implementation of the internal innovation efforts. It will examine topics such as the leadership role in internal innovation processes, the importance of organizational fit as a manager ties the various parts of a business together to support innovation, and how to encourage the participation of employees in the innovation efforts. Finally, Chapter 5 will investigate the evaluation and control process in internal innovation activities. This chapter will examine how managers can determine if they are reaching the desired outcomes, maintaining relevant controls for personnel, and finding the best practices for the organization. The student should have a solid understanding of internal innovation and how to be successful at such efforts after finishing this part of the book.

SOURCES

GlaxoSmithKline annual report. 2009. http://www.gsk.com/investors/build-report.htm.

Whaler, J. 2009. Glaxo's Big Bet on Battling Pandemics. *Wall Street Journal*, (Oct. 9): A1, A16.

CHAPTER **3**

Planning for Innovation

OVERVIEW

This part of the book addresses the issues involved with internal innovation. In this chapter, we focus on how to lay the foundation for internal innovation efforts through planning. There is some overlap in the efforts for both product and process innovation, but there are also differences. This chapter addresses both product and process innovations as well as the following specific issues:

- The pros and cons of internal innovation
- Product innovation and process innovation
- The basic steps in the planning process for innovation
- The technology stages and their possibilities for innovation
- The development of a climate for internal innovation

INTRODUCTION

Innovation is how firms *create* new technology, products, or processes. This text separates innovation, which occurs internally, from situations in which the firm obtains a new technology or process from an external source. Our prior experience with business shows that fundamentally different skills are employed in each activity. To illustrate, the creation of new technology is impacted by the firm's efforts to encourage creativity and teamwork across various disciplines. As a result, it can take time to develop a new product, which can be a disadvantage if there are already competing products in the market or if other firms are close to entering the market. In contrast, if the firm purchases technology from a firm that is already in the market, or that is close to entering the market, a firm can obtain the new technology quickly. The difficulty is that the firm must then integrate that technology or business unit. These are just a few of the differences between the two approaches to technology and its management. Each approach has its benefits and drawbacks, which will be detailed later, but the skills involved in the two activities are different.

Many firms use both approaches—innovate through internal processes and acquire technology from external sources. However, a firm usually concentrates on one area more than the other. As noted, Cisco has both internal and external efforts to obtain technology, but it is best known for its external efforts. In contrast, a firm like Corning also uses internal and external means to obtain technology, but it is best known for its internal innovation efforts. Therefore, in this text, the internal innovation process and the external acquisition of technology are separated.

It is difficult for firms to innovate successfully on a consistent basis because of the complexity of the process. Many issues and components must be balanced to create an environment conducive to innovation. It takes not only the right people but also the right structure and the right reward system, as well as some luck, to succeed. Thus, when a business plans and lays the groundwork for an innovation strategy, it must consider a wide range of variables.

Typically, firms will have a variety of ongoing innovation endeavors, and these efforts are part of the integrated attempts to produce a platform of different products that complement each other. However, to aid in the analysis and understanding of innovation, the text will focus on the development of single products and domains because they are easier to understand and demonstrate. The concern for the integration of the various products into platforms will be addressed in greater detail in Appendix 3.

PLANNING: A COMPLEX PROCESS

Planning for innovation is a complex process in which a wide range of issues must be examined. 3M provides an excellent example of the range of issues a firm must address to promote internal innovation. The company relies heavily on internal innovation. Even while faced with a devastating recession, 3M posted all-time record sales in 2009 of $23.5 billion. This performance is an outcome of 3M's history of R&D investments and continued commitment to innovation and new product development. From 2006 to 2009, one of every $18 of revenue went to research and development.[1]

To achieve this success, 3M has a number of elements in place that encourage individual, group, and organizational innovation. These elements include:

- set stretch targets that push the firm to be innovative in order to meet the new sales and product development targets
- allocate time, space, and resources for employees to explore new ideas
- provide stage resources so that ideas that come forward can receive funding to move to the next stage of development
- encourage cross-functional teams to develop new ideas

Thus, 3M has built an environment that encourages innovation and accepts failure, while having a clear plan for the movement of innovations from development to the market. The result of this total set of activities is the creation of one of the most innovative firms in the world.

This chapter addresses the planning process for innovation, and Chapter 4 will build on that discussion to explore implementing those innovation plans. Chapter 5 will discuss the evaluation and control that need to occur to ensure that internal efforts stay on track and innovative.

TO INNOVATE OR NOT TO INNOVATE

The first concern for internal innovation planning is considering the basic question of whether the firm should innovate or not. Individuals and firms often overlook this question. The innovation strategy has a number of benefits, but it also has drawbacks. The managers in the organization need to balance the benefits and drawbacks and try to find the best fit for the organization. A firm deciding to pursue innovation cannot be naïve about the pros and cons of an innovative strategy.

Factors That Favor Innovation

To determine whether an innovation strategy is appropriate, the firm and its managers need to examine the potential benefits and drawbacks realistically. The benefits include:

1. Greater control of the process and the outcomes
2. Greater understanding of the technology produced and how to apply it
3. Greater ability to potentially develop the next generation of technology
4. Greater profit potential as a first mover

The first three benefits emerge as a result of the creative process associated with innovation. For example, an understanding of how the product or process was developed, and what actions worked and what did not in the development process, provides unique knowledge to the innovative firm. This unique knowledge can help direct the future development of that technology or product because the innovative firm understands the product or process.

For example, the innovative firm better understands what steps are needed to produce the product more efficiently or effectively. Thus, a firm can use the knowledge gained from the innovative process to its benefit in a way that others cannot. These first three points can lead to a competitive advantage for the business because the business can develop internal resources around the technology that are unique and that other firms cannot easily match.

The last point in the four benefits to internal innovation listed above is to be a first mover[2]. The concern for being a first mover is more externally focused than the other three. A **first mover** is a firm that is first to market in some manner. The firm may be first to market with a given product, the first into a given market area, or the first to employ a given technology in a particular fashion. Being first often leads to competitive advantages, such as customer loyalty and brand recognition. These advantages can be difficult for competitors to match or overcome. For example, if a technological product has high **switching costs**—costs to switch from one producer to another—it can make customers very loyal to the first producer because they have already purchased its product. Similarly, the level of comfort and support can make a customer unwilling to switch to a new producer. It was this factor that helped establish IBM's competitive advantage in computer mainframes for many years. Typically, in the 1950s-1980s, businesses that needed computers had top managers who did not necessarily feel comfortable with computers. IBM's high level of support to firms with such managers led to strong customer loyalty because the managers felt very comfortable with IBM.

In some settings, it can be more beneficial to be a follower in an industry. The first mover can spend many resources educating the market about the product. Additionally, the follower can learn from the mistakes of the first mover. In a more extreme example of how first mover advantages can turn, the firm that pioneered the use of the containerized cargo for shipping overseas went bankrupt. Containerized cargo has the benefit of ease of loading and unloading and greater control over shrinkage of goods because the products are sealed in containers. Thus, in the case of containerized cargo, it was followers into the industry that were able to build on the customer education provided by the pioneering firm. Typically, if not a first mover, then a **fast follower**, or a firm that quickly follows the first mover into the market, will perform best in an industry.

It is also important to note that first-mover advantages, like customer loyalty, can be lost. For example, once managers began to feel more comfortable with computers, costs became a greater consideration in purchasing decisions, and customer support was a smaller consideration. As a result, competitors began attacking IBM in the late 1980s. This caused IBM to not only lose its leading position in the computer hardware market but almost pushed the firm into bankruptcy.

Factors That Discourage Innovation

There are also reasons not to embrace internal innovation as a strategy. When deciding whether to pursue this strategy, potential drawbacks need to be considered. They include:

1. The time required for an internal development strategy is greater. If a new product is purchased, it is almost immediately available to the firm. However, if it is developed internally, the process may take years to develop the new processes and to market the new product.
2. There is a greater risk of failure to develop the right product at the right time.
3. Keeping a pipeline of new products and/or processes is difficult at best. There is always the danger that another firm will enter the market first.

These drawbacks can have a significant impact on the strategic posture of an organization. To illustrate the potential drawbacks, consider Christensen's discussion of hydraulic shovels in his book *The Innovator's Dilemma: When New Technologies Cause Great Firms to Fail.*[3] Until the development of hydraulic shovels, the established manufacturers were focused on developing larger mechanical earth-moving equipment designed to move greater and greater amounts of earth. However, many customers were not interested in moving larger amounts of earth; they needed an efficient machine that was compact and flexible. The firm JCB introduced the first backhoe, or hydraulic shovel.

The existing earth-moving firms had relatively large, well-funded innovation efforts. However, their innovation efforts failed to generate the right product at the right time for their market. The established firms had focused only on large mechanical equipment; their innovation effort to respond to the development of the backhoe took time. Thus, most firms were caught by surprise, and many failed. Christensen refers to the ability of a new technology to quickly replace the established thinking on a given domain as a disruptive technology. (In Chapter 2 recall we presented the differentiation between a disruptive technology and a radical technology.) In this case, the disruptive technology made useless much of the extensive investment in innovations by the established firms because they were investing in the wrong technology and product—bigger earth-moving vehicles. The ability of the established firms to respond to this disruptive technology was also slowed because they had to shift their internal innovation efforts to this new perspective.

Therefore, there can be strong negative implications to pursuing an innovation strategy. Managers must clearly understand both the benefits and drawbacks to internal innovation as they move forward. This understanding requires not only technological understanding but also an understanding of the market and the customer. This requires multiple foci as the innovative firm looks for new opportunities. It is important that the "opportunities" can be matched with the firm's strengths and abilities as new products and processes are developed.

TYPES OF INNOVATION

Once the firm determines that internal innovation is an appropriate strategy, it must determine the specific type of innovation to pursue. The planning process is, in general, the same for the different types of innovation, but subtle variations occur, and these are examined next.

Innovations can be classified in a number of ways; however, one of the most common is from the perspective of product and process innovations.

While discussed separately here, these concepts are intertwined, and there is rarely one without some effect on the other. After all, if a new product is developed, the firm needs to develop some way of producing and marketing the product, and this means new processes must also be put in place. Likewise, processes may lead to new interactions within the organization. This, in turn, may lead to new products as different individuals exchange information and ideas within the organization.

Product Innovation

For most organizations, product innovations are the center of their research and development (R&D) efforts. Although R&D can occur in a separate unit of the organization, today it is more often spread throughout the firm. Thus, there is not always a single department or area called R&D. Instead, there are people focused on the goals and objectives established for R&D throughout the organization. In the case of GlaxoSmithKline, they have established research efforts around the world.

No matter how the firm chooses to structure its R&D efforts, it must be clear where it is in the R&D process and the type of innovation efforts that are needed. The types of innovation efforts found in the R&D process include:

1. Basic: pure research and development
2. Applied: new product development
3. Systems integration: product improvement or market expansion

A firm may have any or all of these different types going on at any given time.

Basic Research: Pure Research and Development

Basic research involves the creation of new knowledge. This knowledge can be new to the firm, or it can be an innovation that was unknown before this effort. Basic research is fundamentally risky, but it has the potential to provide great rewards such as leading to new products or ways of doing business. The goal of an innovation strategy is to create value for the firm and its customers. This goal cannot be forgotten even in basic research. Thus, academic institutions, government agencies, and specialized research labs typically focus on basic research because value creation for these entities is typically not determined by monetary profit (See Figure 3.1).

To illustrate, consider the investigation of laser physics. This area was developed from work by Albert Einstein and is based on absorption, spontaneous emission, and stimulated emission of electromagnetic radiation. In the beginning, this research was undertaken to extend our knowledge of how radiation and light interact, not to make products. The basic research in this domain did not immediately lead to new products. Basic science is motivated by the broad curiosity of the researcher, not specific product interests.

Applied Research: New Product Development

A firm then builds on the basic research and conducts applied research. **Applied research** utilizes the new knowledge developed by the basic research to create new products. The new product development can then lead to the

Type of Research	Goals	Examples from Particle Physics
Basic	Pure R&D —create new knowledge	Einstein's work on emission of electromagnetic radiation
Applied	New product development	Laser printers, laser lights, compact discs, laser cutting
Systems Integration	Incremental improvement in products	Laser light shows, improved laser knives for surgery, smaller laser printers

FIGURE **3.1** Types of Research Efforts

firm changing its strategic position in the industry or, at least, changing its potential position in the industry. This should lead to the firm gaining some measure of competitive advantage.

The purpose of applied research is to add value to the firm and its customers in the marketplace. The risks of applying the innovations from this type of research are less than those in basic research. The probability of success and high reward is moderate. To continue our illustration of laser physics, basic research established the foundation with principles that most individuals do not understand, but the applications that have emerged from that technology are numerous and familiar to many—laser printers, compact discs, laser knives used in surgery, barcode scanners, and laser lightshows.

Systems Integration: Product Improvement

This third type of R&D is the most incremental in nature. **Systems integration** is aimed at supporting existing business improvements in established products or opening new markets with an existing product. This type of integration has low risks and rewards associated with it. In fact, most of the risks are negative; not changing can lead to strategic disadvantage. Many firms call this type of innovation **tweaking** because it involves adjusting the ways the firm organizes its existing knowledge to increase its leverage. Systems integration is most concerned with the fit among parts of the organization and how to improve the fit with existing knowledge bases. Thus, medical imaging was applied research that flowed from the basic research on particle physics. Systems integration research occurred, followed by tweaking. This process has led to better laser printers, more spectacular light shows, and better laser knives for surgery.

Which Type of Innovation Efforts Should a Firm Focus Upon?

Which of the three types of research is best for an organization to pursue? The firm should match the type of research it wishes to pursue to its needs and capabilities. One of the key elements in making such a determination is what the firm's competitors are doing. This analysis of strategic position—what is occurring in the industry now and what will occur in the industry in the future—is difficult. However, the analysis is aided by closely monitoring the competition.

The firm must first decide whom to monitor. There is no formula for determining which competitors to monitor. One means to quickly identify the position of the firm and its competitors is a strategic group map. A firm does not compete against everyone in its industry. Instead, it competes against some firms more directly than it does against others. Thus, a firm needs a means to segment its industry into relevant groups so the business can identify which firms to target.

To illustrate, a firm like Nintendo with its product Nintendo DS and DSi has dominated the handheld gaming industry. However, in 2008, LeapFrog—the educational electronic learning leader—decided to challenge with Didj. Nintendo was not directly concerned about LeapFrog until it entered into its gaming territory. For Nintendo, LeapFrog was not part of its strategic group originally—now it is. LeapFrog has changed its strategic group. Depending on how it identifies its product lines, LeapFrog may need to watch two strategic groups, one for educational electronic learning products and one for electronic gaming.[4] A strategic group map helps identify firms that should be monitored most closely. As a result of strategic actions by others, a firm's competitors may be a constantly evolving set of entities.

To better understand who a firm's competitors are, the firm can construct a **strategic group map**. This map reveals the graphic positioning of firms in an industry. To generate such a map, a firm first chooses two axes that represent critical factors in the industry. The factors on these axes can vary based on what the analyst believes is important. However, variables that are highly correlated should be avoided. Thus, factors like cost of product and quality in the same map are avoided. These two factors both increase at approximately the same rate and would not provide much insight. Instead, look for two distinct factors that are relevant to the industry and differentiate segments in the industry. The distinct factors used may (or may not) include either cost or quality but not both; the key is to gain the broadest insight by using factors that are not related.

The firm performing such analysis will then place the competitors in the industry on the map. It should become clear that various firms group together. To illustrate, for the pharmaceutical industry strategic groups could be determined by variables like R&D expenditures (high and low) and specific areas of research (high blood pressure/heart disease, gastrointestinal, etc.).

Once the firms in the same strategic group are determined, the managers study the actions, including the products, processes, and potential actions of each competitor. This information can come through a variety of sources, such as published articles, national associations, analytical reports by industry experts, academic studies, and so on. From the information gathered, managers may learn which firms are investing heavily in areas such as new product research, new processes, and new markets. If a firm is not where others are, it may be at risk. This does not mean that the firm needs to react, but it does need to balance the risks, rewards, and costs to ensure that it is making a conscious choice and that it has plans of actions if a competitor makes a strategic breakthrough.

Process Innovation

The purpose of **process innovation** is to increase the efficiencies or the effectiveness of an organization. Changes in processes require the organization and individuals to adapt to the innovations and this can lead to opposition in the organization. However, if properly applied, process innovations offer the organization and its personnel opportunities to improve the value of the organization and to continue the organization's viability. Thus, process innovations help to improve the output-to-input ratio of the firm.

The most common actions that address process innovation are new product development, restructuring, reengineering, and value destruction. New product development through project management will be dealt with in depth in Appendix 2 at the end of the text. Therefore, the focus here is on the other three types of process innovation.

Restructuring

A major reorganization of a firm is often referred to as restructuring. Recall that GlaxoSmithKline has restructured its R&D unit into eight major focus groups with subunits focused on specific areas. It involves substantive changes in communication and coordination patterns within the organization. Most organizations experience a constant level of small changes in their processes, tasks, and people. However, periodically, the organization needs to undertake a major review of what it does and why. A major reevaluation is commonly caused by events such as:

1. Information is not reaching the proper people to make timely decisions; this leads to slow decision making.
2. Opportunities and threats are missed by the organization.
3. A disruption has occurred in the firm's environment that has caused the firm extreme stress.

The most common restructuring activity in today's organizations is downsizing and reengineering.

Downsizing is a type of restructuring that occurs when a firm either sells some of its units or lays off employees.[5] Although employees generally view these as negative, the impact on the firm depends on the reasons for downsizing and the process that the firm is undertaking in response to those reasons. Processes may range from across the board cuts (not usually effective in the long term) to carefully considering what business the firm wants to pursue and then divesting or selling any asset/business unit that does not meet the newly stated vision of the firm (more time consuming in the short term but more effective in the long term). Many firms that downsize have experienced negative results. Survivors of the downsizing often feel overworked and are uncertain if they might be laid off next. As a result, the expected financial benefits of the restructuring are often not met. The planning for and the goals of downsizing should be extensive and clear.

Reengineering

Reengineering requires fundamental rethinking and radical redesign of work processes.[6] Often firms are using processes in the firm that were developed many years before and are no longer relevant. Reengineering requires the firm to think about each step in their work processes and why does the firm do things the way it does. Bennis and Mische state that reengineering has five specific goals:[7]

- increasing productivity
- optimizing value to shareholders
- achieving quantum results
- consolidating functions
- eliminating unnecessary levels and work

In pursing reengineering the firms should ask several relevant questions. These include:

1. Why is work (your work) performed the way it is?
2. What value is added by this process?
3. How can this work be done better?

From these questions reengineering can help identify processes within the organization that create no value for the firm. These processes may have created value once but are now conducted more from habit than from careful analysis.

In pursuing reengineering many firms gather a diverse group of individuals from the firm. They focus on what their customers obtain from them and why the customers want it. The organization then works backward from that initial point and examines each activity that is part of the production of the good or service. The firm should try to eliminate any activity that no longer provides value to the customers' desired outcome. In this manner of asking fundamental questions, the firm expects to eliminate unproductive and unnecessary activities and develop new ways of improving efficiency.

To illustrate, a firm that manufactured custom long-haul trucks commonly performed a credit check on potential customers before producing the truck. Such a credit check was not difficult but typically took several days to a week to obtain. However, speed is critical to not losing customers in most purchasing situations. The classic advice given to salespeople is that the deal is not done until it is closed. The firm determined that individuals did not casually come in and attempt to buy a custom long-haul truck that costs over $100,000. Therefore, the firm dropped the requirement of running a credit check before starting to produce the truck. The resulting increase in business was far greater than the cost that resulted when a few individuals ultimately did not qualify for the purchase.

Similar questions can be asked even of simple things. For example, why do large firms require a purchase request for small items like a box of pencils? It is cheaper with less paperwork to give each department $25 purchase cards to an office-supply store and let them buy pencils or other small items as they are needed. Reengineering asks these types of questions that large organizations often do not ask.

Choosing to Pursue Process Innovation

Process innovation is difficult to plan and implement. However, the firm should constantly be on the lookout for improvements in systems and processes. The problem with most process innovations is that they require social as well as work design changes. Too often, firms wait to make process changes, such as restructuring and reengineering, until the organization faces a crisis situation. Individuals may be more open to experimenting with new methods in times of crisis. On the other hand, the fear of what may happen next can hurt the efforts to be innovative. Managers and agents of change need to make system and structural evaluation and innovation a continuing part of the organization. This is because the ability to make changes can be far easier if individuals do not feel they are losing their jobs and positions in the process. A firm that is moving forward but making changes has places and new opportunities for those that are displaced by the changes.

INNOVATION PLANNING PROCESS

No matter what type of internal innovation the firm decides to pursue, there are common aspects to the planning process. The first is for the firm to recognize some fundamental dimensions of internal innovation:

1. The firm must acknowledge that the goal of internal innovation is for the firm to outperform its competition. Innovation is not a goal in and of itself but is part of the firm's total strategic effort.
2. Internal innovation is a process that involves many individuals, capabilities, and resources. To illustrate, Art Fry is typically credited with the creation of Post-it Notes at 3M. However, it was Spencer Silver, another 3M employee, who created the low-tack adhesive on the back of Post-it Notes. Silver could not find a use for the adhesive. Art Fry came up with the use for the adhesive after a presentation by Silver. Ultimately, it took both individuals' innovations to produce the successful product—the Post-it Notes.[8] Incremental innovations or product tweaking with Post-it notes is common—for example, 3M now sells pens with little post-its stored in the barrel of the pen to mark important pages.
3. Resources are critical to the innovation process. The way resources are managed and allocated determines the types of innovation actions the organization is capable of undertaking.

Once the firm has recognized these different dimensions of internal development, certain steps need to be pursued in the planning process. Five specific activities are part of the innovation planning: setting vision, mission, goals and objectives, strategies, and tactics.

Determine the Vision

After the organization affirms that it wants to pursue innovation as a strategy, the development of the firm's vision is the initial step. Typically, key leaders in the organization and the board of directors accomplish this process. A **vision** summarizes where a business wants to go and includes an understanding of how technology supports the firm's vision. The vision helps the firm focus its efforts more clearly on what the innovation plan wishes to accomplish.

Therefore, for a firm like Federal Express, the vision is that familiar statement: "When a package positively, absolutely has to be there overnight." This statement lets whoever hears it know the firm is focused on speedy package delivery. The development of the ability to deliver packages overnight was a radical process innovation when it occurred. The vision statement makes the focus of the business clear.

Set the Mission

Once there is a clear vision of what the organization is and the role of technology, the firm then develops its **mission**. This brief statement, usually fewer than sixty words, builds on the firm's vision to specify what it does and how. Thus, a firm's mission is more specific than its vision. The mission is typically developed with far greater involvement of the firm's employees. The goal is to have them not only understand what the mission of the firm is but also buy into that statement and fulfill it.

As noted in Chapter 2, every firm should have a mission statement. The mission statement helps the firm stay focused on what it is attempting to accomplish and how. The "how" leads to the firm's approach to innovation typically being included or reflected in the firm's mission statement. Thus, for a firm like Google, the mission statement is:[9]

> Our mission is to organize the world's information and make it universally accessible and useful.
>
> We believe that the most effective, and ultimately the most profitable, way to accomplish our mission is to put the needs of our users first. We have found that offering a high-quality user experience leads to increased traffic and strong word-of-mouth promotion.

As a result, for Google, the role of innovation and how it fits into the organization is stated clearly, and this provides direction for the efforts of the firm.

A mission statement provides the written purpose to fulfill the strategic intent expressed in the vision. It is easy for firms to lose sight of what they wish to accomplish. The mission statement helps the firm stay focused as it begins to plan and implement its goals and objectives.

Establish Goals and Objectives

After the mission statement, the firm establishes its goals and objectives. These set how the firm will carry out its mission. This step is particularly important for an internal innovation strategy. The goals and objectives specify what innovations, among other things, the firm wishes to accomplish over the short and long term. Therefore, if the firm wishes to develop a new product or process, the goals and objectives of the firm should specify that.

An earlier discussion introduced the concept that an innovation strategy needs to create value for the firm. The innovation strategy is pursued to produce value that contributes to the firm's success. The firm needs to have goals and objectives that ensure the organization is constantly working toward that success. From the goals and objectives, the firm will plan its allocation of resources to help ensure that those goals and objectives can be reached. So what appears like a simple goal—to dominate a given product or geographical area—will lead to a wide variety of other planning efforts to ensure that the goal is reached.

The pharmaceutical industry provides an example of the benefit of having long-term and short-term goals. Pharmaceutical firms will not compete in all categories of drugs but will develop goals and objectives for specific drug categories that will help it succeed at its mission. Thus, a firm may have a long-term goal to dominate an area like sexual dysfunction in males. For a pharmaceutical firm, long term can mean twenty years. To implement this goal, the firm will allocate researchers, equipment, and time to research efforts in this area. The researchers then target specific niches within that broad category where they believe they will have a positive impact. Then they work in those chosen areas and ultimately plan to develop specific drugs in the chosen domain.

Pfizer used this model. The firm had identified sexual dysfunction as a potential growth area they wanted to pursue aggressively. In 1991, the firm patented a heart medicine named Sildenafil. In 1994, one side effect of the drug that was documented in trials—increased blood flow to the penis—began to receive greater attention. This led to a pill form of Sildenafil that was named Viagra. Thus, Pfizer had identified sexual dysfunction as a strategic area they wanted to address. They had people and resources in place to ensure that this goal was reached. These individuals allowed the firm to take advantage of an opportunity from an unexpected domain, a heart medicine. The outcome was the most profitable drug ever developed. Today, Pfizer faces competition from other products like Cialas and Levitar. The result has been a loss in market share from 80 percent in 2000 to 50 percent in 2008. Even more disturbing for Pfizer is the patent for the drug Sildenafil will expire in 2012 and generic drug manufacturers are likely to erode Pfizer's market share further. The firm also had long-term goals, which included expanding the way the product could be used. For example, the company tested a quicker acting nasal version of the product. That did not work as Pfizer had hoped. Now, Pfizer is pushing other products through its developmental pipeline in order to offset the loss of the patent on this very successful product.

Therefore, firms that choose internal innovation need both long-term and short-term goals. These goals and objectives put into action the mission of the firm. The internal innovation process cannot rely on a method that simply waits to see what might happen. Instead, through innovation planning, the firm targets specific areas, and that targeting can lead to success for both short-term and long-term changes.

Set Strategy

Once the organization sets goals and objectives, it needs a strategy to achieve those goals and objectives. There are specific goals and objectives for the different levels of the organization. You will recall from Chapter 2 that in an organization that has many different units the definition of which portfolio of business units the firm wants to compete in defines that firm's corporate strategy. The corporate strategy should be based on the goals and objectives that have been established. Typically, the corporate strategy and the planning associated with it are quite broad.

Thus, each of the individual business units will have a business-level strategy. As the process cascades, it becomes more specific as plans are developed for the operational level or individuals that have specialized functions within a group, such as R&D or marketing. Figure 3.2 summarizes the different levels of strategy and their innovation concerns. The specific strategies for each of these levels will be discussed in greater depth when implementation is examined in Chapter 4.

Specific Tactics and Actions

Once the organization has determined its goals and objectives for each of its different levels, then the actual activities of individuals within the organization necessary to achieve those goals must be specified. Everyone in the firm should be acting in a way that helps the firm reach its goals; its goals should fulfill the mission, and the mission should support the vision. The best set of goals and objectives is useless without a plan of action. It is through operational actions, often called tactical activities, that strategic plans come to life and operational effectiveness is gained.[10] The actions that 3M takes to encourage employees to be innovative (time to explore projects, seed money, etc.) are examples of tactics the firm designed to support a vision of innovation. Figure 3.3 summarizes the steps in the strategic planning process for internal innovation.

Innovation Concerns for Different Levels of Strategy	
Strategic	Coordinate entire organization; Exploit new technologies; Assess external factors; Maximize returns; Determine which technologies and units to emphasize process.
Business	Coordination and implementation of innovation of individual business unit; Manage product and technical innovations; Ensure cost-effectiveness of that unit and relevant technologies.
Operational	Manage change at operational level; Provide training; Assist marketing; Control costs; Supply technical support.

FIGURE **3.2** Levels of Strategy and their Innovation Concerns

Planning

Strategic Vision	Summarizes what the business is about
Mission	Summarizes the firm's vision of itself, its values, and what it values
Goals and Objectives	Establishes how the firm will carry out its mission
Tactics and Actions	Delineates the specific operational activities the firm will undertake to achieve its goals and objectives

FIGURE **3.3** Steps in Planning for Innovation

APPLICATION OF THE PLANNING PROCESS

As noted, the planning process is an effort that builds on itself. The firm cannot determine what strategy to pursue until it is clear where the firm ultimately wants to go. Thus, the planning process cascades, starting very broadly and moving to specific steps that will help the firm accomplish its ultimate goals.

The belief that such a process is critical began with the experience of pioneering firms such as General Electric. The firm started in the 1960s with strategic planning efforts, and that endeavor is, in part, credited for the great success the firm has experienced over the last forty years. Today, the benefits of planning continue to be widely accepted. For example, planning is credited with the ability to turn around IBM. In 1996, IBM was facing serious problems as noted before as customer loyalty from its first mover position deteriorated and the personal computer became the dominant growth engine in the industry. Lou Gerstner decided that the planning process was critical to help the firm turn around. In doing so, he decided that the planning process should have five key characteristics.[11] These were:

- Ensuring that the identification of strategic issues focused outward on economic and technology issues rather than on internal organizational outcomes
- Ensuring that strategic planners focused on the critical implications, risks, and trade-offs inherent in strategic alternatives
- Structuring an ongoing process for top managers to regularly examine fundamental strategic challenges and opportunities
- Connecting strategic planning directly to resource allocation decisions so that implementation issues are considered during the planning process
- Promoting widespread support and involvement in the entire process

The planning process set up by Gerstner led IBM to face the reality of the decline of mainframe computers. As a result, IBM is now a software company led by Samuel Palmisano (an IBM lifer) that follows these five views within the organization and in working with its clients.[12]

- **Hungry for change.** Continually evolve the approach to marketing.
- **Innovative beyond customer imagination.** Market through direct customer collaboration.
- **Globally integrated.** Operate as a global marketing organization.
- **Disruptive by nature.** Encourage renewed and compelling thinking.
- **Genuine, not just generous.** Enable a genuine organization.

When planning started in the 1960s, it typically was a very structured process, and the firm discussed its strategic plan once a year. The outcome usually resulted in a thick business plan. However, too often, that plan tended to be a document that had little impact on the firm's actions. The firm with such a planning process did not focus on the long term; instead, it focused on the crisis of the moment.

Today, the focus is on establishing strategic planning processes that are constant and flexible. The goal is to help the firm identify events that occur in its environment and then to adjust and adapt constantly. For IBM, the outcome of its new approach to planning is an ability to respond, with significant resources, to new opportunities or threats within thirty to ninety days.[13] Clearly, IBM used its 5 views on how to work with clients to change the organization. This flexibility for the firm has continued with the company as it is today—primarily a computer services company rather than a manufacturing firm.

3.1 **REAL WORLD LENS**	**R&D Planning: Ranbaxy**

R&D Planning: Ranbaxy

Ranbaxy Laboratories is an Indian firm that is one of the world's leading producers of medicines for the treatment of tuberculosis and malaria. The largest pharmaceutical firm in India, its research team is made up of more than 350 researchers. However, the firm has established plans to almost double the number of researchers. Previously, Ranbaxy focused on generic drugs that matched the needs of emerging markets. However, the firm has decided to shift its strategic mission to focus on a worldwide pharmaceutical presence. As a result, it knows that it must develop drugs that are branded. It refers to these as new chemical entities. This shift in mission lead by Ranbaxy also led the firm to change its strategic planning process. The firm realizes that to participate in the branded pharmaceuticals market it must reach the world's largest market for such goods: the United States. Therefore, it has built a new research facility that will not only develop new products but will also conduct the human testing required by the FDA. In addition, Ranbaxy had diversified into Europe, Brazil, Russia, and China.

REAL WORLD LENS *(continued)*

In late 2008 the very attractive positioning of Ranbaxy led the Japanese firm Daiichi Sankyo to purchase controlling interest in the Indian firm.

1. What problems do you believe Ranbaxy will encounter because of its change in strategy? How should it plan for such possibilities?

References

Jain, A. 2003. Ranbaxy on a hiring spree, woos US scientist. *Financial Express*. http://www.financialexpress.com/fe_full_story.php?content_id=27629.

Ranbaxy stock analysis. 2003. *India InfoLine. Com* (Jul. 15). http://www.indiainfoline.com/stok/inid/in88.html.

Research and Markets: 2007. The Indian Pharmaceutical market is Predicted to Continue to Grow. *M2 Communications Ltd.* October 9, 2007. http://www.ranbaxy.com/investorinformation/annual_pr2008.aspx

There is no strict formula for how a firm should conduct such a process. Some firms will have frequent meetings, perhaps monthly, in which they discuss changes in the environment, and the actions needed in the firm. Such an approach is particularly appropriate for small technology firms where there is rapid change. In larger technology firms, the large annual planning effort is still typical, but they also need to be sure that there are mechanisms for relatively constant feedback to the organization. In nontechnology firms, whose environmental changes tend to be slower, there still will be the classic annual planning process with little effort to revisit the plan until the next year. For example, in the concrete industry, what will change so much that there is the need to revisit the plan in less than a year? Therefore, the planning process will differ for each firm based on its industry, size, and the type of innovation effort it is attempting.

FACTORS THAT AID INNOVATION PLANNING

As noted at the beginning of the chapter, the management of the innovation process involves a wide range of issues inside the firm. These various activities need to be in place to facilitate and plan for innovation activities. Many of these processes will be examined in greater depth as the implementation of innovation plans is reviewed. However, some of the key areas that managers need to address as they conduct planning will be touched upon briefly. These can be categorized into three broad areas:

1. Creativity
2. Organization-wide issues
3. Political issues

Each will be reviewed in turn.

Creativity

A well-known creative individual such as Dean Kamen, who invented the iBOT wheelchair discussed in Chapter 1, can have a tremendous impact on a firm. For example, Kamen has also had the creativity to invent things like a peritoneal dialysis machine, a new system to treat cutaneous T-cell lymphoma, and the Sterling engine. However, creativity can also occur in teams. The Viagra example earlier in this chapter did not involve one individual but a team of individuals who developed the drug and then recognized other uses for it.

Creativity requires more than simply creating new ideas. From a strategic standpoint, it is vital that individuals not only are creative but that the creative activity within the firm gets translated to the bottom line of the organization. In the 1970s, Xerox held patents on many of the parts that eventually formed the prototype for the personal computer. However, Xerox was focused on photocopying technology and failed to recognize the potential of the creative genius of its personnel at the Palo Alto Research Center (PARC) at least in the case of the personal computer. However, Xerox's PARC is still considered the granddaddy of high-tech research centers having produced such products as Ethernet and laser printing.[14]

Organizations can ensure that creativity is continuous and not sporadic by utilizing certain mechanisms that encourage creativity. Figure 3.4 shows a partial list of tools and techniques that support creative thinking. These tools include:[15]

1. Adventuring
2. Confronting
3. Portfolio of skills

Adventuring	Confronting	Portfolio of Skills
Encouraging employees to try new ideas	Focus on building new skills in employees	Have employees ask "what if"
Experimenting	Encourage cross-functional team activities	Encourage devil's advocates
Making mistakes		Encourage debates by employees

FIGURE **3.4** Innovation Requires Perceptual Challenging

Source: *Perpetual Challenging and Its Components*, "Enhancing Organizational Creativity: The Process of Perpetual Challenging," Constantine Andriopoulos and Andy Lowe. *Management Decision* (2000, Vol. 38, No. 10). © MCB University Press. http://www.emeraldinsight.com/md.htm. Republished with permission, Emerald Group Publishing Limited.

Adventuring

Adventuring occurs when individuals explore areas that are outside their comfort zone and perhaps even that of the organization. Adventuring should have as its goal the generation of new ideas to deal with uncertainty. Experimentation is the most commonly known method of adventuring; however, introspection (working from what is already known) and scenario making (thinking of various paths that innovations take) are other methods of looking for new ideas. The previous example of 3M allowing employees to spend some of their time on projects of their choice is an example of such adventuring. Samsung of Korea does something similar when it sends young, fast-track managers to different parts of the world with the simple instruction to identify new opportunities and learn about that culture so that the firm can learn how to operate better in the specified foreign country.

Confronting

The second way of encouraging innovation is confronting, which is a process that encourages deliberate debates among employees through such techniques as devil's advocate or "what if" questioning. These debates can be conceptual or contextual. An example of a conceptual confrontation is asking the question, "What if a printer is assembled from the side rather than the top?" This question led to a breakthrough in manufacturing process design for printer manufacturers in recent years.

Contextual confrontation is more familiar to people. When faced with a deadline, people tend to be more focused and creative. An obstacle, such as a firm deadline, stimulates innovation as individuals narrow the focus to what needs to be solved now. Another contextual confrontation occurs at conferences where individuals come together to discuss new ideas or products. 3M helps to create such situations by sponsoring conferences, both internal and external to the firm, where new innovative ideas are presented. These conferences encourage the debate and discussion of ideas that confront the established perspective on a given topic.

Portfolio of Skills

The third method for encouraging innovation is through the creation of a portfolio of skills. In this situation, creative employees are encouraged to stretch beyond their normal boundaries into new domains. Thus, while confronting asks employees to enter into debates based on skills they already have, this method encourages creativity by asking employees to learn new skills. This process can lead to an exchange of ideas that might not occur normally. The biggest disadvantage is the danger of creating too diverse a portfolio of skills for any given individual. The use of cross-functional teams represents an effort by 3M and GlaxoSmithKline to build such portfolios. The firm often takes individuals from a variety of disciplines and places them on teams when exploring new domains. In these settings, not only does each team member bring his or her own set of perspectives to the process, but other team members gain fresh insights and new skills that they take back to their various departments.

Organization-wide Issues

Organization-wide issues bring together individual cognitive efforts of several people in the firm. As stressed throughout this book, although there are situations where brilliant individuals develop significant new innovations, in most organizations there are multiple individuals involved in the process. The organizational processes that connect and encourage connections among these individuals are critical to the nature of the innovation planning process. The organizational processes are multidimensional and include:

1. Communication
2. Reward systems
3. Organizational assumptions

Communication

Knowledge management involves "spreading the word" through communication channels about both the needs and the opportunities for the organization. These needs and opportunities should then be integrated into the planning process. To show the impact of communication, consider the prior example of the development of the Post-it Note—communication about an adhesive that had no apparent use led to the development of the product.

Good communication not only makes individuals in the planning process aware of needs and opportunities; it also helps ensure that the organization is working toward a common direction as it performs innovation planning. It is easy for one part of the organization to believe it understands the problem that the organization should address in its planning, while another part of the organization is building a plan that sees the problem from a very different perspective.

The sharing of knowledge can be both formal and informal. Formal communication is required by the organization and includes such things as posting on internal discussion boards or through e-mailing major points of a staff meeting or a weekly report to all key personnel. The organizational structure typically indicates the lines of formal communication. The firm should develop those formal communication processes that ensure everyone in the organization receives the necessary formal information, although managers do not want to bury employees in unnecessary information. Thus, in designing formal information channels the firm wants to design efficient structures and processes that get timely, correct information to appropriate personnel.

Informal communication takes the form of e-mails, phone calls, and face-to-face visits that managers take upon themselves to do. The organization should keep the reality of informal communication in mind and develop mechanisms that encourage building rich networks among individuals across the organization. One benefit of training and development, which draws on individuals from a wide variety of departments and units, is the development of such informal networks. Company softball teams or bowling teams also help build informal networks within the organization.

While it is important to encourage formal and informal knowledge sharing, the organization should be cautious that it does not have an overload of information. The organization should encourage both formal and informal communication while limiting that which clogs the system. Periodically, the organization needs to evaluate the flow of information to be sure that communication (particularly formal) is getting the right information to the right people at the right time to make timely plans and take timely actions.

Reward Systems

Organizations must plan reward systems for individuals and groups that develop and sustain the internal innovation processes. If specific innovations are desired in the organization, individuals must be rewarded for that activity. Too often, managers of an organization reward A, while hoping for B.[16] If a firm desires something specific, individuals must be rewarded for achieving that goal.

Reward structures will be discussed in more detail in the next chapter when implementation of the plan is discussed. At this stage, it suffices to say that as plans for implementing an innovation strategy are developed, the manager must be aware of the role of rewards and not create an internal innovation strategy that the reward structure does not support.

Organizational Assumptions

The assumptions used in the planning process and through the innovation process need to be monitored periodically to be sure that the base assumptions are still true. There should be planned reviews. These will be discussed in more detail in Chapter 5. Toyota is typically considered a great example of a firm that uses MTI effectively. The safety issues that Toyota is facing from 2009 and 2010 clearly illustrate that even admired systems and processes need evaluation and adjustment.[17] There were basic planning assumptions that should have been challenged in Toyota but were not. More about how to develop and monitor the organization processes necessary to avoid such problems using an internal innovation strategy will be presented in Chapters 4 and 5.

Political Processes

Finally, innovation planning is also impacted by the political nature of organizations. Innovation implies change. Change is often seen as threatening to the social infrastructure of an organization. In addition, information is a source of power within an organization. Because intelligence gathered and processed about trends and potential technological challenges becomes information, power is potentially created. The result is that innovation can lead to changes in the power structures and thus the politics in the firm. To illustrate, if a new technology emerges, sales personnel may lose power while engineers and production personnel gain power. Thus, withholding vital information may be empowering for the individual on a short-term basis but may be detrimental to the long-term success of the organization. These shifts are caused by changes outside of the organization but can have a very real impact on how innovation planning occurs in the organization.

However, political actions can be used to benefit the organization. A leader can use power and politics to move the organization in a new direction. Some managers, such as Samuel Plamisano at IBM, and Bill Gates of

Microsoft are known for this ability. Both of these individuals used their power and political clout to make radical changes in their organization's products and processes. These individuals were strong willed and highly visible leaders both in their firms and in the wider public. They used this stature to argue forcibly for their views.

TECHNOLOGY STAGES AND PLANNING

Technology goes through an evolutionary process. The evolutionary stages can simply be seen as start-up, growth, maturity, and aging or decline (these are similar to the S curve stages discussed in Chapter 2). For the firm, the natural progression of the technology along its life cycle requires adjustments in planning during each stage. The division between the end of one stage and the beginning of another is not always clear. However, the manager should be aware of such stages and have planning processes consistent with the life cycle of the technology and the major product(s). Each of the four stages is described next as well as how the planning for innovative activities differs at each stage.

Start-up

Figure 3.5 shows how the focus, information needed, communication, firm power sources, and types of innovation change during the life cycle. For example,

	Start-up	Growth	Maturity	Decline
Focus	Broad	Potential competitors; Monitor	Problem solving; Improvement	Avoiding losses
Information Needed	Science trends; Breakthroughs; Developments	Trends; Competitors' actions	Competitors' actions; Other uses of technology	Market sustainability; Competitors' actions
Communication	Opportunity generation; Informational	Customer/supplier critical	Production processes for efficiency; Trends	Cost/benefit; Trends
Power Sources	Contacts with external groups; Knowledge	Marketing; Procurement	R&D, production; Marketing	Sales; Finance
Key Personnel	Boundary spanners; Technical personnel	Sales & procurement; Middle managers	Production; Marketing; Innovative thinkers	Sales; Cost experts; Strategic decision makers
Strategic Actions	R&D; New product and market development	New product and market development	Restructuring; Market expansion	Reengineering— looking for renewal; Value destruction

FIGURE **3.5** Technology Stage and Contributing Processes

at the beginning of a new technology, basic and applied research are needed. The new scientific trends, breakthroughs, and developments must be closely monitored. At start-up, the firm is seeking an opportunity that will give it some type of sustainable competitive advantage or at least help the organization maintain its competitive position.

An important aspect of planning for innovation is understanding where the power and communication needs are during the various stages. At start-up, external groups are very important. The firm is looking for opportunities in the environment, and contacts with external groups are critical to identify those opportunities and understand how to take advantage of them. Today, firms in the nanotechnology industry spend much of their time and energy externally. They must educate both potential investors and potential users of their products about the promise in the technology. The strategic actions for this stage focus on R&D activities and new product development.

Growth

During the growth stage, the firm consolidates the innovation it has generated as the industry standard as some product designs become dominant. At this stage, some consolidation in the industry begins as many of the start-ups either fail or are purchased by other firms. The organization still interacts with the external environment to establish that its technology is the best. At the same time, the firm focuses internally to ensure that its structures and processes are consistent with its growth. The firm's emphasis in innovation planning shifts to product improvement or adjustment during this stage. This phase of technology requires that the firm improve its product and its production. If the firm cannot establish its efficiency, it will not survive the initial stages of the consolidation.

To illustrate this, one can look to broadband firms. Firms today spend much of their time communicating the need for this technology and its benefits. However, at the same time, the broadband industry is established enough so that price sensitivity and price competition are beginning to occur. Therefore, firms are moving from their external focus to one that is more concentrated on internal efficiency. The result of this greater focus on efficiency in the industry is that the planning process similarly needs to focus on these issues. In the growth stage, the firm is larger and has a greater concern for efficiency. Process innovation becomes more important as the firm positions itself for the mature stage of the technology.

Maturity

As the technology or product moves into the maturity stage, the organizational emphasis moves even more to an efficiency focus. Thus, process innovation planning becomes more important. The consolidation in the industry will continue. Thus, in the PC industry (a more mature industry than broadband), firms were growing but at a much slower rate in the period around 2005. Dell dominated the PC industry because it was the most efficient company, and its direct marketing process was still the most innovative. However, even now Dell is having struggles. The PC industry is struggling as consumers are not updating as often and the new technology of the iPhone and Blackberry is replacing some of the most common uses. In the PC industry now, firms such as Acer are gaining market share with small, lightweight netbooks, while firms such as Dell are struggling to keep pace.

The flow of organizational communication is focused more internally in maturity than in an early-stage industry. There is only limited need in the PC industry to communicate about the product and how it is used. Communication remains critical, but the focus lies with the boundary spanners within the organization—especially those in sales and purchasing. In the planning process, these two domains gain even greater importance.

Aging or Decline

At the end of a technology's life cycle, the focus becomes avoiding losses, and if the organization decides to maintain a position with the technology, then the emphasis in innovation is systems integration and process efficiency. In other words, the firm is looking for synergies to support continuation of the technology.

Therefore, during the decline stage, the internal structures and processes of the firm become even more important. The control of costs is the overriding emphasis. As a result, finance and sales represent the power sources for the firm and will receive a greater focus in the planning effort. The commodity microchip industry clearly illustrates this phase of the life cycle. If the firm decides to leave the technology, it must plan its exit and the destruction or reallocation of the value the firm has in the technology.

3.2

REAL

WORLD

LENS

Government's Role in Innovation

In many international settings, the government plays a critical role in research and development planning. For example, in Korea, there is the national R&D Ministry of Science and Technology. This ministry plays an active role in targeting specific areas of research and development in the nation. Much of the research in other nations such as China follows this type of model. In the United States, the government plays a critical role, providing 30 to 40 percent of the annual expenditures for research and development. However, the model in the United States has been to develop cooperative ventures with the government, education, and/or business as partners. Although there are some government laboratories, most of the expenditures are grants provided to academic institutions and to businesses.

1. How else can government policy affect the ability of the organization to pursue an innovation strategy?

DEVELOPING A CLIMATE FOR INNOVATION

3M is an example of a firm with a variety of programs that promote innovation. However, other firms have tried similar programs and have not been successful with their innovation strategy. In part, other firms have failed because they did not have a climate that supports innovation. If the climate is not suitable, the initial efforts to plan for the innovation will not succeed nor will the later efforts at implementation.

This climate for planning is more likely to develop if the manager keeps in mind several myths about innovation that are widely held but not true. These myths need to be recognized initially if the climate for innovation is to be established. These myths are summarized in Figure 3.6.

One of the most commonly held myths is that ideas are the key to innovation. Ideas are important, but the reality is that there are many brilliant ideas that are never discussed or even presented. Many individuals learn early in their life in the organization not to discuss ideas that sound too different. The individual fears ridicule or rejection; so an excellent idea may go unexplored. In addition, past experience and expectations may lead managers to reject a good idea that is presented for some basic reason such as, "We tried that five years ago and it didn't work," or "We've never done it that way before."

A climate in the organization that encourages the presentation of new ideas and potential innovations must be encouraged if planning for innovation is to succeed. This climate is particularly important in the planning process because it is at this stage that new approaches are typically discussed and integrated into the organization's efforts.

Another myth is that the innovation has a single critical moment. The "eureka" moment with big product discovery is limited. Instead, innovation and the planning for innovation are a long-term process that helps the organization maintain focus while changing and adapting as necessary. For example, flat-panel displays that are so common in large televisions and notebook

Myth	Reality
• Ideas are the engine for innovation. • A good process generates all the innovations needed. • If we have the next big idea, we will be successful. • Through innovation, we can grow our way to prosperity. • A good evaluation method will eliminate bad ideas. • The entrepreneurial firm is the one that will be most successful in the long run.	• There are lots of brilliant ideas that you've never heard of. • Leadership, culture, and process form a three-legged stool. • Develop the discipline to grow ideas. • Objectively analyze opportunities—grow when you are ready. • Don't blind yourself to the human dimension when evaluating ideas. • Seek a balance of behaviors—entrepreneurship is not always the right strategy.

FIGURE **3.6** Myths about Innovation

computers took more than twenty years to develop from discovery of the process to large-scale production of large-screen products. Several factors caused problems and delays in the widespread adoption of the technology. First, making a larger version of the screen than was initially developed proved very difficult to mass-produce. The waves in the screen became prominent as screen size increased, which resulted in the images blurring. This caused general dissatisfaction by consumers. Second, the specialized glass components became extremely fragile when the size was increased. At one time, it was estimated that manufacturer breakage of the glass in the screens was 50 percent. This type of breakage led to the third problem—excessive cost for the final product.[18]

During the 1990s, several major breakthroughs occurred that allowed the development of the larger screens. In the meantime, companies in this industry found small-screen applications that led to profitability. Some of these applications were hand-held devices such as palm cellular telephones, calculators, electronic games, radar screens, and global positioning devices. It took systematic champions who ensured the planning process continued to focus on flat screens and their potential that led ultimately to the large-screen technology. Without these champions, the effort may have been abandoned, and digital, large-screen television would still be a product of the future. Thus, the planning process is not a one-time effort that produces all the answers. Instead, it is a long process where plans are made, changed, and then adapted again. Planning must be an active, living process that incorporates many views and then makes adjustments.

Another myth is that only big ideas are useful to focus on for innovation. Instead, the planning process should include a full range of ideas and efforts. As noted, 3M created Post-it Notes, a simple idea about little notes with sticky material that allow them to adhere to other items without leaving a residue. This product has been very profitable for the firm. The most profitable option may be a small idea or system integration, not a revolutionary change. The planning process needs to be sensitive to this fact and not overlook the small issues.

SUMMARY

This chapter has laid the foundation for the exploration of internal innovation by examining the planning that must underlie such internal efforts. The chapter laid this foundation by helping determine if an internal or external technology focus is most appropriate for the given situation. The chapter then specified different types of innovation. The steps in the planning process were outlined as establishing the vision of the firm, its mission, goals and objectives, strategies, and tactics. How these planning steps need to consider the stage of technology development was also discussed. The means to ensure that a creative process is employed in such planning by the technology-focused firm concluded the chapter.

MANAGERIAL GUIDELINES

To plan for innovation, managers need to follow several guidelines. These guidelines will help managers develop the right climate for success in innovation activities whether the innovation activities are aimed at product or process innovation.

1. Recognize that innovative individuals exist throughout the organization. More ideas flowing through the organization increases the chances of organizational breakthrough. The involvement of a wide range of individuals in the planning process helps to reap the benefits of these innovative individuals.
2. Periodically reconsider how work is being accomplished. Too often, the traditional ways of doing things hurt the emergence of new ideas. Be sure that your processes allow traditions to be examined also.
3. Encourage "turning the prism" to put a new light on problems. Discontinuous thinking fosters new models and paradigms in planning.
4. Ask employees what their biggest aggravations are. These aggravations may be little things, but they may be blocking good ideas in the planning process.
5. Train people to recognize their creative abilities and participate in the planning innovation process. Creativity is like height and weight—everybody has some. Managers need to provide an environment that helps individuals use their innovation for the benefit of the organization.

Guiding Questions

To help the manager establish the innovation planning process, the following checklist should be employed to help ensure the process achieves maximum benefit.

1. As the planning process begins, have the firms' current activities been examined to determine whether they are providing leverage for the firm to maintain a sustainable competitive advantage?
2. In the planning process, are concepts being developed that provide an immediate measurable and discernible competitive advantage?
3. Does the process of innovation add to the firm's value for shareholders and other stakeholders? How? If not, can the planning process be changed to ensure that it does?
4. Do employees feel encouraged about their importance in the innovation activities of the firm, particularly the planning process?
5. Have the leaders of innovation been identified at all levels of the organization? Are these individuals involved in the planning process?
6. Does the result of the planning process indicate what needs to be done next, who needs to do it, and when it needs to be done?
7. How are "newness" and "failure" handled by the firm?

CASE **3.1** THE REAL WORLD

Innovation Planning: Corning

Corning Inc. is well recognized in the United States because it is more than 150 years old. There are very few large American firms that are this age. The key reason that Corning has been able to survive this long is its internal innovation strategies and processes.

The company was founded in 1851 by Amory Houghton Sr. Prior to starting the glass company, Houghton had a varied career from carpenter to trading goods on the docks at Cambridge, Massachusetts.

(continues)

CASE **3.1** (*continued*)

With this entrepreneurial background, he founded the glass company that became Corning Glass. Today, one of the continuing core values of Corning is the entrepreneurial spirit of Houghton. This value continues to influence the firm as Corning constantly seeks out new products to manufacture.

Houghton was an innovator, and this focus continued in Corning after his death. In 1908, Corning established one of the first industrial laboratories in the United States. This early commitment to internal innovation through research and development has produced a number of products that established new domains for Corning. One of the most widely recognized innovations is optical fiber. In 2000, the firm won the President's National Medal of Technology for the development of this innovation. This award is given for significant contribution to the technological life of the country.

Corning Today

Today, Corning is a publicly traded firm with revenues of approximately $3 billion. The firm has two principal divisions: technology segment (53 percent of revenues) and telecommunications (46 percent).

The technology segment is the direct descendant of the firm's initial expertise in glass. For example, Corning recently used its internal research capabilities to develop products like the active matrix crystal display glass used in flat-panel displays for notebook computers and televisions. The firm also has developed ceramic technologies that are used in environmental products for pollution control such as diesel substrate and filters. In addition, Corning continues to manufacture the glass panels for cathode ray televisions.

The expansion into telecommunications is built on the firm's development of fiber optics. However, Corning has expanded from its dominant position in fiber-optic cable to produce the hardware equipment for the telecommunications industry. This includes cable assemblers, optical couplers, splice equipment, and test equipment.

Building an Internal Innovation Foundation

The base for this extensive set of products is Corning's internal research and development efforts. The firm spends approximately 10 percent of its revenues on research and development. Between 1995 and 2000, the firm deepened its commitment to internal development by increasing the number of research and development personnel by 67 percent, to more than 1,500 individuals.

In addition to hiring the individuals to conduct the research and to develop new products, the firm has a culture and process that support innovation. The first part of that process is planning for innovation. Corning has identified three broad areas to focus its future growth on: ceramics, optical fibers, and photonic parts. Strategically, the firm has targeted specific areas in each of these three domains where it wants to

CASE 3.1 *(continued)*

pursue new product innovation. One means it uses to decide what areas to target is offsite meetings with top line and technical managers. These meetings occur every four to six weeks and include discussions about the relevant markets and what actions the firm should take. The firm also interacts extensively with its customers to ensure that it is identifying key movements and product needs in its planning process.

In implementing the innovations that are planned, Corning has developed a unified and systematic approach. The firm uses cross-functional teams that include scientists, engineers, marketing specialists, and others from key domains in the business to work on innovations. The firm also encourages cross-fertilization of ideas through periodic "Growth Days" when different products are showcased. In these settings, a wide variety of individuals gather, listen to a presentation, and learn about and comment on new products and processes that are presented. The firm also has policies that support innovation. For example, an employee bonus can be up to 200 percent of base salary depending on performance and the nature of the contribution.

Corning generates far more ideas and products than it can pursue during any given time. The result is a need for a process to evaluate the different ideas and products so that judgments can be made on which innovations to initiate support to or to continue support of. The ability to justify the product at each step of the process is critical for the team promoting it. The evaluation system is designed to be flexible as well as to avoid the continuation of projects that are not meeting expectations.

Corning's Efforts Bring Results

The result of these innovation efforts is that in the last decade Corning has received over 50 percent of its revenue from products that had been in existence fewer than five years. It is worth noting that Corning's extensive internal development efforts often lead to innovations that do not fit with the company's primary focus.

Now Corning is tapping its ample budget for research and development to turn simple sand into a succession of big products, from heat-resistant glass for railroad lanterns and CorningWare ceramics to optical fiber and LCD screens. Now, even as other manufacturers are pulling back on R&D, *Corning* is pushing ahead to find the next product. The staff at its R&D facility in the Silicon Valley is zeroing in on three areas: improving high-speed communications between computers using optical fiber, adding solar power to handheld devices, and developing better displays for smartphones and laptops.

1. How did Corning address the issues presented in this chapter?
2. What advice would you give to Corning about planning for innovation?
3. If you were a Corning competitor, what would worry you most? How would you compete with them?

(continues)

CASE **3.1** *(continued)*

References

Kotelnikov, V. Business e-coach. Case Study—Corning: Managing innovation through in-company ventures. http://www.1000ventures.com/business_guide/im_internal_startups_cs_corning.html.

McConnon, A. 2008, Corning R&D hits the road. *Business Week*, May 25 86.

Stigson, Bjorn, and Gilson, Jean-Marc. How innovation supports sustainability. 2004. *Chemical Week*, 166 (20): 16–.

Anthony, Scott, and Fuson, Scott. Instilling a culture of innovation. 2004. *Chemical Week*, 166 (32): 16.

CRITICAL THINKING

Relating to Your World

1. In developing an organizational climate that supports innovation, management often overlooks critical issues. This is true especially when setting up and nurturing communication networks. Many good ideas are lost because the right person is not available at the right place at the right time to bring the innovation to fruition. What issues do you believe are critical in planning for the integration and sharing of information in an innovative firm? How do these issues differ for product and process innovations?

2. In this chapter, we discussed several companies and how they have created climates for innovation. At each level of the organization (top, middle, lower), what do you think are the critical issues? What are the potential advantages of an innovative strategy for individuals at each level? What should be the biggest fear at each level?

3. You have done a number of innovative things in your life. Think about things that you have used for something besides its intended purpose. List some of these on a sheet of paper. What can managers do to help individuals capture that same feeling in the work environment?

WWW EXERCISES

1. Use your favorite Internet search engine and find an example of a successful innovation and one that was not so successful. (Finding failures is difficult.) What were the reasons given for success? For failure? Which of these reasons relate to poor planning? What does this tell you about the keys to success in innovation?

2. Do you believe creativity can be learned? Find an article or website that is devoted to capturing creativity in the work environment. What does the author say about individual creativity and group creativity?

3. Find an article or website that provides guidelines for developing an innovative organization. What do you think of the advice given? Compare the advice you find to the advice your classmates find.

AUDIT EXERCISE

In developing plans for an innovative strategy, the top management of the organization must ensure that the strategy helps the organization add value for customers and shareholders plus maintains or builds a competitive advantage for the firm. Because of the uncertainty of most innovative products and/or processes, managers need to develop a systematic approach to assess how time, effort, and resources should be spent and which opportunities will be pursued. McGrath and MacMillan[19] presented the following list of factors for determining the value of an option.

1. Favorability of the demand for the product
2. Factors that could speed adoption
3. Factors that could block the success of an innovation
4. Likelihood of strong competitive response
5. Likelihood that the potential competitive advantage is sustainable
6. Factors within your organization that would allow you to set standards
7. Cost factors in commercialization
8. Resources available for commercialization
9. The level of novelty the innovation captures
10. Cost considerations in development
11. Other opportunities that could be leveraged
12. Potential area where damage might occur

What are the key questions that need to be addressed under each area? What areas covered in this chapter do you believe are most relevant for each of these factors? Explain your answer.

DISCUSSION QUESTIONS

1. What are the key resources that must be managed in planning for innovation? How is the mix of the resources different for product and process innovation?
2. What factors favor innovation as a strategic activity? What are the possible hindrances to pursuing such a strategy?
3. Technology usually goes through a four-stage process. How is innovation planning different for each of these stages?
4. Discuss the difference between newness creation and process changes in the innovation arena. Which do you think is more difficult and why?
5. The discussion of how different governments are involved in the innovation process is a reflection of the culture of those countries. How do you believe the culture of the country and its view of innovation influence the organization's climate for innovation? What should managers take into consideration when planning for innovation based on these factors?
6. How are the innovation processes used at Corning similar to those used at Ranbaxy? How are they different?

PART TWO OPENING CASE: GLAXOSMITHKLINE

1. What are the special planning needs for GSK?
2. What industry trends should a firm like GSK consider in its planning processes?

KEY TERMS

applied research 80

basic research 80

downsizing 83

fast follower 78

first mover 78

mission 86

process innovation 83

reengineering 84

strategic group map 82

switching costs 78

systems integration 81

tweaking 81

vision 86

NOTES

1. 3M's 2008 10-K Annual Report
2. Chandler, A. 1990. The enduring logic of industrial success. *Harvard Business Review* (Mar.–Apr.): 130–140.
3. Christensen, C. 1997. *The Innovator's Dilemma: When New Technologies Cause Great Firms to Fail.* Boston: Harvard Business School Press.
4. Anonymous. 2008. LeapFrog Enterprises, Inc.; LeapFrog Announces First Quarter 2008 Financial Results. *Business & Finance Week.* May 19. 248.
5. Cameron, K., S.Freeman, and A.Mishra. 1991. Best practices in white-collar downsizing: Managing contradictions. *Academy of Management Executive*, 5: 57–73.
6. DuBrin, A. 1995. *Re-engineering Survival Guide.* Cincinnati, OH: Thomson Executive Press.
7. Bennis, W., and M.Mische. 1995. *The 21st Century Organization.* San Francisco: Jossey-Bass.
8. _____, Art Fry and Spencer Silver. Post-it® notes. http://web.mit.edu/invent/iow/frysilver.html.
9. Google 2008 Annual Report.
10. Porter, M. 1996. What is strategy? *Harvard Business Review* (Nov.–Dec.): 61–78.
11. Garr, D. 2000. *IBM Redux: Lou Gerstner and the Business Turnaround of the Decade.* New York: Harper Business.
12. Hennessy, M. (2008). The Enterprise Of the Future. *Research Technology Management*, 51(5), 7–8.
13. Steinberg, N. 2002. IBM's continuous strategic evolution: Constant planning for rapid innovation. http://www.itsma.com/research/abstracts/CS0002 .htm.
14. Bsales, J. and C.Metz. 2007. Today's ideas, Tomorrow's tech: We go behind the scenes at the biggest names in high-tech research, uncovering five projects clever enough to reinvent modern computing. *PC Magazine*, 26(14), 67–72.
15. Andriopoulos, C., and A.Lowe. 2000. Enhancing organizational creativity: The process of perpetual challenging. *Management Decision*, 38 (10): 734–749.
16. Kerr, S. 1995. On the folly of rewarding A, while hoping for B. *Academy of Management Executive*, 9 (1): 7–14 (reprint).
17. Hlavacek, J., C.Maxwell, and J.Williams. 2009. Learn from new product failures. *Research Technology Management*, (Jul.-Aug.): 31–39.
18. Murtha, T., S.Lenway, and J.Hart. 2002. *Managing New Industry Creation: Global Knowledge Formation and Entrepreneurship in High Technology.* Palo Alto, CA: Stanford University Press.
19. McGrath, R., and I.MacMillan. 2000. Assessing technology projects using real options reasoning. *Research Technology Management*, 43 (4): 35–49.

CHAPTER 4

Implementation in Innovation

OVERVIEW

This chapter continues the discussion on innovation by addressing the issues surrounding implementation of the plans discussed in Chapter 3. Once the foundation for internal innovation is laid and plans have been made, it then takes all the employees in the organization, to achieve success. The process of putting new ideas to work is the implementation of innovation. The specific issues addressed in this chapter are:

- Leadership's role in bolstering innovation
- How key personnel engage in the innovation process
- How the organization achieves fit during implementation
- How employees are encouraged to buy into the process of innovation

_____

INTRODUCTION

Once the firm has laid the foundation for an innovation strategy through the planning process, implementation becomes the concern. Implementation efforts must occur whether the goal is a small change in a product or a radical shift in an entire industry. Consistent with the strategic process view of this text, if managers plan well and properly implement the plan, the firm should have successful outcomes. The examination of implementation in this chapter concerns how firms should organize systems, structures, people, and processes into a consistent, synergistic whole to achieve innovative capacity, profitability, and long-term sustainable competitive advantage.

KEY INITIAL QUESTIONS FOR IMPLEMENTATION

The key initial questions in implementation are:

1. What should we be doing now, and what can we do later?
2. What are the time and/or specialized skills required for the prioritized activities?
3. What should be delegated and to whom?

The answers to these questions should emerge from the planning process and provide the foundation for implementation. These answers form the basis for evaluation and control, which will be explored in the next chapter (see Figure 4.1). We now turn to a discussion of each of these questions.

FIGURE **4.1** Key Questions in Implementation

What Should We Be Doing?

A central element in the implementation of an innovation strategy is prioritizing the actions necessary to carry out the strategy. The organization will establish its goals and objectives in its planning process. However, these goals and objectives are not all equal and the manager should prioritize them during the planning process. In the implementation phase, the organization needs to make choices on how to act on these goals and objectives. Some goals and objectives require more effort to achieve, while some goals and objectives are more important than others. There will also be a variety of means to achieve the different goals and objectives, and as a result managers must also determine the best means to obtain the desired results. For example, if the firm has a goal to increase new product sales by some predetermined percentage, the firm can increase demand through advertising or developing new uses for the product. The firm will need to determine which path, or combination of paths, it will pursue to meet its goal of increasing sales.

These various choices will be more effectively and efficiently accomplished if the firm is clear about what actions are critical today, what needs to be done later, and what can be ignored for now. This prioritization should be based on the strategic choices important to the success of the innovation strategy, and timing associated with the various choices.

Prioritizing activities is not done in isolation with an assumption that the implementation choices that are made will never change. Instead, the organization must adapt its prioritization as the environment changes and as the technology moves through various stages. The environmental change may be external or internal to the firm. For example, if a firm is dependent on an individual with a key skill or relationship and that individual leaves the firm, the implementation efforts of the firm may need to change.

Similarly, if external situations change, the firm may also need to change its implementation efforts. The energy price fluctuations in recent years have brought dramatic changes in the R&D efforts of many auto manufacturers. For example, the price escalation in energy has increased the importance of hybrid cars. Prior to the energy price escalation there was some interest in alternative cars, including hybrid cars, solar cars, and hydrogen-fueled cars. However, the expected time frame for such products to be commercially viable was thought to still be far in the future. However, this time frame has changed with the increase in fuel prices and concerns about sustainability because of CO_2 emissions.

To illustrate, the first hybrid SUV that relied on both battery fuel cells and a traditional internal combustion engine was introduced in 2004 by Honda. Other automakers thought Honda had the wrong focus, and instead focused on alternatively powered vehicles—namely solar energy. For example, GM sponsors an annual engineering competition for solar-powered automobiles, but the practical use of solar power may still be years away, if ever commercially viable. However, the rise in energy prices and the concern for green house gases have allowed companies with hybrid cars to be significantly ahead in the marketplace of those firms that focused on solar cars. The demand by consumers is for something that is immediately available, not years in the future.

The growth in demand of the hybrid car is such that in September of 2009, most of the world's major car manufacturers (including Toyota, GM, Ford, Nissan, and Daimler to name a few) issued a joint letter of understanding about the development and market introduction of fuel cell powered, totally electric vehicles. These automobile manufacturers believe that 2015 will mark the point in time where fuel cell powered, totally electric vehicles will outsell traditional gas-powered vehicles. Because each manufacturer will implement its own production and commercial strategies, it is likely that fully electric vehicles powered by fuel cells will happen before 2015 for some manufacturers. Thus today firms are seeking to move beyond hybrid cars to the development of low to zero emission vehicles utilizing hydrogen fuel cells.[1]

While the automobile manufacturers have signed a letter of understanding, that does not mean they do not want to "**leapfrog**" each other. In 2009, Toyota and Honda led in fuel cell technology. However, the other manufacturers are trying to develop new technology that leapfrogs, or skips over the existing generation of products, to introduce a product with significantly new technology. This leapfrogging is similar to what occurred in emergent markets with cell phones. In these markets cell phones were able to leapfrog over the use of landline telephones. A nation with an emerging market economy may never be able to develop extensive landline systems for telephones. Instead, these economies leapfrog by moving directly to cell phone use. The use of cell phones in most emerging economies has, in turn, allowed a variety of other leapfrogging opportunities. For example, the banking for most people in emerging economies in Africa occurs over the cell phone not in traditional buildings.[2]

Therefore, firms must adapt their prioritization of activities as key conditions change. Just as goals must change as situations change, the implementation of a strategy must also adapt. The firm that is unwilling to adapt can find itself in a difficult situation.

What Is the Time and/or Specialized Skill Requirements for Key Activities?

Once the various actions are prioritized, the second question the firm must answer is what are the resources needed to successfully complete these activities? Organizations do not have unlimited resources for the implementation process. Activities higher on the prioritization list should receive the resources. Thus, rather than trying to make everyone in the organization happy by giving them some resources, the organization must target what is critical and focus its efforts on those areas. There are several areas to focus upon as the firm sets priorities about the firm's key activities. These concerns include timing, human resources, and the effective use of existing platforms.

Timing

Once the organization understands the available resources to complete the prioritized activity, it should ensure that those resources are available when and where they are needed by the given projects. Conflict often arises because managers believe their individual projects are the most important, and they may be unwilling to shift personnel and other resources to another task that is part of a strategic goal with a higher priority. This may be particularly true in situations where the manager must meet given performance standards, and the shifting of resources makes meeting those performance objectives more difficult.

Human Resources

Most resources can be purchased by a firm when needed. But human resources are typically developed internally by the organization over time. To have the right number of employees with the desired skills requires the firm to hire today for a task that is required in two or three years. In addition, to maintain the value of that human resource, the company must continue to educate the employee. To illustrate, an engineer may be hired right out of college. Usually, there is a one- to two-year training program initially aimed at teaching the engineer "the business" of the company. With each passing month, the engineer gains value for the firm, but it may be at least a year or two after the engineer is hired before he or she starts "paying off" where the engineer generates more value than costs. If engineers in some fields do not continue to study new developments (usually with company support), the engineers can lose technical competence quickly and become a cost to the firm. Thus, activities such as training and development, and tuition reimbursement, are processes to help the firm protect the value and availability of its human resources. An example from a firm that may not typically be thought of as a business with innovative practices is Mike's Carwash. Mike's has 37 locations in Indiana and Ohio. Remember, innovation has simply to be new to the business/industry. Thus, innovation can occur in many industries that we do not typically associate with the management of technology and innovation. Mike's has earned a reputation for stellar service by rigorously interviewing entry-level job candidates and then reducing turnover through thorough training and great incentives. About two-thirds of the company's employees work part time, but they stay much longer than average for the industry. Because turnover hurts profits and service, the management has developed training programs (including a 10-minute customer service refresher video each week) and provides a tuition reimbursement program that allows workers to pursue other long-term career interests. This longer tenure of employees allows the firm to better decide who would make good managers for each of its locations. All of these processes are innovative and relatively unique in the carwash industry. They have given Mike's a competitive advantage.[3]

The Effective Use of Existing Platforms

When ensuring that the resources necessary to complete a goal or objective are present, the firm does not start from scratch. Instead, firms can leverage existing technology platforms as a means to save resources. For existing products the firm has technological expertise, relationships, and knowledge of distribution channels in place. The firm may be able to apply these resources to more than one product or process. We call these complementary when synergies can be found. The topic of complementary technologies will be examined more closely in Chapter 7, when the implementation of technology acquisition strategy is examined. However, the topic is also of concern to internal development and will be reviewed briefly here.

A firm is a composite of various resources. The ability to have these resources used in different domains within the firm allows synergies to develop. For example, there may be two industrial products that serve different markets, but if the sales of both products occur principally at trade shows, there is potential synergy in the marketing of the products. The firm will have expertise in how to sell at trade shows, and the products may even be sold at the same trade shows. However, it is more difficult to generate synergy if two products appear similar on the surface but require different skill sets to produce or sell. For example, a product that is sold in large box stores, such as Walmart, may appear similar to another retail product that is sold at high-end specialty stores, such as Saks Fifth Avenue. However, the ability to achieve synergies in selling the product at both will be difficult because the two distribution channels are too dissimilar. Even though the products may appear similar in some ways, the quality and nature of the product sold at Saks may be much different from the one sold at Walmart. Thus, it cannot be assumed that there are synergies that can be established.

To illustrate the benefit of technology platforms, recall GlaxoSmithKline has organized around several key areas including prescription drugs and vaccines. The prescription drugs they focus on are in the areas of asthma, malaria, depression, migraines, diabetes, heart failure, digestive conditions, and cancer. Thus, they seek to leverage their expertise in these areas to develop drug improvements faster and to facilitate getting changes through the regulatory system more efficiently. Their reputation with related products in a particular drug line (or platform) should help them in the marketplace also.

What and to Whom to Delegate?

The third question that should be initially asked in implementation is what to delegate and to whom to delegate? Ultimately, there needs to be someone clearly in charge of the various aspects of the implementation process. There are two benefits to **delegation**. First, the person in charge will become the champion for that product during the innovation process. The champion advocates for resources to ensure that the activities required will occur. The second benefit will be explored in greater detail in Chapter 5 as we look at responsibility in evaluation and control. However, when someone is delegated the authority over an activity, it is expected that person will complete that aspect of implementation. Thus, it is also possible to determine who did not follow through if the implementation activity does not occur.

KEY IMPLEMENTATION ISSUES

Once the organization determines the foundation through asking key questions, it needs to address the four critical issues necessary for implementing an innovation strategy. These elements are:

- Leadership
- Engagement
- Extension
- Alignment

Leadership	Engagement
Install supportive systems and policies Create mechanisms for innovation Allocate ample resources to critical activities	Build a knowledge-based culture Training and development Mentoring

What to do now versus later?

What is the timing, human resources, and existing platforms that are necessary or useful?

What should be delegated and to whom?

Alignment	Extension
Tie rewards to achievement Build "fit" Build a capable organization	Develop and share lessons learned Monitor organizational competencies Look for other opportunities

FIGURE **4.2** Four Elements of Implementation

These issues are interrelated and must be coordinated to help the organization achieve the types of innovation it wants. Although all four are essential for product or process innovation, new product development requires more engagement and extension, and process innovation requires more leadership and alignment skills. The text will next examine these four issues and the key actions within each in greater detail (Figure 4.2).

Leadership

In technology-focused firms, one of the key concerns is the leadership of the organization. There is typically a team of key leaders who guide the firm and play a pivotal role in its success. In part, this greater reliance on key leaders occurs because these firms compete in environments that are rapidly changing and evolving.[4] Please note that pivotal leaders can be found at any level of the organization, as well as in any functional or product area.

The leaders of successful innovative companies ensure that there is cooperative behavior among employees, a culture of collaboration, and cross-functional initiatives.[5] Two areas are of particular concern in seeking to ensure there is such collaboration. These are the skill mix of the leadership team and the actions typically taken by the leaders.

Skill Mix

It has been stressed throughout this text that innovation is not accomplished by an individual working alone. Rather, innovation is accomplished most often by a team of individuals with different skills. The need for different skill sets within the firm is evident as the topic of leadership is considered.

<table>
<tr><td>

4.1

REAL

WORLD

LENS

</td><td>

Interpharm Holdings

Interpharm is a manufacturer and distributor of generic drugs. The firm has nineteen pharmaceutical products, representing eleven distinct drugs, some of which are sold at different dosage levels.

</td></tr>
</table>

The firm in May 2003 sold its Atec Group, whose focus was computer operations, in an effort to better focus the firm on its core competency–pharmaceuticals. The cash proceeds from that sale were $3.6 million. The firm then used these proceeds to enhance its growth through internal innovation. The result was an expansion of expenditures on R&D. These expenditures allowed the hiring of new employees and the creation of better facilities. However, the promise of new products was not immediately realized. In 2008, Amneal acquired Interpharm Holdings and will use its outstanding management and financial strength to better realize the potential of Interpharm's product line, to ensure the outcomes of its well defined product pipeline, and to leverage manufacturing capabilities. The development timeframe for pharmaceuticals is very long. The inflow of cash in 2003 helped Interpharm but the firm needed still more resources and time to realize its desired product development. The financial stability of Amneal will help to realize these goals. The goal—research and development of new products–remains the goal for the newly integrated Amneal and Interpharm.

1. Why are new drugs critical to the success of a pharmaceutical company?
2. What do you think about the changes to Interpharm Holdings fortunes in the last decade?
3. Discuss the fit you see developing in Interpharm/Amneal's internal innovation efforts.

References

Anon 2008. Amneal pharmaceuticals to acquire assets of Interpharm Holdings, *The Medical News*, April 24. http://www.news-medical.net/news/2008/04/24/37769.aspx.

Corporate update. Company announces plans for expansion of product line and capacity. 2003. *Drug Week* (Nov. 28): 104–110.

Interpharm Holdings, Inc.: Generics company reports increased revenue. 2004. *Biotech Week* (Jun. 16): 260–265.

To illustrate, Microsoft is now one of the world's leading companies. It is impossible to think of the firm without thinking of Bill Gates, who clearly provided leadership in founding the firm. However, the founding team that created Microsoft involved more than just Bill Gates; it also involved many other individuals including Paul Allen, who played a critical role during the firm's development. Bill Gates is typically most associated with the external marketing efforts of the firm, and Allen was the internal technologist. It took both of them, and many others, for the firm to be successful. Allen has now left Microsoft but he still consults with the firm and advises it.

Occasionally, there is a single individual who plays all key roles in the firm, but more often, a team of individuals is present. It is critical that leadership within the group ensures actions are completed. However, the leader needs to foster flexibility and freedom so others can be creative enough to make the innovation strategy successful. The leadership of a firm that wants to develop technology internally must build an innovation network with 4 critical steps:[6]

1. Connect and link individuals: To be innovative, the leader must find groups of people who have different skills and approaches, link them together, and define what the goals are for the group. This definition should be very general at first.
2. Set boundaries and engage: As the interconnected individuals form a group, the leader's skill set should include the setting of targets of success, a timeframe, and engage the members in the task quickly.
3. Support and govern: The leader's skill set must include the allocation of resources so that success is possible. Actions here include defining administrative support, key knowledge and information inputs, and determining who the key people are for the innovation group's leadership and sponsorship.
4. Manage and track: Establishing performance criteria for individual and group successes is a key leader/manager role. In addition, as individuals come and go in the innovation network, the changes must be managed.

Today, in Microsoft Bill Gates is undertaking new roles—not as active CEO/leader, but rather as guru or idea person for the firm. In addition, he is spending more and more of his time with the Bill and Malinda Gates Foundation, which supports a number of far-ranging activities. Gates sees this period of his life as a "giving-back" time—Microsoft is still important and he still wields great influence, but the wealth created by the success of Microsoft has led him to pursue more external interests in education and environmental concerns. He has been able to make this move because the leadership team he built at Microsoft is so strong.

We often discuss leadership in terms of **formal** and **informal leadership**. That is, does someone's power come from his or her official position in the firm or from other sources such as respect or knowledge? To illustrate, Seymour Cray was one of the founders of Control Data Corporation during the 1950s. He went on to establish Cray Computers in 1972. In 1976, he developed the Cray-1, a large, very powerful, high-performance computer. This computer established the domain of supercomputing. In 1980, he stepped aside as CEO, and in 1981, he relinquished his position as chairman of the board of directors of Cray Computers. These steps by Dr. Cray do not mean he left the firm. Instead, he removed himself from these administrative posts so that he could devote his talents and time to developing the next generation of supercomputers. This resulted in the Cray-2 introduced in 1985. Cray was widely acknowledged as a leader both in the firm and in the industry. Some of this leadership came from his formal role as a founder of a major computer firm. More important, his leadership also came from his knowledge and capabilities.

The power from his knowledge and capabilities was such that later in his life Cray exerted leadership in the firm without holding any high-level, formal position in the firm. In the mid-1980s, he was just a contractor for the firm. However, despite his limited official position he still shaped the organization and the domain of supercomputing. He believed the greatest contribution he could make to the firm was intellectual, and that is where he focused his efforts. Bill Gates hopes to have a similar informal impact on Microsoft.

Leadership Actions

Leaders are successful when they accomplish given tasks and goals. However, a leader does not achieve goals simply by giving verbal or written directives. Instead, leaders take actions to encourage the tasks needed to achieve the goals. Specifically, there are three leadership actions that successful innovation leaders pursue:

1. Create a supportive environment
2. Create mechanisms for innovation
3. Allocate resources

We now examine each of the three leadership actions.

Create a Supportive Environment When leaders encourage innovation in the organization, they must clearly indicate the firm's direction. These efforts should support the vision and mission of the firm. The key leaders then support efforts that move the firm in that direction. Their championing of the implementation effort for innovation must be more than words. This is important for all innovation, but it is critical for product innovation.

To illustrate how such supportive systems and policies can impact a firm, let's look at Microsoft again. The firm has had success in developing new products. However, it is actually much better at taking others' technology and bringing it to market in a more accessible manner. Recall that Microsoft, in its early stages, licensed the MS-DOS operating system to IBM, but it did not invent the concept of the operating system. Similarly, others invented many critical aspects of Microsoft's Windows program, such as the bit-mapped display, and the use of the mouse.

Other firms may create more innovative products, but Microsoft excels in the presence of support activities to ensure that the idea is taken to market in a form that can be used widely by customers. This ability requires the firm to have a variety of procedures and actions that allow it to take ideas and quickly move them to the marketplace. Thus, the leadership of the firm is clear on its goals, and they support those efforts within the firm to position it as a market leader. Microsoft excels at the growth stage of a technology, not at the start-up.

If support systems are to work as desired and aid the innovation process, there must be coordination and integration between the leader and the various units and actions of the organization. It has been found that to coordinate and integrate these various systems, firms need to:

1. Avoid paralysis of analysis
2. Delegate effectively

Too often, technology firms can be paralyzed by analysis. For example, a product team may want to have extensive research data before making a decision. This can include consumer research about potential demand, estimates of sales, and estimated cost structures to justify the expenditure of funds and manufacturing for their idea. Managers cannot rely solely on those numbers when they implement an innovation strategy; leaders ultimately make the decisions and it must involve their judgment. However, there is a tendency by managers to generate an ever-increasing amount of analysis to justify their actions. This can result in over-analysis of a situation. Managers need to make judgments when it comes to technology. Typically, perfect information or clear answers are not available. Instead, the manager should obtain a reasonable amount of information and make a decision.

The law of diminishing returns from economics is relevant when considering how much analysis is appropriate. If the firm keeps running tests and tweaking the planned innovation, the margin of improvement in value and quality is likely to get smaller and smaller. At some point, the tweaking activity costs more than its potential benefits. The cost of analysis is similar. When launching a new product, Microsoft wants that product to be free of errors. However, it is understood that the firm cannot assure buyers that there are absolutely no problems. At some stage, the firm has to put the product on the market. If Microsoft does not act consumers will replace Microsoft's existing product with those of a competitor. The new product from the competitor will most likely have its own imperfections but would still take market share from Microsoft. Thus, Microsoft ultimately must act and address any problems in a new product release through the use of online updates for those who have purchased the product.

The leader also needs to delegate effectively. The leader may support the innovation strategy and hurt the process by being too involved. Managers cannot micromanage each and every project in the organization. This is a disservice to the organization, to the manager, and to the employees. The organization is not getting the leadership it needs from the manager, the manager is not developing leadership skills, and the employees are not getting the growth opportunities they need. Instead, the leader needs to empower subordinates to act on the different parts of the innovation strategy. However, when delegating that authority, the leader needs to indicate clear support for the innovation strategy. This show of support should include the indication that the organization will accept failure. The leader should monitor the process, but trust employees. This will allow employees to learn from their experiences. The leader can then spend time on other activities while employees gain new skills as they are empowered to pursue activities. A research chemist who is promoted to manager of the lab must coordinate the resources and communicate the actions of the team of researchers and technicians. If this new manager continues doing the work of a research chemist he or she will leave the lab without the leadership and advocacy voice it needs to optimize the chances of success.

Create Mechanisms for Innovation A second issue that must be addressed by the firm's leadership is the installation of systems that are supportive of innovation. These systems range from standard operating procedures that affect everyday life in the firm to special programs that encourage and reward the exchange of ideas. Two such special programs may be restructuring and reengineering. A better understanding of who does what for the organization should emerge from restructuring and reengineering efforts. These efforts do not have to be organization-wide but may simply involve redesigning mechanisms for innovation through the better integration and utilization of support and staff functions of the organization.

For example, at the oil product company CITGO the purchasing department was recently credited with helping the production department save hundreds of thousands of dollars annually. The firm had in place policies encouraging the different departments to observe and understand what the other departments in the firm were doing. Through efforts encouraged by this policy, the purchasing department noticed the number of returns of their packaged products had increased significantly. Because the returns occurred across many product lines, the production department had not noticed that the returns were related to one element—improper labeling.

Purchasing, working with the production department and labeling suppliers, was able to spearhead a process innovation that led to changes in how labeling was handled. Rather than having employees look up labels in a paper catalog that identified which label to place on a product, an automated system was developed. Now, when an order is relayed electronically to the plant, a picture of the label and packaging with accompanying information is immediately available online. This has resulted in a significant decrease in the number of returns.

The purchasing department did not make the changes, but their support played a key role in changing how the labeling was carried out. Although the technology employed was known within the organization, it was not being used for this application. The value of the company increased through the cooperative efforts established by the firm. Methods to reward those who use such innovative and cross-departmental efforts to increase efficiency or effectiveness for the firm were already part of the organizational policies. The firm rewarded both the purchasing and production departments for the savings obtained. Thus, the mechanisms were in place to encourage innovation and adaptation inside the firm, and the policies and procedures supported the action.

For many organizations, it is difficult to establish the type of knowledge-exchange atmosphere that encourages the questioning and openness needed to promote innovation. It is easy to say, "This firm wants an open, honest exchange of ideas," but the reality of most organizations is that politics and power have significant influence on the process of idea exchange. However, policies and procedures can be established to help encourage such interactions. It is the leaders of the organization who determine whether these policies are more than words on paper.

TradeKing, a pioneer in incorporating unique Web 2.0 functionality to its site, illustrates a similar ability to create an innovative environment. TradeKing created an online community where traders can interact to share ideas in an open, trusted environment. A key illustration of this openness is that

TradeKing's CEO, Don Montanaro, is one of the few CEOs of a financial services firm with his own blog. Montanaro posts his blog directly on the TradeKing site and his e-mail is clearly listed for others to engage him in dialogue. The CEO ensured direct interaction was not only a written policy but enacted it through his example of openness.[7] The result is an environment in which people both inside and outside the firm feel free to offer insights and suggestions.

Allocate Ample Resources to Critical Activities   The last of the actions for leaders, if implementation of an innovative strategy is to succeed, is the allocation of ample resources for the activities that are desired. This is true in the implementation of any strategy; for innovative strategies, especially new product development, it is even more critical. There must be adequate money, people, and other resources to allow the speculation, trial, and error if the innovation strategy is to be successful. Because innovation, whether radical or continuous, involves a change in the status quo, it requires resources that allow for the inefficiencies of experimentation and risk taking. It also means that resources must be available for those who advocate a different path. The leadership must realize that experimentation, risk taking, and difference of perceptions should not be viewed as frightening but rather as opportunities. Without some type of innovative process, the organization will atrophy. The environment of the firm will change; leaders help allocate resources to move the organization in a desired direction to a desirable outcome.

To test that an atmosphere supportive of innovation has been created leadership needs to review several questions periodically:

1. Is there an open, questioning attitude among the employees?
2. Is the organization avoiding "ruts" that are not questioned?
3. Does everyone believe that their opinion will be listened to and counted?
4. Are there strategic gaps between where we are now and where we want to be? If so, how do we close them?
5. Is there a vision of where the firm wants to be in the competitive environment?

Engagement

Engagement is the second key implementation issue. The central question in engagement is how to get the various entities in an organization moving in the same direction. Culture is the pivotal element in determining the level of engagement within the organization. Microsoft's industry reinvents itself every twelve to twenty-four months. To do this, Microsoft needs a culture where its products are not replaced by competitors'; instead, the firm makes its own products obsolete. To illustrate, Microsoft originally sold MS-DOS operating systems for computers. It then created Windows, which replaced MS-DOS. Now, about every two years, it replaces its current Windows operating system with a new version. This level of adaptability only comes about because of a culture that stresses change and flexibility. Microsoft's employees believe in this culture and daily act in a manner consistent with it. For new product development or product improvement activities, engagement similar to that found at Microsoft is critical if the firm wishes to be successful over a long period of time.

The need for organizational flexibility has been noted by a wide variety of authors. In fact, Richard D'Aveni,[8] who developed the concept of hyper-competition, argues that this adaptability and flexibility are critical to the success of any firm in advanced economies. He argues that as a result of the very fast speed of competition today, if a firm cannot be adaptable and flexible, it will fail. Such adaptability requires the active engagement of employees.

The leaders of an organization should create an internal environment that allows individuals to believe they are part of a system and organization that will allow innovation; that is, they are actively engaged in the firm. If the employees do not believe that they are part of the system and organization, there typically will be problems.

Problems in Engagement

To say that the organization should have strong engagement by the employees is easy. However, in practice, it can be difficult to develop such engagement. Potential problems that need to be overcome include inertia, fear, and complacency.

Inertia If a product works, the tendency is for the firm and its management to continue without disruption. Internal politics in the organization support this tactic because power is established and political systems are understood. If systems are changed through innovative activities, such as restructuring, then those power and political relationships are disrupted. Thus, established forces in the firm will desire inertia but inertia will make strategic actions in support of innovation almost impossible to implement and sustain. Inertia can be overcome with movement toward an open, egalitarian process that encourages direct, candid, future-oriented discussions. Such inertia can occur in a whole industry. Health care records management is a good example. Individual health systems, hospitals, and medical practices all have their own systems and do not want to implement a new system that forces them to change. There is technology that would allow all of these various records to be tied together efficiently. The result would be lower costs and better health care but inertia is part of the problem that prevents the parties from coming together for the change. This inertia has been present for over 10 years, preventing cost savings and better customer service.

Fear Even when the change is perceived as positive, fear can be part of the response. For individuals in the organization, it is a fear of loss—loss of personal power, loss of respect, or loss of a job. Fear is part of the reason there have been no changes to health records. The proposals to generate electronic health care data files for all people would allow physicians and others to know what was happening with any patient at any given time. But fear of change here includes the fear of others being able to see what any doctor does may call some of their decisions into question. Fear is best managed by success. Small successes build positive beliefs about organizational and individual abilities to manage the innovative process. For the organization, the fear of cannibalizing existing products or implementing a flawed process is very real. Because of fear, it is likely that the innovation process will lead to higher levels of conflict, stress, and political activities by employees.

Complacency   Complacency is similar to inertia; however, inertia concerns organizational structure, and complacency impacts the organization's efforts to make broader changes. Such complacency emerges when there is satisfaction with the status quo. "If it ain't broke, don't fix it" should not be heard in an organization undertaking an innovation strategy. There should be a constant process to offset the possibility of complacency.[9] This process involves capturing and locating knowledge and expertise, transferring and sharing knowledge, and enabling individuals to develop new ways of thinking about things. Health care providers in the United States need to examine the successes in other industries and in other countries with recordkeeping technologies. To simply say our health care is the best in the world and not look at the others on how to improve is a sign of complacency.

Overcoming the Problems

The process organizations use to overcome engagement difficulties is change management. Managing the change process helps employees believe they are part of a team where involvement is expected and trust is built. The steps to transforming an organization were developed by Kotter and are presented in Figure 4.3.[10] Three methods can help transform the firm by helping employees believe they are part of the organization and important to the innovation process. These are:

1. Building a knowledge-based culture
2. Training and development
3. Mentoring employees

Each of these is reviewed briefly next.

Building a Knowledge-Based Culture   One of the most difficult items to identify and proactively impact is the culture of the organization. Culture is like art; one knows it when one sees it but it is a qualitative characteristic of an organization that can be hard to define. Over time, individuals develop a

- Establish a sense of urgency
- Form a powerful guiding coalition
- Create a vision
- Communicate the vision
- Empower others to act on the vision
- Plan for and create short-term wins
- Consolidate improvements and produce still more change
- Institutionalize new approaches

FIGURE **4.3** Kotter's Eight Steps to Organizational Transformation framework to interpret events and activities that occur within the firm. This framework is the culture of the organization, and it takes time to develop and is difficult to change. However, the leaders of the firm can encourage the organization to have a change-accepting culture through their own behaviors. The leadership of an organization helps to establish the culture by providing powerful signals of the culture. Thus, if the leadership of the organization seems to embrace change the rest of the organization is more likely to adapt that characteristic into the firm's culture.

The purpose of an innovative strategy is to create and effectively utilize new products and processes.[11] To make this happen, a firm needs to have a culture that encourages the sharing of knowledge within the organization, even though it is sometimes difficult. For example, Xerox wanted to build a database for technicians on how to repair products. At first, the technicians were reluctant to submit tips on how to make repairs because it was not what they normally did. In addition, some feared that if they shared their "special" knowledge, they would become unnecessary. Engineers were asked to enter some tips to encourage technicians to share. In some groups, rewards, including cash and T-shirts, for submitting tips were offered. Today, Xerox's "Eureka" system holds about 70,000 suggestions and saves the company millions of dollars a year in repair costs.[12]

Training and Development It has already been noted that training is critical to ensure that the proper human resources are present in the organization when needed. However, training and development also have the benefit of helping employees grow as professionals and as individuals. Honing skills through training and development is an important key to getting individuals to believe in what the organization is doing and where it is going. If an organization is going to be innovative, then it must renew itself by focusing on new technologies, new processes, and new ways of doing things. The organization cannot do this with employees who have outdated skills.

The more successful skill-development programs in organizations are characterized by:

1. Informed opportunism: Once the organization decides to be innovative, opportunities for individuals to grow and explore should be developed to enhance creativity and knowledge sharing. The training and development must be as dynamic as the changes to which the employees need to respond.

2. Directed empowerment: Development occurs as individuals are placed in charge of the various efforts of the innovation activities in the firm. These new experiences will increase the skills of employees as they are forced to learn new things to meet the needs of their new responsibility. The organization must support these individuals in their new roles by allowing them time to develop the skills and providing the training and advice necessary to meet their new responsibilities.

3. A turning prism: Development also allows individuals to change their view of given information. For example, benchmarking occurs when an organization compares itself to "the best" and makes changes to improve

areas in which it is not performing up to par. The information from firms outside the organization helps the business to see itself in new ways.

Mentoring   Another important means to connect the employee to the firm is through **mentoring**, the direct one-on-one activity between employees in the organization or a system designed to allow two people to learn from each other. Typically, in this setting, one party takes on the principal role of sharing information and guidance with the other employee. Traditionally, this relationship involves an older employee mentoring a new, younger member of the organization in some informal manner. However, some mentoring programs are more formal. The Women's Alliance at Xerox uses mentor-matching software to find mentors for women wishing one. The innovative use of match-making software has helped Xerox build a support network for its women and has led to the formation of other mentoring groups for other affinity groups.[13]

Most people have individuals in their lives who have influenced their approach to problems, how they think, how they do things, and how they solve problems. Organizations that are successful in innovation tend to encourage mentoring among a wide variety of individuals. Mentoring should not occur only in a downward direction with senior employees mentoring junior employees. Instead, it should also occur with junior employees mentoring senior employees. This is particularly true in a technology-focused firm because the newer employees may actually be more knowledgeable about a new technology and its application.

Albert Bandura, a social psychologist, wrote in 1977 that "the capacity to learn by observation enables people to acquire large, integrated patterns of behavior without … tedious trial and error."[14] Mentoring allows individuals in the organization to share lessons learned. By encouraging individuals to engage in mentoring, the organization extends the knowledge base of individuals to others in the organization.

Extension

Extension is the third key implementation issue and is concerned with firms having sufficient knowledge of product and market competencies so new ideas lead to action through a filter of experience. New product development, product improvements, and new market entry all depend on extension processes. Extension requires an organizational memory so that lessons learned in the past can be used in the future. Until 1994, Microsoft had chosen not to compete in the Internet space. This resulted in its late entry into that domain. However, once Microsoft determined that it should enter, it did so aggressively and has been very successful. The firm's competencies were important in allowing it to be aggressive upon entry into the market. From a prior analysis of the domain, Microsoft knew the existing problems and began to design specific means to overcome those liabilities. The end result was that the firm virtually overnight went from not competing in a domain to becoming one of its major competitors.

Thus, extension ensures that product and market competencies are understood sufficiently so that actions that lead to innovation can be taken. Earlier, the development of flat-panel displays was discussed. Smaller products entered the market first because the technology to make bigger displays was not available. A new process for making the glass panels had to be developed before the large screen flat-panel televisions, widely available today, could be produced. The techniques for the glass production for flat screens have now led to new processes in other types of glass manufacturing, resulting in a radical product innovation in a very traditional industry.

There are three key actions that can help generate extensions:

1. Knowledge sharing within the organization
2. Monitoring competencies
3. Looking for new opportunities

We now examine each of these.

Knowledge Sharing

As noted previously, the ability to create a culture where employees share knowledge within the firm is critical in the management of technology. An outcome of that knowledge-sharing culture is that it also aids in the extension of the firm's current products.

Not all innovations make the competencies within the organization obsolete. There are typically many small, continuous innovations that allow the organization to maintain or improve competitive positioning continuously. These smaller changes frequently occur because of shared lessons. A small success in one area then leads to other similar changes that ultimately can have a major impact on the organization.

To illustrate, in the hotel industry, many innovations are extensions tied to renovation. Generally speaking, when a hotel wants to change, it does so by renovating or by changing room designations. Nonsmoking rooms were innovative when they first appeared. Because they were successful, more companies adopted nonsmoking rooms. Some hotels now have nonsmoking floors, and some hotels are now completely nonsmoking. Although these innovations are continuous, they have become more prevalent and will continue to do so in the foreseeable future. They add value for some important stakeholders of the firm.

Such small innovations can then grow into a niche or area of competitive advantage. The difficulty with such extensions is that competitors often match them easily, and the sustainability of their competitive advantage is typically not great.

Monitor Competencies

The organization must be aware of what competencies it possesses. The resource-based view of the firm states the principal means for a firm to gain a competitive advantage is through the skills and knowledge of its employees.[15] The competencies of a firm's employees cannot be easily duplicated by other businesses. This is in contrast to hard assets such as machinery because almost any firm that has the money can buy a similar asset.

The ability to focus the competencies of the organization's employees on a given problem requires a clear picture of what competencies are present in the organization. Therefore, many firms perform a skills inventory to identify competencies that are present or needed. Such a skills inventory is simply a listing of the various training, understandings, and certifications each employee has.

For example, e-business and e-commerce are domains that have come into prominence in recent years. The skills needed to manage in such an environment are different from the skills in more traditional business environments. For example, in e-commerce, the number of mouse clicks that it takes to reach an Internet destination is critical. Such a concept has little relevance in most other retail settings. Walmart understood that retail over the Internet would be different from retail in a store when it set up its e-commerce unit. Now it is creating "new" ways of merging technology and brick and mortar buildings. It is working on a new generation of e-commerce enabled stores. A smaller, "high-efficiency" store is being developed and will include a commitment to pick-up service. There will be drive-through service for individuals who order goods online. The changes in store design and service delivery require Walmart to look at store design in a whole new way.[16] In implementing an innovation strategy, the firm must monitor its competencies and enhance and develop those needed skills throughout the organization.

Look for Other Opportunities

The last element of the extension category of implementation is identifying new opportunities. Often, one innovation leads to a wide set of other opportunities for the same technology. New uses for existing products can lead to a whole new market or to unexpected innovations. Everett Rogers calls this sharing of information **diffusion**.[17] It is a special type of communication that is concerned with the spread of new ideas. The main elements of diffusion are:

1. An innovation or new idea
2. That is communicated through certain channels
3. Over time and in a timely manner
4. Among members of a social system (an organization is a social system)

The amount of information diffusion in a firm greatly influences the innovativeness and rate of adoption in the organization.

Diffusion is based on the organization's willingness to act on opportunities as they become evident. Frequently, the opportunities emerge from the organization's boundary spanners. These are individuals who interact with and gather information from the external environment. These individuals do not restrict themselves to the boundaries of the organization but instead have contacts and interactions across multiple organizations. Prime examples of boundary spanning, or reaching outside the firm, are sales personnel interacting with customers or researchers at a professional association meeting. Interacting with customers or with others working in the

<table>
<tr><td>

4.2

REAL

WORLD

LENS

</td><td>

Meeting Industry Standards

Industry standards that evolve can be a major motivator for new innovation efforts by a business. To illustrate, the worldwide semiconductor technology leaders set the International Technology Roadmap for Semiconductors (ITRS). The road map set standards for firms competing in the industry on requirements and technology timelines within the industry. The road map required that manufacturers like Intel and Texas Instruments meet certain standards, as well as requiring that material manufacturers such as Applied Materials meet certain standards. These standards are updated periodically and a report is issued periodically.

</td></tr>
</table>

The strategic difficulty for many firms was that if they did not meet the standards, they could find themselves at a competitive disadvantage. This difficulty is compounded by the fact that the semiconductor industry moves very rapidly, and if a firm does not develop products that meet the standards internally, it may not be able to buy companies or product lines externally that allow it to meet those standards.

Thus, in materials science, if a firm does not meet the standards, it may find that competitors will. For a look at the updated ITRS standards report, go to http://www.itrs.net.

1. What other external environmental factors could influence the firm's implementation activities? Be specific and explain your answer.

References

Singer, P. 2005. 2004 ITRS updated. *Semiconductor International*, 28 (1): 17–23.

Somekh, S. 2005. Economic reality drives roadmap strategy. *Solid State Technology*, 48 (1): 44–46.

same area promotes knowledge sharing among the parties. This knowledge can provide valuable insight to changes in the competitive environment and to potential opportunities.

To illustrate, knowledge diffusion of new technologies led to the development of sport shoes. Prior to these new technologies, the shoe industry was relatively stable. Typically, the rubber-soled shoe was for professions such as nursing or low-cost products for children. The creation of new technologies then led to the development of shoes designed specifically for athletes, particularly shoes for basketball or track. However, it was soon recognized that the same technology could be extended to produce shoes for a wide range of niches. Quickly soccer shoes, aerobic shoes, and water-polo shoes, to name just a few, were produced. Sport shoes have now become a dominant segment of the shoe industry and have been expanded to shoes for activities such as walking.

Alignment

The last issue critical to implementation of innovation is alignment. It has been noted that the firm should have a wide variety of systems to facilitate an innovation strategy—this is process. However, it is not sufficient simply to have the systems in place. Instead, there must be alignment among the various systems. Thus, **alignment** represents a fit among the systems within the firm as they support the firm's strategy and a fit with the firm's external environment.

Firms need to be creative. This creativity must be targeted and grounded in the financial reality of the firm. It is unrealistic for an organization to assume that it will have unlimited resources. Microsoft has a culture that promotes creativity and innovation; however, there are systems in place to ensure creativity does not mean chaos. The firm uses a single, global financial reporting system. This system makes financial information easy to access by individuals and units that work in the finance function, but information is stored in a central data warehouse also. Thus, financial data are available to everyone in the organization. The system is set up so that most employees do not have to deal with the complex software directly; rather, they use simple web-based menus to view the data. Thus, the means to track activities, their financial aspects, and other ongoing projects in the firm are easily accessible. This results in individual/team knowledge and awareness of available resources for pursuing new ideas, products, and processes.

Ensuring that the organization's systems are aligned is difficult. Too often, the expression "the right hand does not know what the left hand is doing" is heard throughout a firm. Phrases like this indicate there is nonalignment in that organization. If a firm wants to pursue an innovative strategy, then the pieces must fit together to bring the resources to the areas of innovation.

The nature of the fit between the various parts of the organization will differ depending on the importance of the type of innovation undertaken. Thus, different types of innovation have unique needs. In Chapter 2, it was recognized that innovation can be radical or continuous. While there are differences in how to align the systems and structures based on the type of innovation, there are three key activities that are necessary for alignment:

1. Build fit
2. Tie rewards to achievement of goals
3. Organizational structure

Figure 4.4 summarizes these key elements of innovation for both incremental and radical innovations. Each will be examined briefly.

Build Fit

Previously, it was stressed that the systems of the organization must fit together in support of the innovation strategy. However, not only must the systems align, but also the different internal groups must support each other. Innovative companies have a hierarchy of importance for various groups that remains relatively consistent for the innovation process. The internal groups are most important because they allow interaction and idea exchange and need to be aligned in support of each other.[18] External constituencies are

Description of Innovation	Characteristics of the Innovation	Key Fit Elements	Examples
Radical Change	Strategic High uncertainty Wide firm impact	Top management support Multiskilled team External alignment	Major restructuring of firm, product line, or market
Continuous Change	Low to medium uncertainty Low firm impact High team impact	Team coordination imperative Document learning to share with peers	Routine improvements to an existing system or service

FIGURE **4.4** Key Fit Elements in Innovation

also important but for different reasons. The groups that should be considered include:

1. Within the department or division: The makeup of this group is determined on a case-by-case basis. However, the mix of individuals on the team must align with the purpose of the innovative processes and the skill needs for that process.
2. Outside the department or division: The firm should focus on reducing interdepartmental or between team barriers. Learning and knowledge sharing across innovation efforts can increase the effectiveness of the processes.
3. Competing with other firms (external): Although there has been a great deal of research that concludes information from competitors is important, the general consensus is that its importance is often exaggerated. This does not mean it can or should be ignored. It is clearly related to the analysis of environmental fit and the determination of what the firm needs to do to maintain or build competitive advantage. But firms need to make their own path while being aware of what others are doing.
4. Suppliers and customers (external): These groups rank only behind competitors when looking at information gathering and the importance of alignment. This group's importance is highly dependent on the type of innovation and the sophistication of the supplier/customer. Because of these variations, the supplier/ customer ranks lower in importance. Where they are important to alignment with the environment, they tend to be very important. However, their importance is not always a key. Commodity products provide an example where competitors' actions have far more influence on building fit. There is little to differentiate commodity products. Thus, the major competitive advantage that a firm can create is based on providing better fit than others do.

Tie Rewards and Incentives to the Achievement of Goals

If a corporation's goal is to be the innovation leader in new product development, then the systems of the organization should be set up to reward individuals and groups that help the firm achieve that goal. This sounds very simplistic, but frequently, organizations and managers state they want, A, and then reward those who do B. To gauge whether a firm is tying rewards and incentives to innovation, the organization should examine:

1. What happens to creative mavericks in the firm? Are they encouraged or swept out?
2. What happens when someone fails? Is there guidance to new paths and analysis of what happened? Or is there blame and then punishment?
3. How often do employees hear, "We want you to take risks, but you had better not fail to meet your quota"?

Innovation happens because employees know that the company expects it and rewards it. However, just as important for the employees is that leaders in the organization recognize that not all efforts to innovate or improve will be successful. For the innovative strategy to succeed rewards and incentives cannot be solely about last year's performance; rather, there must also be a willingness to reward stretching toward a new product, business model, market, or process. 3M depends on its employees' abilities to create new products. The firm understands that not all of the products will be successful. However, it expects employees to continue trying new ideas. Microsoft similarly expects to develop new products consistently and try new things. The ability of the organization to shift from a pure software development firm to one that competed in the Internet domain almost overnight was, in part, due to a reward system that encouraged individuals to take risks.

Organizational Structure

The structure of the organization is a key to allowing the exploration of new ideas and processes. There is much debate as to whether structure follows strategy, or vice versa. Both can be true. When there are changes in strategy, changes in structure are necessary to adjust to the information-processing needs of the new strategy. The structure of any organization has two basic purposes: communication and coordination. Likewise, the strategy follows the structure because if the ability to communicate and coordinate the changes is not part of the organization, then innovative strategy cannot occur.

Generally speaking, the structure of the organization must be characterized by flexibility and openness if innovation is to take place. Structure rigidity is one of the primary reasons large companies in mature industries have a difficult time being innovative. However, this does not need to be the fate of the organization. The firms that enjoy long-term success and continued leadership have structures that allow new ideas to develop and flow through the organization. The specifics of the structure are dependent on the type and degree of innovation.

As can be seen in Figure 4.5, the degree of innovation (radical or incremental) and the type of innovation (product or process) influence the key elements in the structure. If an organization wants to develop radical new

	Radical	Continuous
Product	"Fuzzy" Front-End Process Flat structure with many communication links Cross-functional approach ingrained Careful monitoring of competitors Most often from R&D unit Most successful: When a separate unit is organized for new product development	Analytical Front-End Process Often emerges from boundary spanners who note competitors' actions Cross-function approach needed Careful monitoring of product improvements by others and emerging substitutes Most successful: When a part of improving product culture
Process	"Fuzzy" Front-End Process More functionally oriented Evaluation team in place—benchmarking group Savings monitoring Most successful: When ideas bubble up the organization	Analytical Front-End Process Functionally oriented Cost evaluation is driver Internal information sources are critical Most successful: When those working in the process are acting as internal suppliers

FIGURE **4.5** Structural Considerations for an Innovation Strategy

products or processes, then the processes at the front end should allow for discontinuous or "fuzzy" thinking. If they do not, then the chances of true radical innovation are reduced significantly. In any firm where innovation is part of the strategy, the amount and type of innovation are related to the innovative capacity of the organization and the industry. Much of that capacity is determined by how well the structure allows the free flow of ideas yet maintains the discipline of relevant environmental scanning and information dissemination.

There are many types of organizational structure. Most of us think of functional or divisional structures. Functional structures have engineers separated from accountants who are also separated from marketing personnel. Divisional structures are divided by geography, product line, customer base, or some similar designation. Most successful innovative companies have hybrid structures that are designed to specifically address the situation facing the firm. Thus, these successful firms pursue a unique structure that is based on the activities in that part of the organization and what works best for the organization's goals. For a small company, the structure can have very simple coordination lines and communication networks. When the organization is larger, the needs for communication and coordination become more complex. The formal structure gives some indication of expected relationships, but it is the culture of the organization that will determine how well the network of people and resources will interact to achieve innovation.

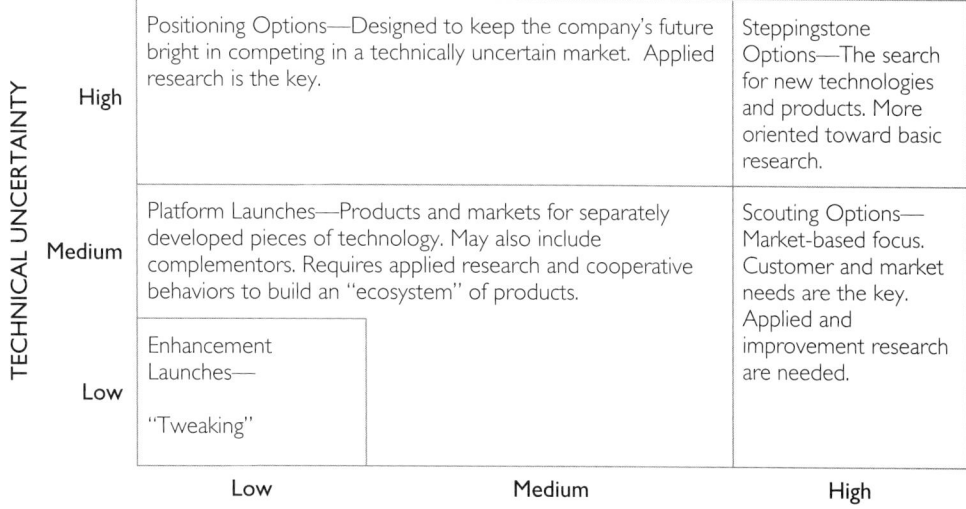

FIGURE **4.6** R&D Portfolio

Source: MacMillan, I. C., and R. G. McGrath. 2002. Crafting R&D project portfolios. *Research Technology Management*, 45 (5): 48.

CRAFTING PORTFOLIOS OF INNOVATION

This chapter has discussed how to make the innovation strategy work. However, there are some mechanics of portfolio management that must be addressed. The leadership must align the products and processes, engage the resources, and provide a basis for innovation extension if the organization is to build and maintain competitive advantage. In Chapter 2, the value chain was presented. This provides a framework for many areas of the organization where innovation takes place. To be successful, the firm needs to develop an R&D portfolio of innovation processes. Figure 4.6 shows one conceptualization of an R&D portfolio.[19] The mix of options is part of the strategic makeup of the firm.

The portfolio of innovative projects can be described as an organizational "ecosystem."[20] The diversity of the portfolio is dependent on the strategic plan and how the implementation keys—leadership, engagement, extension, and alignment—are employed.

SUMMARY

This chapter has addressed the implementation of internal development technology efforts in a firm. There are critical elements in the implementation of such efforts, including the firm's leadership, engagement of key personnel in the process, and the fitting together of the various key elements. The firm may have great strengths in any of the aspects critical to the implementation effort. However, the firm must be able to fit these parts together so that it can produce success for the business.

MANAGERIAL GUIDELINES

The guidelines for implementing an innovative strategy depend on the type of innovation and its goals. In new product development or extension, the guidelines include tactics that depend on characteristics of the new product.[21]

1. Relative advantage: How much better is the product than those with which it is designed to compete?
2. Compatibility with current product usage: How well does the new product fit with the current product usage or end-user activity?
3. Complexity: Will frustration or confusion arise in understanding the innovative product's basic idea?
4. Divisibility and testability: How easily can trial portions of the new product be provided and used?
5. Communicability or observability: How likely is the product to appear where potential users can see it?

Therefore, the managerial guidelines for a new product in an existing market, an old product in a new market, or a new product in a new market include the following:

1. Determine the relative advantage of the product in its market
2. Be aware of the possible influences of the innovative product/ market and how it may affect the previous product/market mix
3. Realize that if the innovation is too complex, then selling it in the organization and to potential customers will be difficult. The leadership should try to negate complexity with clarity of purpose and planning

4. Build awareness through providing samples or demonstrations of the new product's characteristics
5. Make the product visible—Intel Inside is a prime example of a normally invisible product (a computer chip) gaining marketing recognition

The key for innovation is the sharing of information and knowledge. It is hard for the organization to move in a certain direction to a new process/structure if there is no understanding of where, why, and how the organization is going to move. To that end, the pitfalls to avoid in implementation include:

1. Applying process innovation too broadly; the proper scope of the process to be changed should be appropriate to the goal.
2. Overlooking people challenges associated with the changes.
3. Being overconfident about the ability to lead the change process.
4. Forcing changes to coexist with some existing "sacred" element of the organizational structure or process.
5. Avoiding elimination of processes and people just to prove something is happening.
6. Relying on technology to fix all systemic problems.
7. Failing to communicate.

To succeed, process changes should be based on a broad vision of fit and action, understanding the processes and methodologies of the innovation strategy being undertaken, and using the best people available to champion the cause of change.

Guiding Questions

Figure 4.7 presents a checklist of questions for leaders to examine to be sure innovation is still the focus. If the firm wants to be successful in implementing an innovative strategy, it needs to know the strengths and weaknesses in each of these areas. There is no one best profile for innovation, but alignment of the vision, leadership, processes, and resources is critical to engaging in this strategy to extend the firm beyond its current limits. Once the firm knows where it stands in these areas, it can begin the process of evaluating why and controlling activities to enhance innovation.

Vision	Leadership
Does the team have a clearly articulated vision, mission, or set of objectives?	Is excellence of central importance to the team?
Does everyone share the vision (objectives)?	How does the manager monitor and improve performance levels?
Is the vision (objectives) clearly stated?	Are all team members committed to excellence?
Did everyone participate in creating it (them)?	Does the leader encourage open idea exchanges?
Is this vision (objectives) attainable?	
Processes	Resources
Do team members share information fully?	Does management support new ideas?
Do all team members participate in decision making?	Do team members support new ideas by giving time, cooperation, and resources?
Are team members comfortable proposing new ideas?	Does the team leader offer practical help and resources for the development of new ideas?
Are team members able to challenge standard practice?	What happens when a new idea fails?
Is there a climate of trust within the group?	

FIGURE **4.7** Checklist for Innovation Strategy

CASE **4.1** THE REAL WORLD
Daikin Industries

Daikin Industries is one of the world's largest manufacturers of air-conditioning equipment (88 percent of the firm's sales come from air-conditioning-related products). The corporation controls approximately 20 percent of the world market and had revenues estimated at approximately $23 billion in revenue in 2009. Research and development has been a critical part of the growth and success of this Japanese firm. Today, the firm is working on new innovations in areas such as air-conditioning systems that clean the air of pollutants, adjust air humidity, and rid the air of odors.

The implementation processes for the firm's innovation efforts are a key part of the firm's success. Daikin is a worldwide firm. This, in part, explains why structurally the firm has established separate units that focus purely on research in electronics and mechanical engineering. These units are quasi-independent to encourage maximum creativity. The work between these and other units of the firm is coordinated by the R&D planning department. In contrast, the firm has established its chemicals research area as part of the dominant unit focused on those activities. This unit is located principally in the United States. Thus, the firm structurally has responded to different markets in different ways to ensure that the maximum amount of appropriate creativity is encouraged. Where there is

(continues)

CASE **4.1** *(continued)*

an environment that encourages creativity, the firm incorporates its research and development efforts in the unit. However, if the culture of the unit or country in which it is located is not as supportive, the research efforts become headquartered in quasi-independent units.

Daikin also encourages maximum creativity by not specifying in extensive detail the role of different units and individuals in those units. The corporation allows each unit and the employees within those units to look for new opportunities and be creative.

Consistent with its organizational flexibility, the firm develops what it calls a "culture of continuous action." The belief is that the firm must act and pursue opportunities when they occur. Daikin's view is that analysis is critical, but analysis cannot paralyze the firm. Again, culturally, the firm has clearly differentiated itself from many Japanese firms where quick action is not always a predominant value.

Within each unit, the corporation encourages a flat structure that has maximum participation by employees. The teams within these units typically are cross-functional and include individuals from various disciplines within the firm.

1. This vignette illustrates some of the differences that must occur in a worldwide firm as it conducts its R&D effort. What are some other differences you might expect?
2. How might these differences vary between a British firm and a Japanese firm?

References

Chilling out. 2000. *New Scientist*, 165 (2223): 15–18.
Daikin Industries web page. http://www.daikin.com/.

CRITICAL THINKING EXERCISES

Relating to Your World

1. We have organized this chapter by activities that need to take place to successfully implement an innovative strategy. There are many ways to characterize a leader. If you are seeking an innovative leader for an organization, what do you think are the ten most important characteristics for the leader to have? How will you determine if an individual has those characteristics? Discuss how executives can ensure a successful managerial fit.
2. We discussed radical versus continuous innovation in Chapter 3. What area of implementation do you believe is most affected by the type of innovation? Why? What are the advantages and disadvantages of working with each of these types of innovations?
3. When you made a radical change in your life—leaving home, getting married, moving to a new city—what kinds of emotions and feelings did you experience? Was there stress associated with the change? As you reflect on that change, what lessons did you learn that you can use to help others who are facing changes in the workplace from the implementation of new technologies? What did you learn about yourself and your reaction to innovation and change implementation?

WWW EXERCISES

1. Use your favorite job listing Internet site. Look for postings for e-commerce managers. Find at least ten and summarize the qualifications listed. How does this list compare to the one you developed in Relating to Your World exercise 1? What does this tell you about what we believe we need in personnel and what we seek when looking for employees?

2. Choose a company that you believe is innovative. Go to their website and find how they describe innovative processes in their organization. How does what you learned in this chapter align with what the company does?

3. Find an article or website that provides guidelines for implementing innovation in organizations. What do you think of the advice given? Compare the advice you find to the advice your classmates find.

AUDIT EXERCISE

Figure 4.7 has a checklist for innovative organizations to consider. How could you use this checklist to determine an organization's readiness to implement an innovative strategy? Be specific about how you would use the checklist. Identify at least one other major area that should be examined. What questions should be asked in that area?

DISCUSSION QUESTIONS

1. Three questions that link innovation planning and implementation were discussed at the beginning of this chapter. How does the answer to each of these questions affect the organization's readiness for implementation?

2. What are the four key issues that should be addressed in implementing an innovation strategy? Briefly define and review the issues.

3. How does implementation differ for product innovation and process innovation? Which of the four key issues do you believe is most affected? Explain.

4. How do you design a knowledge-based culture using the training and development processes? What type of innovation is this in a successful high-tech firm? In a stable firm with little external change? Explain.

5. "Meeting industry standards" reflects the interaction between evaluation and implementation. The plans of a firm are affected by industry standards. Using Figure 4.1 and Real World Lens 4.2, discuss how the industry standards affect the firm's approach to the three boxes (planning, key needs, and implementation).

6. How are the innovation processes used at GlaxoSmithKline similar to those used at Daikin? How are they different?

PART TWO OPENING CASE: GLAXOSMITHKLINE

1. Based on GSK's past performance, what do you believe are the critical implementation issues for GSK? Justify your answer.

2. With the restructuring of 2008, GSK made some fundamental changes in its alignment and fit processes. How might these changes affect the other critical areas of implementation?

KEY TERMS

alignment 127
delegation 112
diffusion 125

formal leadership 115
informal leadership 115
leapfrog 110

mentoring 123

NOTES

1. http://www.greencarcongress.com/2009/09/automakers-fcv-20090909.html.
2. Ewing, J. 2007. Upwardly Mobile in Africa: How basic cell phones are sparking economic hope and growth in emerging—and even non-emerging—nations. *Business Week*, (4052): 64–71.
3. Jones, T. 2008. Inner Strength: Mike's Express Carwash builds success through its employees. *Modern Care Care*, (Apr.): 46–53.
4. Armbrecht et al. 2001. Knowledge management in research and development. *Research Technology Management*, 44 (4): 28–48.
5. Stone, F. 2004. Deconstructing silos and supporting collaboration. *Employment Relations Today*, 31 (1): 11–18.
6. Barsh, J., M. Capozzi, and J. Davidson. 2008. Leadership and innovation. *The McKinsey Quarterly*, 1, 37–47.
7. Anonymous. 2008. TradeKing; TradeKing Receives 'Official Honoree' Distinction by 12th Annual Webby Awards. *Business & Finance Week*, (Apr. 28): 122.
8. D'Aveni, R., with R. Gunther. 1995. *Hypercompetitive Rivalries*. New York: Free Press.
9. Buckingham, M., and C. Coffman. 1999. *First Break All the Rules: What the World's Greatest Managers Do Differently*. New York: Simon & Schuster.
10. Kotter, J. 1995. Leading change: Why transformation efforts fail. *Harvard Business Review* (Mar.–Apr.): 59–67.
11. James, W. 2002. Best HR practices for today's innovation management. *Research Technology Management*, 45 (1): 57–60.
12. Thurm, S. 2006. Companies struggle to pass on knowledge that workers acquire. *Wall Street Journal* (Eastern edition), January 23. B1.
13. Carvin, B. 2009. The great mentor. *T & D*, January, 46–50.
14. Bandura, A. 1977. *Social Learning Theory*. New York: General Learning Press.
15. Barney, J. 1991. Firm resources and sustained competitive advantage. *Journal of Management*, 17 (1): 99–120.
16. Birchall, J. 2009. Order online and pick up at the drive-through; Kmart and its US rivals are starting to experiment with hybrid shopping outlet. *Financial Times*, (Jan. 28): 14.
17. Rogers, E. 1995. *Diffusion of Innovations* (4th ed.). New York: Free Press.
18. Linton, J. 2000. The role of relationships and reciprocity in implementation of process innovation. *Engineering Management Journal*, 12 (3): 34–38.
19. MacMillan, I., and R. McGrath. 2002. Crafting R&D project portfolios. *Research Technology Management*, 45 (5): 48–59.
20. Cusumano, M., and A. Gawer. 2002. The elements of platform leadership. *MIT Sloan Management Review*, 43 (3): 51–58.
21. Crawford, C., and C. DiBenedetto. 2003. *New Products Management* (7th ed.). Philadelphia: McGraw-Hill.

CHAPTER 5

Evaluation and Control in Innovation

OVERVIEW

This chapter addresses the evaluation and control processes when a firm focuses on an internal innovation strategy. Evaluation and control are ongoing processes in the firm. Initially, the manager evaluates the current environment (internal and external) and future trends to determine if there is a need for change in the strategy, goals, or implementation processes. Control is the next step in the innovation process as the manager makes adjustments in the strategy and goals or the organizational actions as a response to the evaluation. This adjustment process involves making decisions about the firm's plans for the future. Some of the specific issues related to evaluation and control for innovation are:

- Determining if the firm is achieving the desired outcomes

- Doing periodic gap analysis

- Ensuring appropriate controls for personnel: financial, strategic, and cultural

- Designing a support structure for evaluation and control processes

- Finding and sharing best practices, including the establishment of quality programs

---

INTRODUCTION

Once a firm chooses internal innovation as a strategy to gain competitive advantage, it must determine what it hopes to accomplish (plan), then seek to achieve those innovation goals (implementation), and ultimately, determine if it is making progress toward those goals (evaluation). If it is not meeting those goals, it must make the necessary changes to move in the desired direction (control). Therefore, evaluation and control test whether the organization's goals are being met and if not, making changes necessary to achieve the goals or amend the goals. The organization cannot assume that its goals and actions will automatically lead to success or that its goals and objectives will remain relevant over time.

As noted in Chapter 1, too often, those managing technology and innovation do not undertake sufficient evaluation and control activities in the innovation process. Instead, the joy of creating the new technology is the focus of the individuals in the firm, not whether the technology is successful in accomplishing the firm's strategic goals. Individuals may produce a product or process that makes significant technological advancement, but if the firm's goal is to earn a given level of profit or obtain a certain market position and the product does not achieve that, it is not a success. Thus, the organization must monitor the outcomes of its goals and performance, making adjustments as needed. Evaluation and control efforts should be part of the ongoing processes within the firm.

EVALUATION AND CONTROL PROCESSES

One of the key reasons for evaluation and control is that the firm's environment will change over time. As the environment changes, the assumptions on which the goals were based, and ultimately the goals themselves, can become irrelevant. For example, when a competitor develops a new, better way of creating a product that results in either a less expensive or improved product, the firm cannot continue with the same goals and actions it established before the change in the competitive environment. If it does, the firm can find itself at a disadvantage in the marketplace, and this disadvantage can lead to failure.

It is tempting to believe that firms will adapt as their competitive environment changes. However, there are numerous industries, such as printing, steel, and automobile manufacturing, where new processes were introduced, and established competitors ignored the changes. Ignoring change can lead to a firm's decline. To illustrate how dramatic environmental changes can be, and how firms must respond, consider the over-the-counter drug industry. In the 1980s, an individual tampered with packages of Tylenol. This person bought several bottles of Tylenol, emptied the contents of several capsules from each bottle, and refilled those capsules with poison. The individual then returned the bottles to stores where unsuspecting consumers bought the tampered-with product. Several consumers were poisoned by the tainted capsules. One outcome of the 1980s Tylenol poisonings was a new packaging design for most over-the-counter medications. These new packages include the wrapped tops and foil sealed bottles used today. Firms that did not adapt quickly, in a year or

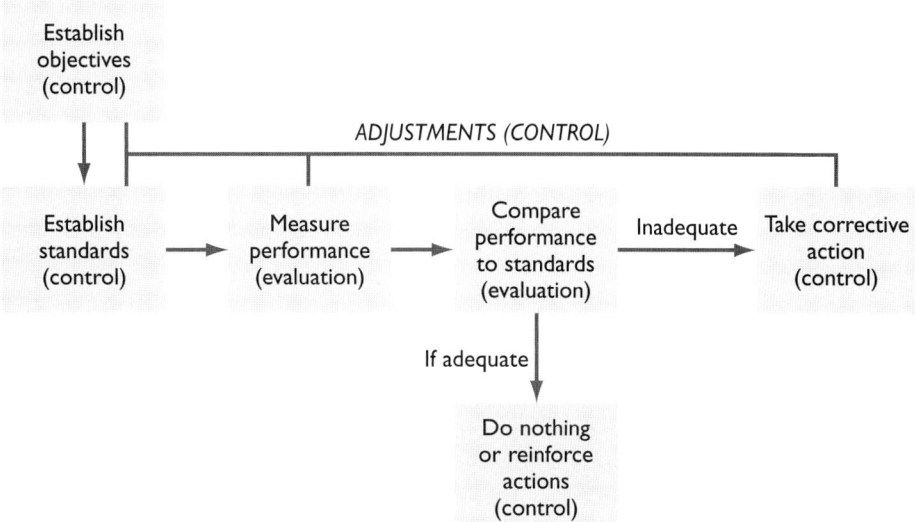

FIGURE **5.1** The Evaluation and Control Process

less, found themselves at a serious competitive disadvantage in the over-the-counter medicine market.

Figure 5.1 summarizes the elements of evaluation and control. This figure illustrates the most commonly used control method: **cybernetic control**. The term *cybernetic control* comes from the biological sciences and refers to the nervous system that serves as a means to provide feedback to the brain. This feedback allows the body to make changes in actions: if you touch a hot dish the pain tells the brain to withdraw the hand. In Figure 5.1, you should see a similar type of feedback in which the firm has information that allows it to make changes in its actions. This model is the foundation of what is discussed in this chapter.

Evaluation and control processes are commonly combined in a business. They are combined because they are intertwined processes in which the firm asks itself how it is doing and what does it need to change to be successful. These questions should be asked often and in connection with each other. However, in learning about evaluation and control, it is helpful to separate the two concepts and examine them separately so that each can be understood better.

EVALUATION

Evaluation of a firm's actions is built around three key questions:

1. Where are we now in comparison with where we want to be?
2. What lies ahead that can affect us either positively or negatively?
3. Where will we end up if we continue on this path?

Answering these questions produces what is referred to as a **gap analysis**. The organization is looking for any potential gap between what it wants to occur, what actually has occurred, and what is likely to occur. A gap analysis is particularly important for internal innovation. As noted in Chapter 3, the time frame for completion of internal innovation is generally longer than for acquisition. If a gap is not addressed early in the internal innovation process, the firm may deviate from its goals to the point that its performance is negatively impacted, and there is no immediate way to address the firm's problems. In contrast, if problems are addressed early, the firm can make the changes necessary to still achieve its goals. Each step in the gap analysis is examined next.

Where Are We Now?

The first step in the gap analysis is to determine if the innovation strategy is working; that is, where are we compared with where we want to be with this innovation process. This is an evaluation of how the firm is doing right now in moving from its current position to its future position. In the planning process, the organization establishes goals and objectives. As discussed in Chapter 3, these goals and objectives typically include both short-term and long-term time frames. The long-term goals and objectives may take years to attain. However, even for an objective that is five years away, there are immediate steps—short-term goals and objectives—that must occur to achieve the long-term goals. The firm should judge its current status against both its short-term goals and objectives and its progress toward long-term goals and objectives.

There are several key evaluations that a technological firm should conduct to understand its current status. These include:

1. Strategic environment evaluation: The firm should examine the strategic direction of the industry and the major competitors' strategic actions in that industry. This segment of the evaluation should include Porter's five-forces analysis, which was discussed in Chapter 2, as well as analysis of other issues that might affect the industry.
2. External environment evaluation: As noted before, not only are customers critical to the firm's success but so are the competitors of the business. The firm needs to examine if the customers are satisfied. Another external concern is the networks with external constituencies such as suppliers and regulators. Are these networks being maintained or built if not already present? This analysis should include both the quality and number of external linkages.
3. Information systems evaluation: Information systems of the firm include the general system for communication processes that allow it to get the right information to the right team/person at the right time to make a good decision. Does the firm have these processes? Is the information being processed by the system appropriate? Earlier, in Chapter 4, paralysis by analysis was discussed. Information systems evaluation helps alleviate this danger by ensuring that the right information in the right amounts is being generated and it gets to the right person on time.
4. Structural analysis evaluation: The firm should evaluate if the structures and processes in the organization allow the emergence of useful innovations. Early coordination and involvement across the organization help

this process. Because the structure of the organization should enhance coordination and communication, this approach enables individuals to explore products and processes early and often.

Although each of these key evaluations should be present in the organization, usually one or two will be more important.

In conducting its evaluation of current status, the firm needs a richer examination than simply comparing its goals to outcomes. The manager should periodically examine if the goal is still valid. To illustrate, a biotechnology firm may set a financial goal of 5 percent growth per year. If the firm underestimated the growth in the market and the entire market is growing at 25 percent per year, meeting the goal of 5 percent growth per year may actually mean the firm is not performing well. Similarly, the business may have a plan to dominate a given product domain over the next five years. However, the given domain can easily disappear within the five years. Thus, dominating a disappearing domain may be a reflection of inability to respond. For example, if in its five-year strategic plan a firm had a goal to dominate the development of banner advertising on the Internet, its goals may need to be revisited. Banner advertising firms typically have lost money, and today, other types of advertising have emerged such as pop-up advertisements and interstitial advertisements that appear as the user moves from one story to another. Thus, a firm may be dominating an industry, but if that industry is no longer viable or is likely to face emergence of new products that consumers prefer, it does little good to continue with such goals. Evaluation must be more involved than simply checking to see if the goal and objectives of the firm are being met.

The organization must determine how much variation it is willing to accept in its goals. There may also be variation in the goal set and the results achieved. This variation may be small. For instance, if sales increased by only 4.9 percent, not the desired 5 percent for the quarter, is that a material difference? The organization will need to determine not only what variation is acceptable but if it is acceptable for all goals and objectives. Variation in some goals may not be critical. However, in others, slight variation can bring negative results. Therefore, in evaluating the gap, the managers of the organization need to use their judgment, not simply rely on a yes–no decision.

The gap analysis proposed here is not unique but is similar to many other tools used by practitioners in business. For example, David Norton and Robert Kaplan[1] developed a procedure called a Balanced Scorecard. In this method, it is argued that the firm needs to look beyond financial measures to other metrics related to customers, business processes, and organizational learning. The organization looks not only at goals but also at how close they are to the goals and the value of the goals relative to the firm's strategy. This method argues for a rich analytical process similar to what is presented here.

What Lies Ahead?

The second step in the gap analysis is the evaluation of what lies ahead that can affect the firm either positively or negatively. This information is obtained by scanning the environment for changes. The environmental issues discussed in Chapter 2 come into play to answer this question. As was discussed in that chapter, the environment shapes the goals that are generated for the firm. However, the firm must also evaluate how the environment might change because of the firm's actions. If the firm is a manufacturer of computers for consumers and the economy appears to be slowing, the firm may experience a downturn. The purchase of a computer by a consumer is highly elastic. Changes in the economy can affect sales, and if the economy slows down, consumers will put off the purchase of a newer computer until they feel more financially secure. Therefore, the computer firm must monitor the economic outlook and respond accordingly.

Although an organization can be innovative and build a new product or develop a new process, the market will determine if the product or process will have a useful life. If competitors introduce a new product, the evaluation of the future outlook for another firm's product(s) may radically change. To illustrate, most college students today have never owned an electric typewriter or a cassette player—standard equipment for students only thirty years ago. Improvements in manufacturing processes, like robotics, make it possible to produce electric typewriters and cassette players at a lower cost than in the 1970s. However, the demand for electric typewriters and cassette players is very, very low despite the fact that the items that could be produced today are of the highest quality and lowest cost in their history. Often, companies do not look for future trends or competitor actions and find themselves improving products and processes that will soon be obsolete.

The types of measures that the future evaluation should consider in the innovative strategy can be classified into three broad categories.

1. Measures of specific outcomes that have been identified to produce competitive advantage in the future: Examples of these include the number of patents, number of new products developed, and new process technologies put into place. These measures indicate efficiency in innovative processes and effectiveness in creating and responding to environmental changes.
2. Measures of outcomes that impact future competitiveness: The primary ways to do this are through customer satisfaction surveys, measurements of quality in inputs and outputs, and change in the number of returns because they all indicate the potential for future repeat customers. The three areas of concern are strategic integration across the value chain, customer needs, and supplier orientation.[2]
3. Measures of future strategic capabilities: This includes items such as reputation, and growth in revenue. These measures indicate the strategic capability of the firm is being enhanced or at least maintained.[3]

Progress in these categories indicates the firm's potential for positive future outcomes. If the firm can build a sustainable competitive advantage through inimitable resources (e.g., patents) or build capabilities that are ahead of competitors, then the firm has a better chance for long-term success. This is, of course, dependent on the firm pursuing the correct course of action. Corning is such a company. The firm is over 150 years old and still leads in glass product innovations. One of the measures it tracks for understanding strategic performance is patents and related products. Corning developed optical fiber and is still a dominant player in this field. However, the firm is constantly updating and building on this older product with new related products and patents. Recently, Corning announced that it has cracked the green light problem (the light frequency needed to make pictures from small projectors clearer) in such cable—next it hopes to take this advance and create a sharp video projection system that fits into a cell phone[4]—a product that will create significant new revenue streams when perfected.

Where Will We End Up If We Continue on This Path?

The last question is a *fundamental* evaluation of whether a different direction for the firm is needed. The firm should ask periodically, "Where are we likely to end up if we continue on this path and is it where we thought when we developed the plan?" Such an evaluation does not happen as often as the comparison between current performance and the stated goals and objectives of the firm. However, the evaluation process periodically needs to look for different opportunities or paths that may have emerged recently or that previous evaluations did not reveal. The firm should not assume it is on the right path simply because it met the current goal. A periodic introspection most often confirms the goals and plans of the firm; however, it can open new horizons that have not been considered (or even known) previously.

For example, 3M initially started out as Minnesota Mining & Manufacturing. Its primary business was mining grit for the sandpaper it produced. It developed Scotch tape as a result of experimentation on glues for making better sandpaper. The founders of 3M never imagined the firm that exists today. Instead, different generations of managers adjusted the plan, increased the innovative culture, and shifted the organization toward new directions. 3M evolved into a firm that makes tape products, medical equipment, and, of course, Post-it notes among other things. Each time the company made a determination that there were other opportunities that had greater potential for the firm, it shifted its direction.

Henry Mintzberg refers to this as an emergent strategy.[5] In technology-related firms, such emergent strategies are particularly critical. Imagine a firm like Apple Computer. No one predicted that the firm's most promising and profitable product would be an iPod for music and not its computers that were so dominant within the firm for many years. The managers at Apple had the capability to recognize an unexpected success and take advantage of this emergent opportunity. The release of the firm's iPad reflects this same emergent strategy.

The minivan provides another example of the importance of recognizing such emergent opportunities during the firm's evaluation. The idea of the minivan was originally developed and rejected by Ford Motor Company. Lee Iacocca was at Ford Motor Company at the time of the rejection. He recognized the potential of the minivan, and he took that rejected idea with him to Chrysler when he was asked to manage that troubled firm. There, the idea was shared and developed. Thus, what Ford viewed as a detour from its plan, Chrysler treated as a new, innovative product that resulted in one of the most profitable automobiles for Chrysler and significantly helped rescue the firm from bankruptcy.

An organization can benefit if it can recognize an emergent opportunity. However, problems can also come to an organization that is not oriented toward taking advantage of emergent opportunities. The organizational context at Ford was not supportive of the minivan innovation, whereas the context at Chrysler was. Chrysler knew from its evaluation system that they were in a major crisis. Because of this recognition, the firm was willing to look further for innovative ideas. The organizational context and how supportive and/or desperate it is will determine if the firm will succeed in identifying such emergent opportunities when it conducts its evaluation.

Too often, individuals and organizations look at deviation from the plan as a negative. When most people encounter a detour sign while driving, they do

---

5.1 REAL WORLD LENS

What Can Go Wrong?

The evaluation and control process in research and development is critical. Consider the following facts:

- It takes one hundred research ideas to generate about ten development projects.
- Only two of those ten projects will make it to commercialization.
- Only one of the two will make money.
- In Great Britain and the United States, about half of the development money is spent on projects that never reach the market.

The result of such development patterns is that firms such as Microsoft will spend over $3 million on development costs for each patent it files. There are numerous stories of firms that failed at research and development processes. For example, in developing a new aircraft engine, one test the engine must pass is the chicken test. As you might expect, an aircraft engine must be able to have a bird hit the engine, and the blades in that jet engine must remain structurally sound. The bird cannot destroy the engine if it accidently strikes it. If a jet engine cannot pass this test, the ability of the jet to take off safely is very limited. One of the major aircraft manufacturers in the world is Rolls-Royce. When developing a new engine, the firm assumed that its new carbon fiber blades could withstand the strike of a bird. They did not when ultimately tested. The result was that much of the engineering and development efforts prior to this point were wasted.

1. What lesson should Rolls-Royce have learned?
2. What suggested changes in the evaluation process would you make?

Reference

Parker, K., and M. Mainelli. 2001. Great mistake in technology commercialization. *Strategic Change*, 10: 383–390.

not envision an opportunity to see more of the community. Rather, the detour is viewed as an inconvenience. The same frequently happens with strategy. However, strategy is not always deliberate or formulated. Typically, there are elements of emergent strategies in successful firms. That is, successful firms are able to obtain a competitive advantage by having the capability to see new opportunities, develop relevant knowledge, and then ensure that knowledge is shared throughout the organization.[6] It is the capability of the managers to see opportunities and shift the organization through emergent strategies that often determines success.

CONTROL

After the organization has evaluated its performance along the three dimensions cited, the firm next answers the question, "What changes need to be made and where?" The decision to make changes or not is the beginning of the control process. This question also completes the key set of questions for the innovation strategy process. Figure 5.2 summarizes the key questions for each stage of the strategy process.

Types of Control

There are a number of ways to classify the types of control in an organization. For firms undertaking an innovation strategy, the control mechanisms that are available can be classified into three principal types:

- Financial
- Strategic
- Cultural

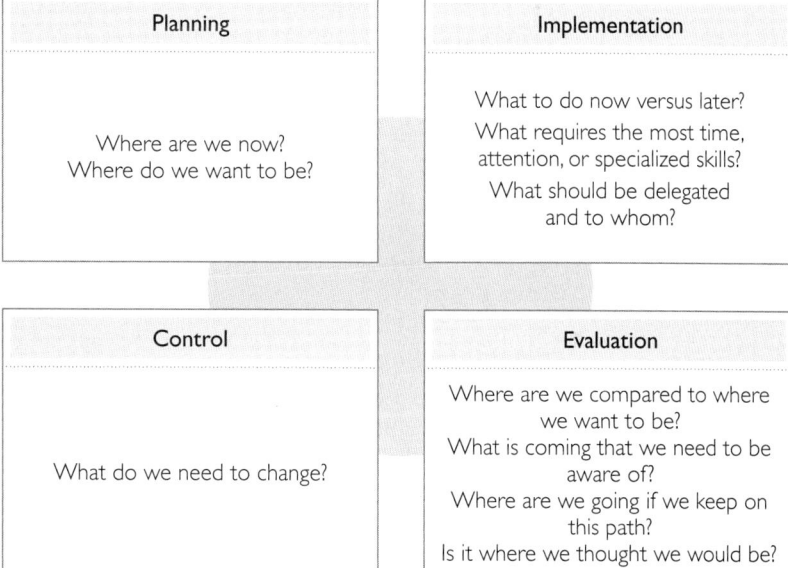

Planning	Implementation
Where are we now? Where do we want to be?	What to do now versus later? What requires the most time, attention, or specialized skills? What should be delegated and to whom?
Control	**Evaluation**
What do we need to change?	Where are we compared to where we want to be? What is coming that we need to be aware of? Where are we going if we keep on this path? Is it where we thought we would be?

FIGURE **5.2** Questions for Each Strategic Phase of Innovation

A firm will use a combination of these controls, although usually one will dominate. Each will be reviewed next.

Financial Controls

Financial controls focus on gaps between the desired financial outcomes and those actually produced by the firm. Therefore, the firm will have goals such as sales growth, profit, and expenses. Those goals can be short term (e.g., the financial performance of the firm this quarter) or long term (two or more years). The goals are then compared to actual outcomes for the time period. If a gap is identified, the firm employs methods to improve those financial results directly. For example, if the firm is not meeting profit goals, it may decide to cut costs. The quantitative nature of the financial measures makes it the easiest to evaluate and correct. However, the use of financial controls alone is usually inadequate. Financial controls simply show whether that particular financial goal is met. It does not address why the goal was or was not met nor does it allow a rich examination of the firm's activities.

Strategic Controls

Strategic controls focus on the firm's meeting of strategic goals. Recall that typical strategic goals can include: to be the market leader for a given product, to be first to market with a new product, or to be viewed as the most innovative firm in the industry. The measures of such domains are more qualitative and difficult to measure. However, for the long-term success of a business, strategic controls are critical. For example, if financial controls, such as profit, are the only measure of success used, there is little incentive to spend money on research and development, which is an expense that has only long-term impact.

However, R&D investment is critical to the long-term health of a technology-based firm. Therefore, innovative firms need strategic controls to ensure that the appropriate actions are taken today to help the firm tomorrow, even if the short-term outcome of these actions is costly. The creation of a future income stream is dependent on the pipeline of innovation that comes from the R&D activities in a firm pursuing an innovation strategy. In using strategic controls, the firm looks for gaps in its strategic goals and outcomes and then addresses them through strategic adjustments. These adjustments typically involve a substantial commitment of resources and may include actions such as expanding the firm's innovation efforts to develop a new product line or a new process within the firm.

Cultural Controls

The last type of control is perhaps the most important for a technology-focused firm: cultural controls. Cultural controls refer to the ability to have individuals act in a manner desired within the firm due to the culture that exists in the firm. Recall from the discussion of implementation in Chapter 4 that the culture of the organization aids the implementation of an innovation strategy. For example, an innovative culture helps ensure that information is shared, risks are taken within the organization, and actions critical to the success of the innovation effort are implemented. The culture of the organization can

	Advantages	Disadvantages	Examples
Financial	Quantitative Easy to interpret	Analysis paralysis is possible Can become narrow Internally focused	% profit increase from new products
Strategic	Sets direction More qualitative Fits environment	Hard to justify based on some financials Can lose sight of where the firm is	Increase market share
Cultural	Very behavioral Very qualitative	Requires managers to be involved on a more personal level	Value enhancement

FIGURE **5.3** Types of Controls: Advantages and Disadvantages

have a similar positive impact when trying to address problems that have been identified by gap analysis.

The strength of cultural controls comes from the relationships within the firm that encourage individuals to act in a certain way. The need for creative thinking requires an environment that encourages risk taking and supports change. However, the firm must also stay focused on creating value while taking risks and being creative. Cultural controls help ensure that the reality of value creation in an environment that nurtures creativity remains a focus for the firm's actions. Cultural controls are typically the least quantifiable types of controls used. Figure 5.3 summarizes the different aspects of each of these different types of control.

Illustration of the Use of the Different Controls

3M has long held that it wants to generate 30 percent of its revenues from products created in the last four years. This is a clear, measurable, financial standard. This standard can be compared to actual revenue sources and recently 3M has not met this goal. Because a gap has been identified between the standard and actual revenues, changes are being made to the existing structure to meet the desired financial goal.[7]

For example, the strategic goal for the firm is to be the industry leader in a variety of domains such as adhesive technologies. This goal has some elements that can be quantified; such as which firms have the greatest market share in the industry. However, interpreting the results of the quantifiable measures is a subjective judgment. If a gap in strategic goals is found, the strategic controls will move the firm to meet those strategic goals.

Finally, 3M wants to be innovative and creative with a high percentage of its revenue coming from new products. The fact that the individuals at the firm believe in and encourage each other to meet this goal is an example of cultural control. The cultural controls are the hardest to implement but have the most impact when implemented properly. Getting the employees to believe in a common goal, encouraging each other to act in certain ways to achieve that goal, and perhaps isolating those who do not pursue such goals are difficult to achieve. However, if the firm's culture supports internal innovation (like 3M's), the resulting organization has the norms and processes in place to have the greatest potential for sustained competitive advantage. Therefore, cultural controls are particularly powerful for encouraging creativity and innovation but are the hardest to create and enforce.

Organizational Levels and Control Factors

If a gap is identified between goals and performance, then adjustments must be made within the organization. The most common adjustments that may occur include:[8]

1. Rethinking business processes by trying to be more cross-functional
2. Looking for improvements by redesigning the processes the firm uses for its internal innovation
3. Empowering those involved in the innovative process to make key decisions
4. Determining if the goals do not match the capabilities of the firm and develop new goals

Making the changes necessary in any of these can be very difficult. As has been noted, organizations can be inflexible and unwilling to recognize the problems at hand. In part, this is why it is critical that the evaluation and control processes of the firm involve the entire organization. There must be support for the control mechanisms and making the necessary changes when gaps exist. Control is a responsibility throughout the firm. Each level of the organization has a different role, and the nature of control is different in each. Figure 5.4 illustrates the management decision level and the group responsible for exerting control at that level. Each of these levels of control will be reviewed in turn. Please note that just as in planning, the effective control process requires that each level be in sync with the levels above and below it.

Boards of Directors

At publicly traded firms, shareholders elect individuals to supervise the managers of the firms. These elected individuals make up the board of directors and they are accountable for control processes in the organization. These individuals have a fiduciary responsibility to the shareholders—that is, a legal responsibility to ensure that the best interests of the shareholders are served by the organization.

The board of directors sets the vision of the organization and provides an important frame for the control mechanisms in the firm. How the organization views change and how it acts and reacts to changes emerge from control mechanisms. The board should not typically be involved with the minor differences between expected and actual results. Instead, these individuals focus on the major gaps that may occur and address whether a radically different

FIGURE **5.4** Level of Organization and Control Concern

direction for the firm is needed. Specifically, the board should be involved with the following types of actions that impact control:[9]

1. Approving the financial support of the major new product and/or process development projects
2. Understanding the number of ways changes in products and processes can affect the firm's strategies and position in its environment and how changes in competitors' domains impact the firm
3. Appreciating and monitoring the risks and rewards created by technology and shifts in technology
4. Questioning technological trend analyses
5. Providing a strategic vision and focus for reviewing investments in technology
6. Using technology to enhance the value of information presentation and understanding, such as financial analyses, risk assessments, and market forecasts
7. Demanding timely updates on the progress of all major projects
8. Ensuring that the desired actions are taken during the control process by firing or hiring the CEO and other top managers

Top Management

Using the vision that the board approves as a base, the top management of the organization sets into motion the control actions and processes of the firm. While the top management will address some of the same strategic issues as the board, they do not usurp the board's position but rather reinforce it. The top management enacts the strategic choices of the board. Thus, they are the principal source of establishing the mission of the organization concerning control systems and processes.

To this end, top management must answer a series of questions when determining what innovation-based projects to undertake. The answers to these questions determine what will be presented to the board, how it is presented to the board, and what recommendations about strategic direction will be made to the board. These questions are:

1. Is the proposed innovation activity consistent with the long-term strategy of the firm?
2. How does what the firm is doing enhance the value of the firm today? How will it enhance the value in the future?
3. What is the cost/benefit? If the costs are great in the short-term, can the firm afford to wait for the payoff?
4. What happens if the technology is leapfrogged? Can the firm recover?
5. What projects should be discontinued?
6. How innovative is the firm's climate and culture? How can innovative thinking and activities be increased?

Divisional Managers and Team Leaders

The direct implementation of the control mission set by the top management of the firm is the responsibility of the divisional managers and team leaders. Thus, the divisional managers determine how to enact the mission and what control actions are needed when, where, and how to successfully complete the control process. These strategies on how the control system will operate then lead to more direct operational actions.

The team leaders, under the guidance of the division leaders, establish further operational details of the control system. These details begin to establish the actions that will take place and include setting the specific goals and objectives that the control system is to achieve. The questions to be addressed include:

1. What specific actions will take place?
2. What kind of knowledge (training) needs to be in place to implement the proposed innovation?
3. Do you have the right people on the team and in the department that is responsible for implementing the innovation?
4. How will you tell if you are making progress toward the desired outcomes?
5. Do individuals on the innovation implementation team understand what the expectations are?
6. Do the team members who are making changes in their work environment believe the organization will help them in the change process?

Department Managers and Team Members

The operational details established by managers and team leaders then lead to the actual tactical actions by department managers and team members. At this level of the organization, the key control questions include:

1. Is the product we are producing meeting criteria for success—within the firm and with customers?
2. Are productivity rates improving with the innovation? Are the improvements at the desired level?

3. Do those whose work is directly affected by changes in process and product understand the "new way of doing things"?
4. What are the key success factors for successful implementation and are they being tracked appropriately?

In many organizations, a key unit at the department level that directly affects evaluation and control is the quality control department. In association with other managers and team members, quality control personnel provide the tools, processes, and mechanisms for day-to-day evaluation and control by department managers and team members. These tools, processes, and mechanisms provide the data that help the firm analyze and understand where improvements are needed and how the improvements might be implemented. Because the quality expectations for technology-focused goods are typically high quality control is an essential part of a strategic approach to innovation.

Quality control departments typically establish quantitative measures to determine the acceptability of different aspects of a product. However, they may also conduct broader evaluations about whether the firm is making the right decision on a range of topics that can impact excellence. For example, the superiority of a firm's output is often dependent on the quality of inputs. The expression "garbage in, garbage out" is true for data, manufacturing, and quality control processes. If the inputs are poor, there is little hope that the output will be acceptable. Thus, today, most quality control departments monitor the firm's total system—inputs, transformations, and outputs. Because of the critical role that quality plays in a strategic approach to innovation, we shall examine the topic in greater detail.

Overview of Quality Issues

In large measure, the view that quality is critical to the firm is due to the work of Edwards Deming.[10] Deming developed the foundations for his approach to quality while working for AT&T. At the time he worked at one of AT&T's manufacturing facilities, there was one supervisor for every two employees in the plant. Dr. Deming felt that there had to be a better way to produce excellence. He developed a quality system for business based on this experience; however, he had limited success convincing firms in the United States to use his methods because the radical change was difficult for many United States managers to accept the critical role of quality in firm strategic success.

Deming had a doctorate in statistics. Following World War II, he was hired to conduct the census in Japan. The Japanese had been world-class producers of military items prior to the war, but their consumer goods were considered shoddy. Because the constitution imposed on Japan following the war prohibited manufacturing military equipment, Japanese firms needed to develop world-class consumer goods to survive economically. Deming convinced the keiretsu in Japan—the vertically and horizontally integrated firms that dominate the Japanese economy—to adopt his system of quality, and the highly integrated nature of business in Japan led to its widespread adoption.

It was not until the early 1980s that quality control became a substantive issue in the United States. At that time, it was widely perceived that because of the superiority of their goods, the Japanese would dominate the world economy; the United States was losing its competitive advantage in the worldwide market. After studying Japanese organizations, United States business leaders discovered the key to Japanese success was the quality of their products, and that a United States citizen had developed the system the Japanese were using. The result was the quick and widespread adoption of quality efforts in the United States. Today, quality is assumed to be critical, and there are many different proponents of quality with many different approaches. This has led most firms to develop their own unique quality control programs that match their unique setting. However, a review of the fourteen points that Deming originally proposed shows that many of these concepts are still central in today's quality programs. The points and a brief interpretation (in parentheses) of each are as follows:

Deming's Fourteen Points of Quality Control[11]

1. Create constancy of purpose. (Have a long-term view.)
2. Adopt a new philosophy. (Have a substantive commitment to the quality program.)
3. Cease dependence on inspection. (Limit the variance between any two items produced by initially doing it correctly.)
4. Move toward a single supplier. (Using one supplier results in less variance in inputs, which limits variance in outputs.)
5. Improve constantly. (Quality is a continuous process.)
6. Institute leadership. (The problems that arise are not due to the workers but instead to the system in which they work. The key is to change the system, not the workers who operate it.)
7. Institute training. (Show employees how to perform their jobs; continuing education is critical.)
8. Drive out fear. (Encourage change by limiting fear so that employees will try new processes and identify things that need to be changed.)
9. Break down barriers between departments. (Provide high-quality outputs to other departments within the firm. A high-quality product will be provided to the external customer only if high quality is maintained within the firm.)
10. Eliminate slogans. (The goal is to change the system in which the individuals work. A slogan to do better will have no impact if the system is the same.)
11. Eliminate management by objective. (To focus exclusively on numbers misses the strategic changes and actions that are needed. A long-term focus on quality requires more than just a short-term number fixation.)
12. Remove barriers to pride of ownership. (Employees need to understand how their work impacts the final output of the firm.)
13. Institute education. (Training and education as to the quality process are important.)
14. The transformation is everyone's job. (Quality must be everyone's goal if the firm is to produce quality outputs.)

Premier Kayak—Gap Analysis of Current Performance

The current online booking system typically employed by kayaking tour companies does not prevent the overbooking of trips and lessons. Premier Kayak was no different and its system only recorded if someone booked an activity. Overbooking resulted in too many customers for the number of kayaks available for the tour or to safely teach the lesson by the allotted instructors.

This has proven to be ineffective, as employees frequently have to reschedule the customers or, even worse, refund the money because the client is upset or unable to do the tour at a different time because of their schedule. For the company and its workers, this is embarrassing and often results in having to deal with very irritated customers. The outcome also placed a strain on the managers to have to change schedules, move boats from different locations, and deal with customers who are not happy.

An alternative to the simple online booking system that is wide spread in this industry is the antiquated paper reservations system where people call in and the firm lists the type of tour or surf lesson, the time of the tour, which ones have booked, the customers going on that particular tour, their contact information (a cell phone number), and the guide for the tour. This book is kept at the office and staff of the firm answer phones when they can and book tours, providing customers with information about services, and taking the payment information from the customers.

The information in the manual system is of limited use since the information in the book is not accessible to all of the managers, kayak guides, firm secretary, and customers at the same time, and from different locations. This results in many problems such as guides not showing up for scheduled tours, guides confusing the meet times, customers confusing the meet times, customers showing up to the wrong location, guides showing up at the wrong locations, and at times overbooking tours. Once again, employees have to reschedule or refund the money. Similar to the online booking system, the paper system is flawed in that a limited number of people can see it at a given time. This results in mistakes that can potentially cause a loss in customers and potential business.

Premier Kayak of North Carolina sought to overcome the problems of online systems and paper systems. They also want a system that is developed to enhance communication among employees. To develop this ideal system they employed a methodology in which the firm generates a model that captures all of the process for making a reservation. From this model they then sought to develop a registration system that is seamless where customers can schedule tours and pay for the tours on line. The system also gives Premier Kayaks the necessary information about the scheduled tours and customers signed up at each location. The system created a central point of reference for any staff member to refer to in the

(continues)

REAL WORLD LENS *(continued)*

event of an inquiry. This system also has a history feature that allows staff to check for previous reservations and possible errors plus limits the chance of overbooking since the system limits the number of individuals that can sign up for a given activity.

1. What gaps do you see that the resulting system focused on?
2. What was the key innovation that Premier Kayak's reservation system implemented to avoid making the reservation system a painful process for customers?

Reference

Martin, F., H. Hall, A. Blakely M. Gayford, and E. Gunter. 2009. The HPT model applied to a kayak company's registration process. *Performance Improvement*, 48 (3) 26–35.

IMPLEMENTING EVALUATION AND CONTROL

As noted at the beginning of the chapter, evaluation and control are a continuous process in the firm. Thus, as a practical matter, evaluation and control will impact the planning and implementation phases of the innovation process in the firm. Figure 5.5 illustrates these interactions and demonstrates that each of these strategic concerns builds on the prior one as analysis and control actions occur.

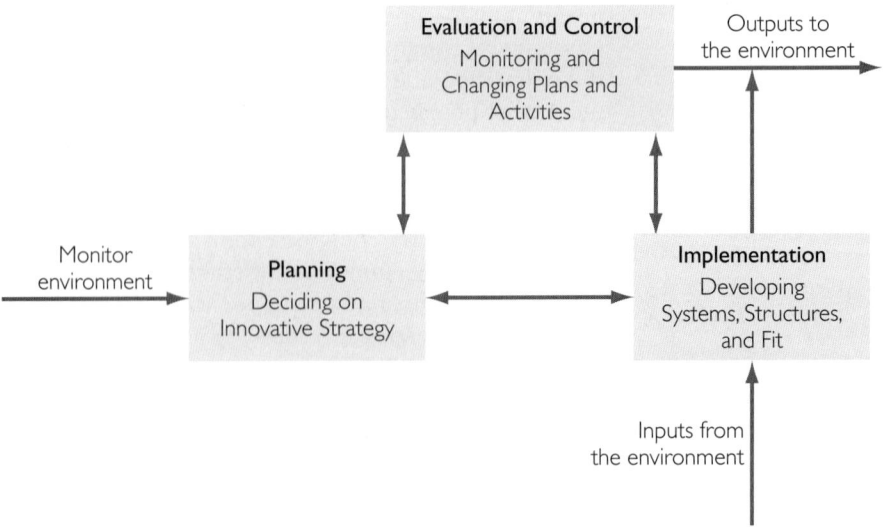

FIGURE **5.5** Interaction of Strategic Concerns

Strategic Elements	Analyze	Act
Planning	What are the key issues for our environment? What are our competitors doing?	Develop template for organizational actions Set measurable objectives for innovation
Implementation	What do we do now? What can we do later? What actions are needed to close the gaps?	Set up structure to support strategy Test progress against the plan
Evaluation and Control	Are we measuring what we are interested in? How do we make adjustments?	Develop means of analysis and information sharing Make adjustments

FIGURE **5.6** Strategic Process Questions

The key questions for each of the three aspects of the strategic process are summarized in Figure 5.6. The integration of the three domains—planning, implementation, and evaluation and control—will now be discussed.

Integrating Evaluation and Control: Planning

The analysis of the organization's environment should occur as the goals and objectives for the innovative strategy are developed. The firm should conduct this process of monitoring the environment regularly. This requires that the organization examine its environment to be sure that there are no major changes among competitors and no changes in external variables looming on the horizon that could affect the firm. As noted previously, the strategy of the technology-focused firm is not developed in isolation. The goals and objectives of the organization should be developed in light of potential actions and reactions by competitors.

As discussed in Chapter 3, the planning stage should result in outputs such as a mission and measurable goals and objectives. All of the goals and objectives should be action oriented and help achieve the mission of the firm. These goals then form the basis for the comparison conducted during evaluation.

Integrating Evaluation and Control: Implementation

The concept of strategic fit becomes critical in the integration of implementation and evaluation and control. If the pieces of the organization do not fit together, then it is unlikely that the organization will reach its goals. In other words, the gap will increase rather than shrink during the evaluation and control effort if there is not a strategic fit. This will result in less efficiency and less effectiveness. A means to analyze a gap, once it is identified, is to examine the inputs into implementation and the outputs from the processes that are being used. This should produce insights into the nature of the gap and what corrections are needed.

In general, the actions that may be required to fill such a gap are easy to state and difficult to accomplish. For example, structures, processes, and procedures need to support innovation. If the goal is to get a product to market within the next ninety days, then the processes and structures to achieve the goal must be in place. Often, managers make the decision to throw more resources at a problem; however, this does not always address the real issues or needs that exist. Most of the problems associated with innovation failure are not technical but rather the softer implementation issues that deal with how the process is set up and managed. Thus, when a gap is identified through evaluation, the implementation of the control aspects must be carefully designed and implemented to ensure that the actions taken produce the desired outcome.

Often, there is too little focus on matching the goals with the activities; senior management adopts "newness" without verifying that the innovation will help the organization achieve its goals. There are several critical decisions that must be made during implementation to ensure the firm does not get mired in an innovative strategy that leads to disaster.[12]

1. Ensure that management support and project goals are clearly understood and in the forefront of the organizing effort.
2. Assess the skills required so that the people needed to meet the project objectives are involved with the process.
3. Set up the infrastructure for the most likely scenario, but have contingencies for other strong possibilities.
4. Don't underestimate the influence of the internal culture and the external environment.
5. Set up a monitoring system that is an integral part of implementation and addresses potential technical and market risks. The approach taken in implementing any strategy or any project to support a strategy should contain the monitoring of potential risks.

The issue of organizational structure and its impact on evaluation and control merit several specific comments.

Organizational Structure

One particular concern during the integration of evaluation and control with the implementation process is the organizational structure that the firm will employ. The purpose of structure is to indicate lines of communication and coordination, and as such, structure should change as processes, products, and systems change. Typically, as the organization evolves and grows it develops a more complex organizational structure. For example, a new organization generally employs a simple structure. In this setting, the founder or senior manager is usually involved in all major decisions. There is limited specialization among the workers and an easy flow of information throughout the organization. Thus, in the evaluation of performance, everyone's actions are clear and known by the CEO, which makes evaluation more direct. Similarly, this structure allows rapid responses within the organization so control responses can be very quick. Although the directness and simplicity have many appealing features, this structure can only be used by very small firms. If large firms try it, the result is likely to be chaotic. The resulting problems can result in the firm implementing a functional structure.

In a functional structure workers are typically divided into professional domains such as accounting, marketing, or manufacturing. This structure allows the organization to separate the employees into efficient groupings as the number of employees increases. However, these groupings make communication between different types of employees more difficult and can result in the employees focusing more on their individual professional unit than on the entire firm. Thus, as the structure grows more complex, so does the nature of the evaluation and control that is needed.

The organization that continues to grow may find that it is in a number of different businesses. As a result, it will then develop a **divisional structure** in which each business unit has its own top manager and its own function-based groupings of employees. This is done to build the flexibility of smaller firms into the structure of a larger firm. The larger a firm becomes, the more likely it is to become slower in decision making and information processing. If a corporation grows large enough, it can have numerous divisions that perform a wide variety of businesses. For example, General Electric has over sixty different businesses; each could be a separate division. However, that many divisions are too hard for the CEO to manage. Therefore, GE organizes its businesses into eight major segments: Aircraft Engines, Commercial Finance, Consumer Finance, Consumer Products, Equipment Management, Insurance, Entertainment, and Power Systems. These eight areas are referred to as **strategic business units (SBUs)** that provide the basis of the organizational structure where divisions are organized into coherent groups. The head of each SBU reports directly to the CEO. The heads of the sixty various businesses report to the head of their respective SBU. The creation of SBUs can be an efficient means to better structure the business, but it creates another layer of management in the organization and can make it even more difficult to judge performance and implement controls.

As organizations grow larger, their evaluation and control become more difficult. These larger organizations seek out structures that allow them to manage people and processes more efficiently. The resulting structures make it more difficult to develop effective firm wide evaluation and control mechanisms because the nature of each group's contribution is not as clear. In addition, the changes needed when a gap between goals and outcomes is found are harder to define and implement.

Size is not the only consideration in deciding when and how structure should be changed to better monitor the organization. Other considerations include level of centralization, level of standardization, how formalized the firm is, and how many linkages are formed in the organization. For truly innovative firms, an ambidextrous approach to structure[13] is usually best. This approach allows the firm to have two different types of structure—it is a hybrid of the more mechanistic functional structure and the more organic matrix or network structure. The purpose of the structure is to have multiple approaches to provide space for pursuing short-term efficiencies while exploring the opportunities of long-term innovation.

Validating the Evaluation and Control Process

Evaluation and control should not be treated as a static process that never changes. Instead, it is a process that must change as the environment changes. The environmental change may be due to competitors' actions, changes in industry trends, or changes in the broad economy. To ensure that the evaluation and control effort is looking at the correct issues, the firm needs to ask two questions periodically.

1. Are we measuring what we are interested in?
2. How do we make adjustments?

In taking actions to implement these questions, the first difficulty is to develop a successful system that measures what the organization wants to measure. It helps if the organization has clearly stated goals. With clearly stated goals, the ability to develop relevant measures becomes easier because the organization can focus on those specific items and think creatively rather than having to think too broadly.

However, even with clear goals, it is difficult to know if what is measured is the key to success. For example, two projects may have only slightly different resource requirements with very different potential outcomes. However, the resources needed are just one part of the equation. If the organization only evaluates the resources needed, then the project with the most positive potential may be overlooked.

Managers must verify that adjustments need to be made and then determine which adjustments are needed. The required adjustments should emerge from the analysis of metrics and performance gaps plus the experience of managers. This requires developing and sharing knowledge within the organization. Innovative firms tend to have mechanisms in place that encourage sharing knowledge and information. Most of these mechanisms are relatively informal and do not follow strict hierarchical lines and boundaries. Thus, while firms such as Microsoft and 3M are clearly innovative, they also have well-established control mechanisms. However, perhaps more important, the culture in both organizations encourages active knowledge and information sharing so that the evaluation and control process is constant and occurs more seamlessly than in many other organizations.

OTHER CONCERNS IN EVALUATION AND CONTROL

There are a number of other issues that organizations should consider as they develop their evaluation and control systems. These include organizational focus, concern for value creation, and benchmarking for best practices. Each of these is examined next.

Organizational Focus

One concern for large organizations in the evaluation and control process is determining whether the organization is accomplishing its overall goals. A single business or an SBU in a firm like General Electric may do very well, but the total organization's performance is what stakeholders are concerned about. For large organizations, it is important that the output of the total organization is kept in mind and that the success of any unit or function is in concert with what is best for the entire organization. Evaluation and control play a critical part in ensuring that the performance of the total organization remains the focus. This occurs by making measures such as total corporate performance or subunit value contribution to the firm part of the evaluation process.

Creation of Value

It has been stressed before, but it merits repeating, that the creation of value must be remembered in all activities of the organization, including evaluation and control. For small organizations, the concern for value creation is typically clear and commonly stressed. However, for large organizations, it can be more difficult. As noted in the discussion of structure, some divisions can become more focused on their own success than on that of the entire organization. There is a need to stress the creation of value for the entire organization in the evaluation and control process at both levels. Larger firms need to monitor divisions to be sure the divisional goals are aligned with corporate goals.

In an innovative firm, the creation of value is the accumulation of effort that produces new, applicable ideas. These new ideas are then put into practice (process) or into the marketplace (product). The return that is sought may be a direct benefit to the bottom line of the organization, or it may enhance other areas of potential value.

Indirect value to other stakeholders is also important, but harder to measure and evaluate. Individual firms need to consciously determine how much emphasis they wish to place on such value creation to these other stakeholders and judge their actions accordingly. Some firms put considerable emphasis on such indirect value creation. For example, McDonald's has always had an objective of being socially responsible. The company saw a need for housing the families of critically ill children and developed an innovative idea—Ronald McDonald houses. These facilities do not contribute directly to McDonald's product line or efficiency processes, but McDonald's used its innovative nature to identify a need, develop a solution, and implement the idea. The result is ongoing positive publicity—an indirect value. The indirect value to McDonald's is hard to measure; for the families of critically ill children, it is priceless.

Benchmarking Internally & Externally

Finally, large organizations need to ensure that they are employing best practices and monitoring the environment for changes. Large organizations can set goals that are based on outdated perceptions. The inability to share information easily throughout the organization encourages such behavior because each team, department, or division can become focused on its own success and not on that of the entire organization. Consistently benchmarking the firm against other leading firms in the industry, or along some specific dimension for firms in other industries, provides valuable insight into how the firm is actually performing and what the future might look like. The ability to take information and act on it when a gap is identified benefits the firm.

To illustrate, IBM was not keeping up with competitors' changes as well as the emergence of new competitive opportunities when personal computers were being developed in the 1980s. IBM began to decline, not because it was not benchmarking internal processes, but rather because it was not monitoring the environment and benchmarking strategic concerns. Detecting a paradigm shift is one of the most difficult things for a large, successful company to do. Success tends to breed contentment with the status quo and inertia; IBM was "Big Blue" and the leader in computer hardware manufacturing at that time. It was busy tweaking its products and processes while the environment was changing radically. In fact, in the 1970s, IBM was a firm others benchmarked their internal processes against. The belief that because the firm is successful today it will be successful tomorrow pervades corporate management and stakeholders.

However, what competitors are doing and the assignment of value to those efforts are key analyses that also need to be performed. The analysis of information from the environment leads to a greater understanding of where new opportunities and threats will emerge in that environment (Figure 5.7). If IBM had benchmarked both internal and external factors, it might have recognized the changes earlier. However, IBM has learned from its past. IBM is now largely a service company. In 2005, IBM sold its money-losing, $10 billion PC business to Lenovo. The Chinese company hopes to build a prominent global brand with IBM's ThinkPad, which has long been a favorite of executives and business travelers. Furthermore, in 2008, IBM licensed its x86 server technology to the same firm. The result is that today IBM is largely a computer-services company and no longer a computer manufacturer, but is doing well.[14] Today, IBM is being used as a benchmark again. Other firms are seeking to follow IBM's path. For example, Dell bought Perot Systems in its effort to become less dependent on hardware and more of a service firm.

Key Areas to Consider	Importance
Organizational Focus	Must remember for a large organization, it is the total organization performance that matters.
Creation of Value	Accumulation of effort to produce new ideas that add value in the marketplace or for stakeholders is the goal.
Best Practices	New technologies and competitive opportunities can develop.

FIGURE **5.7** Key Areas and Their Importance

SUMMARY

This chapter has addressed evaluation and control when a firm pursues an internal innovation strategy. Once the firm has established its goals, it needs to evaluate whether it is reaching those goals and, if not, make changes either in the goals or in its actions. The evaluation between the firm's goals and achievements are referred to as a gap analysis. In the evaluation process, it is important that the firm be open to new emergent types of opportunities and take advantage of them. The controls that the firm uses to make changes in its actions are categorized as financial, strategic, and cultural. All three types of control will be present in the firm, although one type or another will dominate. Critical parts of the control process include the role of board of director members and top management of the firm. The structure of the firm will also impact the ultimate nature of the evaluation and control process. Key activities that can aid the evaluation and control process include quality management and benchmarking. The evaluation and control system is important to the value and performance of the innovation strategy.

MANAGERIAL GUIDELINES

In conducting the evaluation and control process managers cannot forget the human element in these activities. Several guidelines that are useful in this regard include:

1. A balance among various approaches in evaluation and control. In other words, there should be more than one type of measurement, and critical issues should be examined from multiple perspectives. No one perspective should dominate, no matter how successful that perspective has been in the past.
2. Involvement of top management. Top management is one of the most important boundary spanning groups in an organization. These individuals should be interacting with all major stakeholders and be able to provide a balancing approach for meeting the needs of the different groups.
3. A focus on the ends. Too often, organizations get caught up in the "right way" or the "way things have always been done before" syndrome. If the goals are clear, then movement toward the goal is important. Learning takes place when better ways emerge, but trying

a different means to the goal is how such improvements can be tested. There is nothing like a drop-dead date to motivate creative thinking about ways to achieve the goal more quickly.
4. Objectives should be specific and understood. Quite simply, if the organization does not know where it is going, it will be very difficult to get there.
5. Feedback is an integral part of the innovation process; it is not a once-a-year occurrence. Many organizations have annual reviews, and some have quarterly reviews. Most organizations and managers do not examine the lessons learned from the last project, do not provide positive feedback, and do not make sure feedback is acted upon. In other words, in most organizations, time is spent measuring outcomes, but then little is done with the information gathered. A successful evaluation and control system collects information, verifies it, makes decisions based on it, and spurs activities to improve the processes and products of the organization.

Guiding Questions

1. Is the evaluation and control process future oriented? Too often, evaluation becomes a criticism of what has gone wrong without building an effort for the future.

2. Are costs and benefits being properly weighed against each other? The bottom-line orientation of many organizations precludes short-term costs with long-term benefits. The lack of long-term orientation can be fatal for the firm.

3. Are a number of innovative ideas being generated? The more ideas that are generated and reviewed, the more likely the firm will find a viable solution for its future.

4. Is the proper amount of funding available to the best potential projects? Underfunding is the biggest drawback to successful innovation strategies.

5. Are the product and process features that are being implemented appropriate and necessary?

6. Do the people in the organization "feel good" about what the expectations, rewards, and outcomes are? The culture of the organization must be positive to enhance creative, innovative activities.

CRITICAL THINKING

Relating to Your World

1. This chapter defined and discussed gap analysis. This type of analysis requires looking to where the organization had planned to be and comparing that to where it actually is. The gap between the two is the area that needs to be addressed. What do you believe are the key characteristics that the organization's plan must have if a gap analysis is to be successful? What must managers guard against in performing a gap analysis? If there is a gap, how will you determine the next step in the evaluation and control process?

2. If you are part of an organization that has an innovative strategy, what are the key inputs from the environment that you need to implement the strategy? What steps will you take to be sure the inputs exceed minimum acceptable levels? What will you do to ensure your output to the environment is acceptable? How are these two periods of evaluation and control the same and how are they different?

3. As a student, you are evaluated periodically during the semester and at the end of the semester. Do you ever change the process that you use in studying in hopes of affecting the outcome? Give an example. When were these adjustments most successful? What characterized the successful adjustments? The unsuccessful ones? Do these characteristics apply to an innovative organization? Why or why not?

CASE **5.1** THE REAL WORLD
IOL Chemicals and Pharmaceuticals Ltd. (IOLC)

IOLC was a bulk chemical producer prior to 1997. At that time, the company chose to shift its focus from bulk chemicals to specialty chemicals. The result was that the R&D function at the firm needed to shift radically. In addition, the firm's system for evaluating and controlling the R&D process needed to change.

The firm's evaluation process maintained the prior efforts such as review meetings and periodic reviews of individual project performance. However, the timing of these reviews has changed. Previously, the projects were all reviewed fairly systematically. Now those projects that are more

CASE **5.1** *(continued)*

time sensitive receive faster evaluation. The firm has also introduced a computer-based system to maintain the records of projects and ensure that accurate monitoring occurs.

Consistent with the shift to a greater focus on research, IOLC moved its research department from a cost center to a profit center. Thus, the firm began interfirm transfers; units in the business could use the outcome of the research unit, but they would be charged a price for those efforts. These units could also choose to use an outside research services in some settings.

An outcome of the shift to a profit center approach was that performance targets had to be revised. The means that the research was presented as a cost to the firm required that the costs and their allocation be much clearer and transparent. One aspect of this change in accounting was that the firm also began to set targets for individual employees and their supervisors. The performance targets allowed not only much clearer accounting but also much clearer evaluation of performance by each employee.

IOLC then also moved to more frequent financial reporting. Rather than quarterly reporting, the firm moved to monthly financial reporting to judge its performance more accurately. The quarterly monitoring of performance was acceptable in the bulk chemical industry, which does not have quick changes in its environment. However, specialty chemicals can have more rapid changes and need to be monitored more closely.

The firm also had to generate new methods to evaluate which projects to pursue. The evaluation of which bulk commodity chemicals to produce was not an elaborate process. Instead, the demand was predicated in a relatively straightforward method based on past demand, and production was conducted accordingly. In contrast, which specialty chemicals to produce is a process much more dependent on the expected cost, evaluating the value of relationships, and other uses for the firm's equipment at that time.

The outcome of this change was that the firm's total sales have more than doubled, and IOLC has become a viable innovator in the chemical industry through its R&D efforts. It is one of the growing Indian firms who plan to expand the market for its stock to increase its global presence. In May, 2008, IOLC announced a $10 million securities offering worldwide to fund these expansion efforts.

1. What were the key evaluation and control processes that IOLC most likely used? Why did you pick the processes you chose?
2. What steps should IOLC take to continue its success?

References

http://organicind.com/aboutus/index.html

Divakar, K. 2000. R&D at Indian Organic Chemicals Limited (IOLC) from a service to an enterprise: A case study. *R&D Management*, 30 (4): 341–348.

Nagaraju, R. 2008. IOL Chemicals board oks raising up to $22.3 mln. *Reuters India*, May 5.

WWW EXERCISES

1. Use your favorite Internet search engine and find an example of a successful evaluation and control process in an R&D firm. What attributes are described as contributing to the success of the evaluation and control process? What factors impact the choices the organization makes in designing its evaluation process? What factors help it to be successful? What do you think about the long-term viability of the process the firm uses?

2. Find a site that discusses the Balanced Scorecard approach to evaluation and control. This method is considered a relatively new technology or approach to evaluating organizational performance. In addition, some control processes are embedded in it. What do you think of the measurements suggested, the process itself, and the potential outcomes? What are the strengths and weaknesses of the Balanced Scorecard method?

3. Find an article or website that provides guidelines for the evaluation and control of an innovative strategy. What do you think of the advice given? Compare the advice you find to the advice your classmates find.

AUDIT EXERCISE

In developing an evaluation and control process, there are many elements that need to be considered. In the chapter, we briefly mentioned Kaplan and Norton's Balanced Scorecard approach to evaluation. Basically, the BSC approach requires that the firm identify its strategic issues and keys and then develop an evaluation process that examines all of them in a balanced manner. The approach was designed to encourage firms to look beyond financial outcomes and take a more balanced review of performance.

1. Many profit-making firms have a strategy map that is tied to four elements: financial results, customers, human resources, and sustainability. If you developed a scorecard that included these areas, what would be your goal in each area, and what measures would you use? What would be your rationale? Figure 5.8 is an example of one segment of a Balanced Scorecard for Novo Nordisk.

Critical Success Factor (CSF) for Delivery Scheduling Program	Rationale	Key Performance Indicator (KPI)	Who Is Responsible?
Customer Relations	On-time delivery can be a competitive advantage	Number of customer complaints	Sales Manager
Contractor (trucking company) Relations	Gained efficiencies mean better performance	Utility of the program for managing timing of deliveries	Director of Operations
Social Responsibility	Save resources by being more energy efficient	Comparison of fuel usage	

FIGURE **5.8** Balanced Scorecard Template

2. What other areas might a profit-making firm include?

3. How would a nonprofit organization use the Balanced Scorecard?

DISCUSSION QUESTIONS

1. Often in management, we discuss evaluation and control as if they are one process. What are the differences between the two? Should they be so closely linked?
2. How will you determine the success of a new product? How will you determine the success of a new manufacturing process? How do the time factors for evaluation differ between product and process innovations?
3. How are evaluation and control different for each stage of the strategic process (planning, implementation, evaluation, and control)? How are they similar?
4. Mechanisms that help control activities in the organization are discussed in this chapter.

Are there other mechanisms that could be used? How do you think innovation processes differ in an organization where the control is bureaucratic and rule-bound in nature versus one where cultural controls are used?
5. How did the evaluation processes at Rolls-Royce differ from those at IOLC? What structural factors do you believe caused these differences?
6. Given the statistics in the beginning of the Rolls-Royce case, why do you think organizations continue to invest heavily in innovative strategies? What are the advantages and disadvantages? What can the organization do to strengthen the advantages?

PART TWO OPENING CASE: GLAXOSMITHKLINE

1. What are the special evaluation needs for a company such as GlaxoSmithKline? What characteristics of GSK do you believe have the most influence on how well it

evaluates progress toward stated innovation goals?
2. What kinds of control systems do you suggest GSK employ? Explain.

KEY TERMS

cybernetic control 139
divisional structure 157

gap analysis 139

strategic business units (SBUs) 157

NOTES

1. Kaplan, R., and D. Norton. 1996. *The Balanced Scorecard: Translating Strategy into Action.* Cambridge, MA: Harvard Business School Press.
2. Chenhall, R. 2005. Integrative strategic performance measurement systems, strategic alignment of manufacturing, learning and strategic outcomes: An exploratory study. *Accounting, Organizations and Society.* 30 (5), 395–422.
3. Marino, K. 1996. Developing consensus on firm competencies and capabilities. *Academy of Management Executive* (Aug.): 6–16.
4. Corcoran, E. 2009. Getting a green light. *Forbes,* June 8. Retrieved from: http://www.corning.com/news_center/in_the_news.aspx
5. Mintzberg, H., and J. Waters. 1985. Of strategies, deliberate and emergent. *Strategic Management Journal,* 6 (3): 257–272.
6. Guimaraes, T., J. Borges-Andrade, M. Machado, and M. Vargas. 2001. Forecasting core competencies in an R&D environment. *R&D Management,* 31 (3): 249–255.
7. Anon. 2009. When Leaders Seek Leaders. *Chief Executive.* Jan/Feb, Iss. 238, 40.

8. Terziovski, M. 2002. Differentiators between high and low performing manufacturing firms: An empirical study. *International Journal of Manufacturing Technology and Management*, 4 (5): 356–371.

9. Faletti, P. 1990. Technology demands directors' attention. *Bottomline* (Nov.): 35–37.

10. Deming, E. 1986. *Out of Crises*. Cambridge, MA: MIT Press.

11. Ibid.

12. Connell, J., G. Edgar, B. Olex, and R. Scholl. 2001. Troubling successes and good failures: Successful new product development requires five critical factors. *Engineering Management Journal*, 13 (4): 35–39.

13. Tushman, M. L., and C. A. O'Reilly III. 1996. Ambidextrous organizations: Managing evolutionary and revolutionary change. *California Management Review*, 38 (4): 8–31.

14. Hamm, S. K. Hall 2008. The quest to design the ultimate portable PC: The superslim ThinkPad 300 is Lenovo's bid for leadership in the high-stakes world of laptops. *Business Week*. 4072, 43–48.

External Strategy

ACER GROUP: A FAMILY OF BRANDS

Part Two of the text examined how firms develop technology through internal innovation processes. Firms can also obtain innovative capability and new technologies externally through strategic processes such as alliances, joint ventures, and mergers/acquisitions. These processes have the benefit of being much quicker than internal development, but the drawbacks include issues such as integration of the different businesses in mergers/acquisitions or managing the relationship between firms in an alliance. Thus, just as with the internal innovation strategy, a firm must evaluate the pros and cons of the different strategic choices involved in obtaining technology externally and then decide what is best for the business in its given context and environment.

Acer Group is a relatively young company, established in 1976. The Acer Group is a family of four brands—Acer, Gateway, Packard Bell, and eMachines. The multi-brand strategy of Acer allows each brand to target different customer needs in the worldwide personal computer market. Acer has grown to be the third-largest maker of personal computers (second largest in notebooks) and in 2008 had revenues exceeding $16 billion. This Taiwanese firm has established itself as a global player in the PC market. How it got there is through innovative use of alliances and acquisitions.

Acer Group: The Firm's History

Acer was founded in 1976 as Multitech. The focus of Multitech was on trade and product design. Just three years later, Multitech designed Taiwan's first mass-produced computer product. The focus from the start was on a product for export—Taiwan is such a small market the firm knew it needed to make a global footprint in the computer market. Multitech, which became Acer in 1987, has a long-term mission "to allow anyone to use and benefit from technology." They have built their reputation on development and manufacture of sophisticated, intuitive, easy-to-use products.

Early Innovations

When Multitech first started it relied on internal new-product development. The PC market was young and the founders saw many opportunities. Acer holds more patents than any other Taiwanese-based corporation and Taiwan firms account for 70 percent of global computer hardware manufacturing. When Acer beat IBM to the market with 32 bit PCs in 1986, it signaled the beginning of the end for IBM's PC business. In 1990, Acer changed its strategic orientation. It became more externally focused in its innovation activities.

External Technology Development

In 1990, shortly after becoming Acer, the firm's attitude towards external innovation efforts changed. During this year Acer acquired Altos Peripherals. This marked the beginning of two decades in which Acer pursued its goal of becoming a major global competitor numerous alliances and acquisitions. Below is a list of several of the more important external innovation efforts:

- 1996—Acer signs a reciprocal patent licensing agreement with IBM, Intel, and Texas Instruments allowing use of each other's patented technology.
- 1997—Acer acquires Texas Instruments' mobile computing unit.
- 1999—Acer Group and IBM form a 7-year procurement and technology alliance.
- 2000—Acer spins off its manufacturing operation to focus on developing technologically advanced, user-friendly solutions.
- 2007—Acer merges with Gateway Inc.
- 2008—Acer merges with Packard Bell Inc.

Becoming a Global Competitor

While Acer was changing its business model from internal innovations as well as evolving from a manufacturing company to a development and marketing firm, it continued to spread its global footprint. It did this through various partnerships and by developing innovative products with its partners and within its own R&D areas. For example, in 2003, Acer launched the Empowering Technology Platform to meld hardware, software, and service to provide end-to-end technologies to customers. In 2008, the Aspire One was launched as the company's first mobile Internet device. In addition, Acer made a strong move into the high-end gaming market with the Aspire Predator series.

These steps were designed to enhance and strengthen Acer's global position. Acer's product range includes PC notebooks and netbooks, desktop computers, storage systems, peripheral devices, LCD televisions, and e-business solutions. The firm is number one in a number of markets with various products. For example, the Europe, Middle East, and Africa (EMEA) market is a stronghold for Acer's mobile-computing solutions. Acer is the largest supplier of LCD televisions in Western Europe. Acer is first in the notebook market in Italy, Spain, Austria, Holland, Switzerland, Russia, Belgium, Denmark, Hungary, Poland, and the Slovakian Republic.

In the United States and Canada, Acer is making its mark through its Channel Business Model (CBM). It developed this model as it expanded beyond Taiwan and continued to improve it as Acer divested its manufacturing facilities. This model allows Acer to be flexible in adapting to global IT market trends. CBM involves collaboration with partners and suppliers to develop and market top-tier products and services. In 2003, they used this model to even co-brand a notebook computer with Ferrari, the Italian carmaker.

Recently, (2009) Acer unveiled the Acer F900 and M900 smartphones at the Mobile World Congress. They began by shipping to channel partners in EMEA and Asia. These products have a relatively large 3.8" wide VGA display, a 3.75G HSPA connectivity for high-speed data transfer, and are the introductory products with Acer's new widget-based user interface that provides easy navigation with vivid 3D animation. The acquisition of Packard Bell was a key to Acer's entrance into this market with this advanced product. Today, Acer continues to use external methods to obtain technology to grow its influence in the global computer hardware and software markets.

Overview of Part Three

This section of the text (Chapters 6, 7, and 8) will explore the methods used to obtain technology externally. These methods include alliances that may be formal such as joint ventures as well as informal alliances that are as simple as a verbal agreement between firms. The section will also explore mergers and acquisitions, the most complex and expensive ways to obtain technology externally. This section will start with the aspects of planning for external efforts to obtain technology in Chapter 6. This chapter will also define many aspects of the external efforts and explain the differences between them. Chapter 7 will focus on the implementation of such efforts. Particular attention will be given to efforts to integrate the results of the planned actions into the existing firm. Finally, Chapter 8 will examine evaluation and control for such external methods. Evaluation and control processes are particularly critical because external means to obtain technology have often not met expected performance goals.

SOURCES

Anonymous. 2009. Acer Showcases Multi-brand Products at Computex 2009 including Aspire Timeline Notebook, Aspire One Netbook, Aspire All-In-One PC.

JCN Newswire - Japan Corporate News Network, Tokyo. June 3. www.acer-group.com

CHAPTER 6

Planning for Obtaining Technology

OVERVIEW

This chapter lays the foundation for the examination of firms obtaining technology externally. The firm may use alliances or acquisitions to obtain technology. Each has its benefits and risks. The evidence is that the firm can overcome these risks by thorough planning. Therefore, this chapter focuses on how the firm should plan to obtain technology externally. The specific material examined in this chapter includes:

- Reasons to acquire technology

- Alliances

- Mergers and acquisitions

- The importance of goals and due diligence

_____

INTRODUCTION

A firm may choose to change its technology mix by means other than internal innovation. This may happen for a variety of reasons such as:

1. The firm's product line is quickly falling behind that of its competitors.
2. A new competitor enters or is about to enter the market that will change the dynamics of the industry.
3. The firm discovers its processes are not as efficient and/or effective as those of its competitors.
4. The firm believes its current products or processes are not going to be successful in the future.

These problems can be solved by internal innovation, but internal responses take time and specialized capabilities. The firm may not have the required time, the capabilities, or the desire to solve these problems through internal innovation. As a result, the firm will look externally to obtain the needed technology to address its problems. The external options available to the firm include forming an alliance that may be informal—such as a simple understanding between two firms—or as formal as a joint venture with another firm that possesses the desired technology or capabilities. Alternatively, the firm can obtain that technology either by buying or merging with a firm that possesses the desired technology. Through each of these external methods, the firm can gain access to the desired new technology or learn new capabilities to produce it. However, each of these options has different benefits and risks that the firm seeking the new technology must consider.

To illustrate, Cisco Systems, which is known for its networking equipment, in 2009 formed a joint venture with EMC, a maker of storage equipment and software. The new joint venture is called Acadia and will sell data center equipment to businesses. The new venture will focus on designing and building systems that rely on virtualization technology that can help customers to create a more flexible technology infrastructure and lower their capital spending costs. For Cisco, the venture will allow the firm to move further into the computer hardware business.[1] Thus, it allows Cisco to accomplish point 2 above—enable the firm to enter a new market and change its dynamics. This chapter examines planning for externally focused technology acquisition efforts. The key element in such planning is initially identifying the exact nature of the activity that will occur. Therefore, this chapter principally presents the benefits and risks associated with the various externally focused methods of obtaining innovation and technology. The chapter initially discusses alliances and then mergers and acquisitions. The last section of the chapter addresses how to plan for externally focused acquisition of innovation. This chapter explores the planning associated with such external methods to obtain technology. As you will remember, Chapter 3 outlined the actual planning process. This chapter does not repeat that but instead focuses on the wide range of issues that managers need to address as they plan for technology-based alliances, mergers, and acquisitions.

ALLIANCES

A **strategic alliance** is a partnership of two or more corporations or business units to achieve strategically significant objectives that are mutually beneficial. The nature of the partnership can vary widely. The costs and level of commitment for each type of partnership can also vary widely, but they are typically less than in a merger or acquisition. One aspect of an alliance that mergers and acquisitions do not have is monitoring costs. These are costs that arise as each firm monitors the partnership to assure that its goals are accomplished as expected and that negative consequences from the partnership are avoided. Thus, to be successful a strategic alliance should involve a relationship in which the strategic advantages outweigh the **transaction costs** involved with the alliance; the transaction costs are the costs of conducting and maintaining the alliance. The choice of what type of alliance depends on what type creates more benefits than costs, as well as other factors such as the learning that occurs in the alliances.

Most often, when individuals think of alliances, they think of joint ventures, or formal agreements between two or more firms where a new separate entity is created. However, joint ventures are just one form of alliance. Alliances can be differentiated along several dimensions. These include level of formality, duration, and location. Each of these dimensions is discussed next.

Formal versus Informal Alliances

One way to classify alliances is by the degree of formality. This is critical because formality can determine the costs and risks involved with the alliance. The formality of alliances can be conceptualized as a continuum with joint ventures anchoring the end that is more formal and informal alliances with no formal documentation anchoring the other end. We begin our discussion with more formal alliances and move to those that are less formal.

In **joint ventures,** two or more firms combine equity to form a new third entity. The level of equity can vary from very small amounts to large multi-million dollar investments. The amount contributed by each party will not necessarily be equal. It is common to have very detailed agreements covering what each party is to provide, what each can expect, and how each is to operate in the joint venture.

To illustrate, General Electric Co., the world's biggest jet-engine maker, has formed a joint venture with China Aviation Industry to sell avionics systems and services for new aircraft. GE and AVIC will start the venture's operations in 2010. The plan is that the venture will export products from both its base in China and from the United States. GE's China Technology Center in Shanghai also will house a commercial aviation center, and 200 jobs will be added in the United States. The 50-50 joint venture will focus on electronic systems and integration services for newly-designed commercial aircraft. The venture's first activity will be bidding on the single-aisle C919, a 168-seat plane being marketed by China's government-controlled Commercial Aircraft Corporation. The C919 is designed to compete with the best-selling 737 model from Boeing Co. and the A320 from Airbus.[2]

Another type of formal alliance is a franchise agreement. In a **franchise agreement**, a contract is established between the company (franchisor) and the individual who buys the business unit (franchisee) to sell a given product or conduct business under the company's trademark. The contract between these two parties typically specifies the time period and geographical region where the franchisee has the right to conduct these activities. Franchising is one of the most rapidly growing business arrangements in the global economy. The franchisor usually provides extensive direction on how the franchise is to be operated. In addition, the franchisor commonly sets standards for behavior by the franchisee in the contract. Failure to follow these standards may result in loss of the franchise. The contract also typically requires not only payment by the franchisee of an initial fee to buy the franchise but also a continuing royalty. One result of the strong direction and expertise that the franchisor provides is that the success of franchisees is commonly much higher than most other types of start-up businesses.

An example of a technology franchisor is Fast-teks On-site Computer services. This franchise requires a cash investment of $19,000 to $40,000, with a total investment of $34,000 to $60,000 by the time the franchisee buys the necessary equipment. Fast-teks is one of the fastest growing on-site businesses and residential computer services firms in North America. It offers business owners and residential customers a one-call solution for their entire computer needs. The franchisee works from home and that eliminates overhead associated with a store location. The franchisee receives marketing support, advisors from the corporation, operating systems for the franchise, and access to purchase goods through purchasing cooperatives.[3] The franchisor receives an initial payment and fees for as long as the franchisee has the business. Therefore, franchisor and franchisee are dependent on each other for their success. This interdependence is the principal reason for the great success of franchising.

Alliances that are intermediate in their formality would still have signed agreements between the parties, but there would be less required of the parties in such an alliance than in a joint venture. Examples of intermediate alliances include consortia and licensing agreements. **Consortia** are characterized by several organizations joining together to share expertise and funding for developing, gathering, and distributing new knowledge. The Oklahoma State University Web Handling Research Center is an example of a consortium. This group of fifteen different industrial partners, three departments in the OSU College of Engineering, and several government agencies combines expertise and funds to sponsor research on web handling technology. The term *web* as used here does not refer to the World Wide Web but instead describes high-technology manufacturing of thin materials that are processed in a continuous, flexible strip form. Web materials cover a broad spectrum from extremely thin plastics to paper, textiles, metals, and composites.[4] The knowledge gained through the consortium has helped decrease the number of defects in web-manufactured materials and has reduced losses for the industrial partners. This consortium, which has been in existence since 1986, has an excellent record of positive payoffs for its industrial partners.

6.1

R E A L

W O R L D

L E N S

Choosing a Franchise

There is a wide range of franchises. Here are some common steps in the process of choosing a franchise.

1. Get general information on the franchise and on the firm by requesting a packet from the franchisor.
2. Investigate the Uniform Franchise Offering Circular. This is a document required from all franchisors. It should have information on:
   a. The history of the franchise, its officers, and directors
   b. A list of all costs and fees
   c. The obligations of all parties
   d. A list of relevant litigation
   e. The success rate of units in the system
   f. Audited financial statements
   g. A list of existing franchisees
   h. A copy of actual franchise agreement
3. Call and visit existing franchisees other than those suggested by the franchisor to see what their experience is.
4. Examine the nature of supports given:
   a. Training programs
   b. Opening support
   c. Marketing programs
   d. Relations between franchisor/franchisee
   e. Actual level of investment

What do you believe are the five key factors that would influence your selection of a franchisor? Why did you choose those five?

In a **licensing agreement**, one firm agrees to pay another firm for the right either to manufacture or to sell a product. The firm selling the right to this product typically loses the right to control various aspects of the product, such as pricing and how the product is marketed, when produced, or sold by the licensee. Thus, there is commonly an agreement between the parties, but the contract only specifies what is to be sold and what the licensee is to receive for licensing the product.

An illustration of the value of licensing agreements was the videotape battle between Sony with its Betamax and JVC with its VHS technology. Betamax was introduced nearly eighteen months before VHS, and Sony was then the larger and stronger company. However, VHS technology was ultimately successful because the tape was longer, and entire movies could be recorded. However, beyond that fact, JVC actively licensed its VHS tape technology to firms such as Matsushita, Hitachi, and Mitsubishi to promote the technology. Sony did not license its Betamax technology. Therefore, despite the fact that it was not the first mover, JVC with its VHS technology was the winner because of its alliances through licensing agreements that it generated to support the technology.

Another type of alliance that is intermediate in formality is **subcontracting** of activities to other firms. These activities may or may not be high value-adding activities to the business, but the activities outsourced typically will not be where the firm's competitive advantage is built. The nature of the interdependence between the contracting firm and the subcontractor will vary with each setting. For example, today firms such as American Express subcontract their computer networks and related support systems to firms such as IBM. Computer systems are clearly important to a firm such as American Express. Thus, the interdependence between American Express and IBM is probably greater than the nature of their contract may indicate.

Informal alliances have the least written about them because there is the least documentation of these activities. In an informal alliance, two firms agree to support each other's activities in some manner. These firms may begin this support without formal agreements either to promote a given product or to aid each other in some way. The agreements are strictly informal. There are few legal protections or means to enforce these agreements. However, if firm A is small and has limited product offerings, it may agree with firm B to refer customers to B if A's customers need particular products that it does not have. Firm B then also informally agrees to refer its customers with needs that B cannot serve to firm A.

The power of such informal alliances can be seen in the growth of Linux, an operating system that replicates the form and function of a UNIX system but is not derived from licensed source code. Rather, it was developed by a group of computer code experts led by Linus Torvalds of Finland. The source code is freely available, enabling the technically astute to alter and amend the system; it also means that there are many freely available utilities and specialist drivers available on the Internet. Around the Linux operating system there has developed an informal alliance of individuals and firms committed to its use and improvement.

Duration of an Alliance

Another way to differentiate alliances is by how long they are expected to last. In considering duration, analysts should recognize that either long- or short-duration alliances may occur in either high- or low-formality situations. For example, as noted above, a subcontracting agreement will involve a written agreement, but the nature of the relationship between the parties will be less extensive than in a joint venture. The duration of a subcontracting agreement may be very short. In contrast, an informal agreement may have no written documents, but the working relationship may continue for years. Thus, formality and duration are not correlated.

Several points should be noted about the duration and formality of alliances. First, the more formal the relationship, the greater should be the detail in the alliance agreement. The documents that specify the obligations and benefits of the alliance also typically specify the duration. Second, the negotiations to develop these agreements often involve long and arduous efforts. Therefore, most firms seek to ensure that the resulting alliance effort will last long enough to provide benefits that will outweigh the initial cost of developing the agreement. A third factor is that the less formal the agreement, the easier it is to abandon if the environment changes or the costs are greater than expected. Thus, duration becomes

Type of Linkage	Duration	Advantages (Strategic Implications)	Disadvantages (Transaction Costs)
Joint Ventures	Long term	Specifies contribution and obligations; New entity created to specifically carry out activity	Strategic drift; Culture clash possible within new entity
Franchise Agreements	Usually long term	Known technology; Long payoff for already developed technology	Contract and monitoring costs
Consortia	Term mixed—payoff determines	Expertise, standards, and share funding	Knowledge leakage
Licensing Agreements	Defined term of agreement	Technology acquisition	Contract costs and contract constraints
Subcontracts	Exists as long as contract is in force	Cost and risk reduced – ideally allows world-class firms in that area to conduct the activity	Little control over quality issues – difficult to enforce even if in contract
Informal Understandings	Exists as long as either party finds a benefit	Opens new opportunities to the parties	Easily disbanded and not enforceable

FIGURE **6.1** Alliances for Technology Acquisition

less certain as the level of informality increases. Figure 6.1 summarizes many of the benefits and drawbacks for the various types of alliances and their duration.

Subcontracting

The prior section defines and briefly discusses subcontracting. This is a type of alliance that is growing in importance as firms seek to position themselves strategically. We include it here under duration because duration is a critical issue for this type of alliance. Many firms subcontract activities that, though important, are not central to their missions such as human resources and customer services through call centers. They subcontract these activities to firms that specialize in these areas. In addition, firms subcontract some technical services because of the difficulty in keeping up-to-date with innovations in the area. While the contracts often have specified time periods, the "duration" of some subcontracts is hard to end.

The firms that provide the outsourced activity are specialized firms with developed technologies that allow them to be more efficient and effective. These firms are better at processing information about new laws and knowledge about their specific areas than the firm that is contracting for their services. Finally, the operation of economies of scale can provide significant benefits to the firm offering the subcontracting services. Although quite controversial when outsourcing to international firms, the results have typically been positive. Overall, subcontracting allows the firm to specialize in domains that are critical to its competitive advantage while other firms that have the expertise conduct other activities. To illustrate Tracker Networks (Pty) Ltd, a leading supplier of vehicle recovery and monitoring systems in South Africa, has subcontracted IBM for the new Managed Continuity and Work Area Recovery Solutions. This is fully outsourced management for applications whose continuous protection and potential recovery is critical. These services will help Tracker Networks to manage its IT infrastructure more efficiently, lower the overall cost of IT operations, minimize business disruption, and implement a superior disaster recovery strategy. The services have reaped benefits, most recently after a fire occurred at one of Tracker's facilities. Using IBM services, Tracker was able to recover critical applications and reroute communications and employees, while maintaining its central business function—recovering stolen and hijacked vehicles.[5]

There can be problems with subcontracting. One is finding specialists to do the activity. The resources spent in the search could be used to improve internal systems, making the subcontract unnecessary. There are also costs in building a relationship with a subcontractor. Many aspects of a contract cannot be detailed and must be developed by experiencing different situations. During the contract, there is a significant need to communicate and work with the party doing the outsourced activity. Additionally, a firm may find itself at a competitive disadvantage because subcontracting the activity resulted in the firm's inability to conduct the activities internally.

Expectations are not always met in subcontracting alliances. To illustrate, in 2007, Mattel had its second major recall of Chinese-made toys in a month. Chinese suppliers were given instructions about what kind of paint they were to use for toymaking; however, the highly competitive environment makes following the rules difficult for the subcontractor. As a result, the subcontractor used cheaper but more dangerous materials without Mattel's knowledge. This ultimately led to the recall. Mattel and other manufacturers are now looking for ways to get the Chinese government to establish an effective regulatory system and guarantee product safety for the subcontracted manufacturing. Mattel finds itself caught in a no-win situation with its customers—poor product quality or higher prices.[6]

Despite these difficulties, the level of subcontracting in the United States continues to expand. As noted in Figure 6.1, the duration of the subcontracting relationship is the length of a contract, but this typically understates the duration; once a given task has been outsourced, it is difficult and expensive to reestablish those skills internally. Thus, firms need to be aware of the potential drawbacks to subcontracting and recognize that once they give up the ability to perform a function, it is difficult to reclaim it.

Location: Domestic versus International Alliances

A third way of classifying alliances is by the locations of the partners. International alliances include challenges such as languages, trusting relationships, and cultural norms, but they also represent opportunities for developing new products and markets. The international alliance, like the domestic alliance, must be based on mutual strategic fit, participant risk and reward, and potential synergy.[7] The reasons for developing international rather than domestic alliances can vary widely. To illustrate a few that are currently in the news, the United States government policy in the early 21st century concerning embryonic stem cell research forced many scientists and organizations to build R&D facilities or enter into joint ventures in Europe (especially Great Britain) to pursue development of potential uses of stem cells. However, in 2009, there was a change in United States government policy and more stem cell research is being developed in the United States.[8] Another common reason to form international alliances is to access centers of innovation for a given industry. For example, many international firms seek out alliances in the Silicon Valley to access cutting-edge technologies that are being developed there. Alternatively, many appliance makers seek out alliances in Japan because most cutting-edge technology for appliances is developed in that nation. Firms in certain countries often seek out international alliances to lower costs. Many United States technology companies have set up alliances with firms in India for this reason.

The move from a domestic to an international firm is a four-stage process for most companies. The firm begins with a technology or idea that enjoys regional success in which it often sells directly or through an agent in the international market. The firm then may use licenses or informal alliances to move more substantially to an international market position. Partial integration of the firm into international operations through an equity investment of a joint venture may occur next. The final stage is global concentration and integration through international mergers or acquisitions.[9] Understanding the value of internationalization efforts involves answering two key questions.[10]

- Is the alliance (arrangement) really adding value to the technological abilities of the firm?
- Is the alliance (or merger/acquisition) contributing to the internationalization of the firm?

Because of the rapid growth in international alliances in recent years, the reasons for pursuing this strategy will be examined in more detail.

Reasons for International Alliances

A wide variety of reasons for international alliances has been proposed. Bruce Kogut is a leading scholar of international business, and he has sought to bring consistency to the reasons for alliances. He argued that they could be summarized into three broad categories:[11]

- Organizational learning
- Cost savings
- Strategic behavior

Organizational learning in international alliances ideally occurs as firms seek to gain knowledge about products, processes, or markets from their alliance partners. The amount of learning through any alliance depends on three factors: (1) the intent to learn, (2) the receptivity to new information, and

(3) the transparency of the partnering firm. There are issues surrounding learning from alliances that managers should consider as they move forward. The technology (product or process) is defined by its competitive significance to the firm(s) as well as its complexity. The more complex the technology, the more difficult it is to learn in an international alliance. As a result, it is more likely that a firm will use subcontracting or acquisition if the technology is too complex.

Learning in international alliances is also impacted by firms in the alliance having unique sets of competencies, unique corporate cultures, and different management styles. Learning occurs as the individuals and teams within each firm combine their knowledge. The goal is to retain the best practices rather than have each side compete to maintain its status quo. As a result, the alliance structure, and expected results, should vary according to the type of learning desired.

International alliances can also be pursued to allow cost savings. As noted before, the cost savings from operating in a foreign market can be substantial because of lower labor costs. For example, it is estimated that IT costs can be cut by up to 70 percent when they are outsourced to India from the United States.[12] The international alliance allows a firm to combine with another firm with expertise in a given market to access those cost savings. The earlier Tracker example demonstrates that there can also be cost savings from an international alliance simply because the expertise of the firms working together results in reduced costs and better service. The ability to cut costs due to greater expertise is why our opening case firm Acer from relatively low cost Taiwan chose to have AT&T from high cost United States develop its unified global network based on Internet Protocol; a move that saved Acer an estimated $15 million over the life of the contract.[13] Thus, even firms known for being low cost such as Acer help ensure this position by seeking out others to outsource activities that can help lower their costs. It should be recognized that there are transaction costs associated with forming an international alliance. As an alliance grows, the costs and risks associated with managing it can also increase. These costs rise even more as the operations are internationalized. As a result, ensuring that there is value created in such international efforts is critical.

International alliances may also be pursued for strategic reasons. A competitor may have entered a given market or geographical region. A firm may wish to match its competitor's actions. However, it may not wish to commit the level of resources necessary to purchase another firm or to start new international operations to match the competitor. This decision not to commit may be because the firm is not sure that the competitor is right or because resources are lacking within the firm. The result is that the firm can enter into an alliance with another firm to develop the strategic response. This strategic choice is cheaper because two firms will share the costs of the activity. Additionally, the long-term commitment is less because the alliance can be abandoned. In the meantime, new knowledge and technologies may be developed because the firms have more access to each other's strengths.

Concerns in Alliances

The major concerns for a firm trying to acquire technology through building an alliance can be summarized as:

1. Finding the proper partner: This is critical for the success of the alliance.
2. Dealing with the ambiguities of the relationship: Because it is not necessarily a permanent, well-defined relationship, the alliance may have unexpected political problems.
3. Discovering that the partnering firms lack a shared vision: A firm could discover that what it thought it was going to gain from the alliance will not actually materialize.
4. Getting the timing right: Both parties must be able to respond when needed, but financial or other strategic concerns may interfere with fulfillment of the alliance agreement.
5. Communicating effectively and efficiently between the alliance partners.
6. Protecting intellectual property: The organization should recognize the potential for loss of knowledge that constitutes a part of the strategic advantage of the firm.
7. Measuring real costs and profits from the alliance: A firm needs to do a realistic cost/benefit analysis. Many organizations find that the original analysis of alliance benefits and costs was not realistic.

In reviewing these concerns, it can be seen that the firm must have a focused strategic goal if it wants an alliance to be successful. Before the firm can identify a proper partner or understand the costs of what it is attempting to do, it needs to have a realistic set of goals for the alliance and an understanding of what the partner firm should bring to the alliance.

Once these issues are defined, the firm can develop an understanding of the costs and determine which firm will make a good partner. Understanding the each firms' needs and ensuring that those needs are met takes time and effort by both parties. There will be ambiguities in these efforts, but a good faith effort must be made to ensure that the alliance meets its goals. This effort typically is built on effective and efficient communication among the parties. However, it occasionally happens that the alliance partners may appear to share the same goals for the alliance, but in fact, there is no shared vision. In these cases, the firm must act quickly to build a shared vision or exit the alliance. Without common goals, it is difficult to build a successful alliance. Finally, the firm must ensure that while it partners with another firm in an alliance, it does not eliminate its own competitive advantages. Many firms will enter a technology alliance simply to learn about a technology that they wish to have themselves. Once they have accomplished this, the firm leaves the alliance and begins to employ or produce the technology on its own. Thus, the firm must ensure that shared information will not provide a competitive advantage to a competitor later. In many ways, the ability to avoid such situations relates to the need to identify the proper partner and to share a common vision with that partner. The firm must also understand the true costs and profits from the alliance. Without this information, sound judgment on the effectiveness of the alliance is not possible.

The critical questions that the partners entering a formal or informal alliance should ask are:[14]

1. Do we have clear goals and expectations? (Even with such clarity, partners also need to recognize that conflicts will happen, but clear goals and expectations reduce the potential for conflict.)
2. What is each member of the alliance responsible for, and what does each bring to the alliance?
3. Will each member of the alliance promise to develop solutions that will solve the problems and needs of the members?
4. Will members promise to meet the goals of the alliance?
5. Who will be responsible if the solutions fail? What compensation will be offered?
6. What are the conditions for dissolution of the alliance?
7. How will disputes be resolved?

MERGERS AND ACQUISITIONS

Another external means for obtaining technology is through mergers and acquisitions. Mergers and acquisitions are not mere linkages; rather, they are permanent changes to the structure of the firms involved. A merger is a transaction involving two or more corporations in which only one permanent corporation survives. An acquisition, on the other hand, is the purchase of a company that is completely absorbed as a subsidiary or division of the acquiring firm.

The terms *merger* and *acquisition* are commonly used interchangeably. However, they represent very different types of business activities. An **acquisition** refers to the outright purchase of a firm or some part of that firm. The result of an acquisition is that the purchasing firm typically remains the dominant force in the newly combined business. Although the purchasing firm does not always remain dominant, an acquisition often results in its managers filling critical management roles. In contrast, a **merger** occurs when two firms combine as relative equals. As a practical matter, it is often difficult to differentiate mergers from acquisitions. As a result, the terms are often used together, *merger/acquisition.*

Mergers and Acquisitions of Technology

In recent years, the number and dollar amounts of mergers and acquisitions in the United States and around the world have soared. For example, in the three-year period from 2004 through 2007, there were over $6 trillion worth of mergers and acquisitions worldwide. However, statistically, these business ventures have a poor record of performance. Consistently, over the past decade, approximately 60 percent of mergers/acquisitions have failed or significantly underperformed within three years of the deal.[15] The perception of the difficulty in achieving success in a merger or acquisition is the reason, when a merger or acquisition is announced, you will typically find that the stock of the acquired firm will increase in value whereas that of the acquiring firm decreases.[16]

If the record is so poor, why do firms continue to pursue acquisition of technology through merger or acquisition? There are a number of benefits that continue to promote mergers and acquisition as they concern technology. First, in contrast to alliances, there is ownership of the business involved and its technology, so the risk of losing proprietary knowledge is reduced. Second, economies of scope and economies of scale may be increased, and thus, costs are reduced. Third, the merger or acquisition may fill the need for innovations that the internal systems of the firm have not or cannot produce but that are needed to increase or at least maintain market share.

However, evidence points to another motivation for mergers and acquisitions that may influence the decision even if managers do not recognize it. There is evidence that the salary of managers will increase as the size of the organization grows. As a result, it is widely argued that one of the principal motivations for mergers and acquisitions is that it may enhance the reputation of the manager as well as increase the manager's financial return.[17] The underlying theory for this belief is called agency theory.

Historically, **agency theory** comes from the recognition that those who own firms and manage them are now separated. Thus, the agents (active managers) may act in their own best interest rather than that of the firm. Thus, self-interest motivations promote growth of organizations through mergers and acquisitions because the managers can obtain higher salaries with larger or more complex organizations.[18]

Strategic Reasons for Mergers or Acquisitions

Mergers and acquisitions can allow a firm to accomplish a variety of strategic goals. The merger/acquisition could allow the participating firms to:

- Enter a market quickly or increase speed to market
- Avoid the costs and risks of new product development
- Gain market power
- Acquire knowledge

Enter a Market Quickly

The merger or acquisition of a business allows the acquiring firm to gain immediate access to the technology, customers, distribution channels, and/or the geographical areas of the acquired firm. If there are barriers to entry, such as existing high levels of customer loyalty, distribution channels where wholesalers are unwilling to take on new firms, or a geographical area where the best retail locations are already taken, a firm can circumvent such restrictions by acquiring one of the existing competitors.

When technology is a critical factor, the ability to enter a market quickly is frequently very important. Customer loyalty may be established quickly, and if the firm is not in a particular market, it will quickly lose any potential to gain that customer loyalty. The ability of a firm to gain such customer loyalty encourages the firm to be a **first mover**, or first competitor in a domain. If there was no such customer loyalty to be obtained, then the acquiring firm could move slower and consider developing the product internally.

To illustrate, in 2005, eBay bought Skype for $2.5 billion. The argument was that it would allow eBay to move into voice-driven auctions. This would create a first mover advantage for eBay in this domain. It was also part of eBay's strategy to move toward developing applications that could be used by smartphones like the iPhone to bid in eBay auctions. Unfortunately, the synergistic results were never found and by 2009, the firm was looking to spin Skype off in an IPO as a separate firm.[19]

It is possible for a firm to be a second mover through a merger or acquisition. **Second movers** move quickly into a market after the first mover. If there is low customer loyalty, the second mover can be in a profitable strategic position. For example, the firms that were second movers with new technology in tracking and handling containerized cargo ultimately replaced the first mover. These firms learned from the mistakes of the first mover and saved money by not having to educate the customer, a cost the first mover incurred.

The quick access to customers obtained through an acquisition has both offensive and defensive strategic uses for a firm. If competitors are permitted to service one aspect of a critical customer's needs, they begin to gain access to that customer. This initial access can lead to opportunities to service other aspects of that customer's needs. Thus, a firm may not have a viable product in a given business area, so it will want to acquire an existing business quickly to limit any opening for competitors to their customer base—a defensive move. Alternatively, a technology firm can often gain initial access to a customer base it does not serve through the acquisition of a firm that provides a product used by those customers—an offensive move. For example, in 1999, Dell Computer purchased ConvergeNet Technologies Inc. to gain the technology for sharing storage on a SAN (storage area network) with existing equipment and to expand its customer base.

Avoid the Costs and Risks of New Product Development

The acquisition of ConvergeNet had another benefit for Dell. A merger/acquisition can be used to control the costs and risks of doing research and development. The expense of developing a new product can be quite high. Large sums of money may be invested and no viable product developed. A merger or acquisition of an existing firm or product means that the acquiring firm does not have to conduct that R&D; instead, it has already been conducted by the firm that was merged or acquired. The creative process is also one that can be difficult to initiate. Nick Allen, vice president and research director at the research firm GartnerGroup Inc., Stamford, Conn., described the benefits to Dell from the acquisition as, "I think Dell looked at that and felt if it doesn't have its own R&D, it can't succeed in the storage or even server space. Dell bought an R&D arm."[20] This approach for Dell to buy into a market by buying established products is also consistent with Dell's R&D budget, which is one of the lowest in the industry on a base of percentage of sales.

Gain Market Power

Market power occurs when a firm has enough market share to shape that market's actions, and it can be a strategic motivation for a merger or acquisition. Market power can be seen in the ability to direct issues such as pricing for that product in the industry. This occurs because the dominant position of the firm becomes such that other smaller firms follow its pricing lead. Thus, if the firm with market power lowers prices, other smaller firms will be forced to do likewise. The market power of a firm can also dictate the behavior of suppliers. If a firm becomes the dominant firm, the supplier may need to conduct business with that firm to remain competitive.

To illustrate, British Petroleum since 1998 has made a number of acquisitions such as Amoco, Arco, Burmah Castrol, and Veba Oil. These acquisitions helped BP to expand and become one of the supermajor oil companies. There are six such firms in the world. These firms exclude state controlled firm and possess 5 percent of the world's oil reserves. The other 95 percent are held by state owned firms such as ARAMCO in Saudi Arabia. These six firms help to drive the world's petroleum industry as they are considered the most innovative and also dominate the free market area of the industry.[21]

Acquire Knowledge

Finally, a firm may pursue a merger or acquisition to gain knowledge about a particular technology. The most valuable commodity for a firm is knowledge. If a firm does not have a viable understanding of a domain, it can purchase a firm with individuals that possess that knowledge. However, one risk of such a strategic effort is that if "acquired people" are the key to gaining such knowledge, those individuals may leave during or after the acquisition. The methods to reduce this risk will be discussed in the next chapter as we examine implementation.

Types of Mergers and Acquisitions

Mergers and acquisitions can be differentiated by two key characteristics:

- Related versus unrelated
- Horizontal versus vertical

Figure 6.2 summarizes the different types of mergers and acquisitions.

Related versus Unrelated

The relatedness of the merger or acquisition concerns how the skills and abilities of the acquiring and acquired firms match each other. Firms that possess similar businesses or skills are referred to as related mergers and acquisitions. Firms that have very dissimilar businesses or skills are referred to as unrelated. This does not mean that the firms have to be similar in all aspects of the two businesses to be related. Instead, the concern is whether the areas that are critical to the two firms are similar. Thus, firms may appear to come from two very distinct industries, but if both firms rely on similar skills, such as marketing, and if both firms employ those skills in a similar manner, the merger or acquisition can be considered related.

To illustrate the difference between related and unrelated acquisition, eBay successfully acquired PayPal, which is closely related to its core business. Both firms had similar technological foundations and approach to marketing on the internet. The result was that it was easier to integrate the two firms. However, since then eBay has made several purchases of fast-growing companies with high margins that were unrelated and have not worked out well.[22] The skill

	Goals	Examples of Desired Outcomes
Horizontal	Learn new skills Gain ground on competitors	Improvements in manufacturing or marketing Reach critical size
Vertical	Access new technology Gain ground on competitors	Upstream or downstream control Cost reduction; Improve quality
Related	Learn new skills Gain ground on competitors	New customers Marketing or manufacturing improvements
Unrelated (most difficult)	Access to new technology Learn new skills	New products, processes, markets Risk diffusion, new customers/suppliers

FIGURE **6.2** Types of Mergers and Acquisitions

sets of the firms in this setting did not match those of eBay and the result was that it was hard to integrate the firms or to share value between them.

In general, firms that pursue related mergers or acquisitions perform better than those that pursue unrelated mergers or acquisitions.[23] When there are similar skills or products acquired, it is easier to understand how to integrate the two businesses (or how to integrate a purchased product line if it was only a partial acquisition). It is also easier to understand how to manage the acquired firm or to place new management in the acquired firm if incumbent managers leave the business. However, for technology acquisitions, the biggest potential opportunities come from constrained acquisitions—that is, acquisitions in the same broad domain of technology but which are unrelated to their current concerns. In this situation, the acquired firm may possess a skill or access to a new market or product line that the acquiring firm does not possess but believes it needs to maintain or enhance competitiveness.[24] Nortel acquired Pingtel Corporation in 2009. Nortel wanted to bring Pingtel's R&D capabilities into its organization. The plan is to use those capabilities to further develop software-based solutions that are more advanced than Nortel can accomplish with its current resources. Nortel expects to speed the development of new information technology centered channels to market.[25]

Horizontal versus Vertical

Mergers and acquisitions can also be classified as horizontal or vertical. **Horizontal mergers and acquisitions** occur when the acquired and acquiring firms are in the same industry. The focus, in contrast to relatedness, is not on skills possessed by the firm but on the actual industry in which the two firms compete. If the products supplied by the merged firms are produced within the same industry, it is a horizontal merger or acquisition. An example of a horizontal merger is the combination of Conoco and Phillips Petroleum in 2002. The two petrochemical firms were in the same industry and combined to become one of the largest oil companies in the world. The two firms were so close in activities that it was estimated that $750 million in cost savings could be generated by eliminating duplications in areas such as accounting.

However, if one firm is a supplier or customer of the other, it is considered a **vertical merger or acquisition**. To illustrate, in 2002, Baxter Healthcorp paid over $300 million to buy ESI Lederle. Baxter is a healthcare company that manufactures a wide range of health-related products. ESI sold injectable drugs. In describing the acquisition, Baxter said:

> *The vertical integration of ESI Lederle's manufacturing capabilities provide a reliable, cost-effective, quality manufacturing source of small volume parenterals, vials and ampules not currently available elsewhere within Baxter.*[26]

Both horizontal and vertical mergers and acquisitions can have a positive impact on a firm. A technology-focused firm's needs should dictate the type of merger/acquisition that is employed. Each type of merger/acquisition (horizontal and vertical) brings different technological benefits to the firm. But as we shall soon discuss, each has different impacts on the planning process.

Whether a merger or acquisition is vertical or horizontal or related or unrelated is not always clear by a surface examination. For example, an oil company may purchase a coal mining company. If the oil company is acquiring another form of energy, the acquisition may be considered horizontal and unrelated. There are different skills and markets for the products. However, if the oil company is doing research and development in coal gasification, then the acquisition of the mining company is vertical and related because it is designed to ensure a ready supply of raw materials. It is important that a firm

---

6.2

REAL

WORLD

LENS

KeySpan Energy

KeySpan Energy is a Brooklyn-based utility company that recently bought Eastern Enterprises, which owned five different gas utilities. KeySpan chose to purchase the firm so that full economies of scale could be obtained. Once that decision was made, the firm began to develop further plans. In terms of IT, three specific steps took place.

1. Assess the current state of IT for both companies, including issues such as staffing, business activities they support, and technology infrastructures used.
2. Create a vision of the post merger business and its IT needs. An end target should be established as the firm begins its planning for the merger, not a moving target.
3. Develop the integration plan. Typically, multiple tasks must take place at the same time. Having different identifiable pieces whose performance can be measured helps allow the firm to be successful.

REAL WORLD LENS *(continued)*

Now, KeySpan has been acquired by the British power company, National Grid. This takeover was designed to do two things. First, National Grid wanted more information about automatic metering. Second, National Grid wanted to spread risk internationally. Some of the new British policies, such as closing coal-fired generation plants, have led National Grid to seek geographic diversification to lessen its risk.

Process and communication were critical to KeySpan and are also critical to National Grid. Therefore, having a clear understanding of where the firm wanted to go and why becomes central in the planning process. Without that clear understanding, the firm cannot fully develop its planning. You will see more on this case in Chapter 8.

1. Most of the models that we have presented include more than three aspects or steps. What do you think of KeySpan's approach? What lessons learned by KeySpan should National Grid heed?
2. How does the degree of relatedness affect the complexity of the acquisition process? Please be specific.

References

Anonymous. 2007. KeySpan Corp.: New York Regulator Approves Acquisition by National Grid. *Wall Street Journal*. (Eastern edition). August, A.12.

Korman, R. 2001. Why acquisitions can turn bitter. *ENR: Engineering News-Record*, 247 (7): 15.

Taub, S. 2002. KeySpan under the gun. *CFO.com* (Apr. 8). http://www.cfo.com. www.nationalgrid.com

wishing to obtain technology externally plan carefully the reasons for acquiring the technology and understand the skills of each firm.

PLANNING THE ACQUISITION OF TECHNOLOGY

Until this point, the focus of this chapter has been on the different types of activities that a firm can pursue to obtain technology externally. The choice of which external method to use will influence the planning process of the given business. This section and the next provide an understanding of what impact the choice of external method has on planning. Figure 6.3 indicates the two critical areas that need to be examined initially in planning. How these issues are implemented is addressed in the next chapter, Chapter 7.

Goals

A key part of the planning process is that the organization's goals should lead to the choice of the external method for obtaining technology and innovation. Each of the methods for externally obtaining technology and innovation has

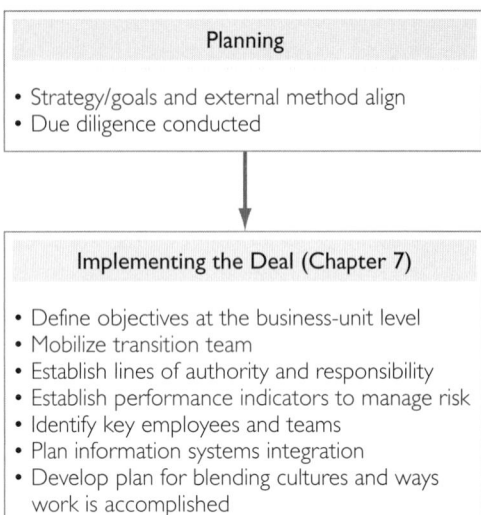

FIGURE **6.3** Planning the Deal and Its Implementation

different benefits and goals. However, if the firm establishes a goal of dominating a new market, it is likely that purchasing a firm through a merger or acquisition is more appropriate than is an informal alliance. Therefore, the firm's planning process in choosing an external method should consider the goals of the firm rather than the goals being shaped by the method chosen.

If the method chosen matches the goals of the firm, then the firm should next ask whether the action *will create value for the organization*. If the answer is no, then the process should stop immediately. Too often in planning an alliance, a firm gets caught up in the possibilities and not the realities.

Due Diligence

Due diligence is the investigation the firm conducts as it begins to develop its planning. In its **due diligence**, the firm investigates all aspects of the firm that may be acquired to ensure that the target is actually as it is perceived. It is

1. **Strategic Diagnosis**
   a. Environmental assessment
   b. Market and technology positioning evaluation
2. **Formulating Strategy**
   a. Technology inventory
   b. Profile current and future technologies
   c. Chart technological requirements
   d. Determine inter- versus intra-firm abilities
3. **Determining Goals**
   a. Appropriation of technology—internal or external
   b. Deployment of technology in product or process
4. **Strategic diagnosis and adjustments, as needed**

FIGURE **6.4** Due Diligence Analyses

possible to find that either the finances or the strategic position of the firm is not as desired through an in-depth examination. Due diligence should include the various analyses presented in Figure 6.4. Each of these requires different information. The firm should prepare as in-depth an analysis of each of these concerns as possible. These analyses build on each other. Because due diligence is an evaluation process, it will be discussed in far greater detail in Chapter 8.

Once the firm has gathered a variety of information, it should integrate the knowledge into its planning process. The firm must be willing to determine that it should not proceed with externally obtaining the innovation or technology if due diligence proves that the effort will not produce the desired results. A decision tree illustrating the use of this information is presented in Figure 6.5.

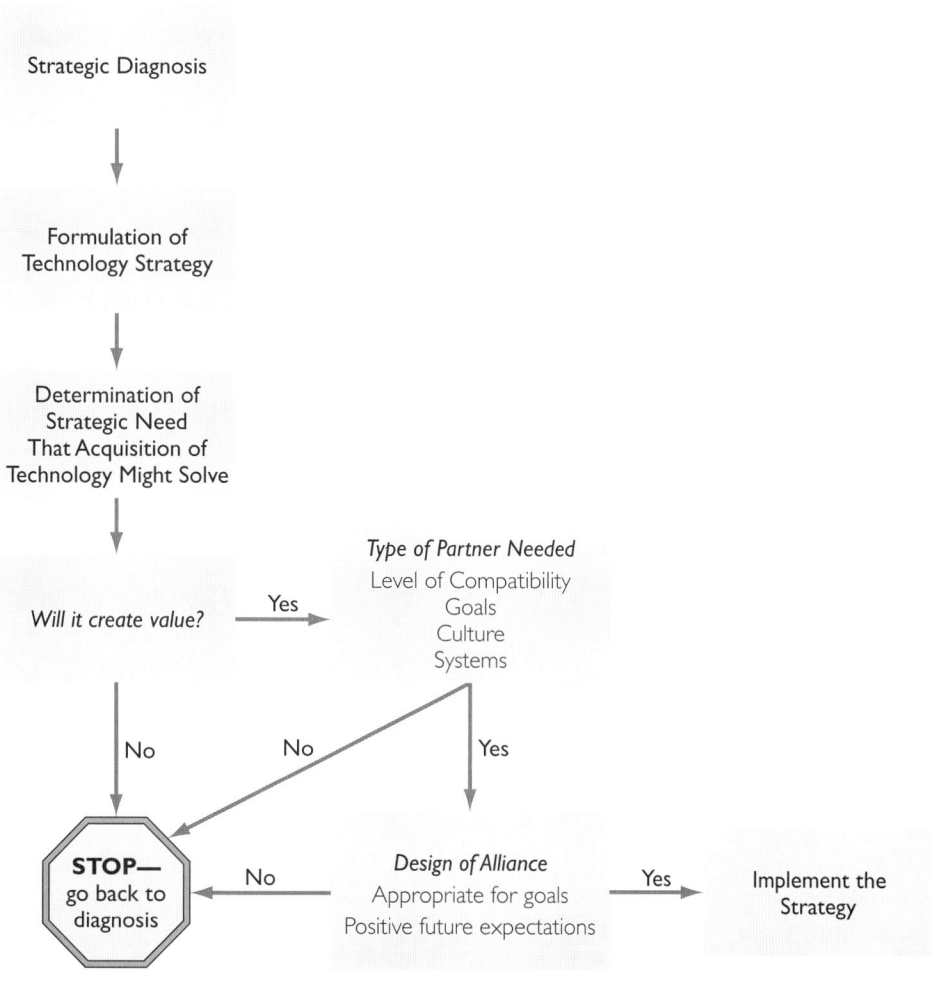

FIGURE **6.5** Decision Tree for Acquisition Technology

Major Mistakes to Avoid

In developing plans for the acquisition of technology, there are several mistakes to avoid. These mistakes emerge time after time when organizations try to merge systems, structures, people, policies, and so forth in alliances or acquisitions. They can be summarized as follows:[27]

1. There is not enough study of existing systems. Most of the energy is spent looking for synergies, not for possible breakdowns or incompatibilities.
2. There is an overemphasis on the needs of the larger, stronger partner. The needs of the smaller or acquired partner must be put into the equation.
3. The timetables for the blending of the organization are unreasonable.
4. Insufficient resources are put into the alliance planning and implementation processes.
5. The decision makers get enamored with making the deal work rather than making the alliance work for the good of all the stakeholders.
6. There is an overemphasis on "we must be together in all things." Sometimes it is better to keep some systems separate, especially in non-ownership situations.

SUMMARY

This chapter has described many of the basic elements of planning for obtaining technology externally. The methods involved can include either alliances or mergers/acquisitions. Alliances can be differentiated by formality, duration, or location. Mergers and acquisitions can be differentiated by relatedness or whether they are vertical or horizontal. The critical issue with all of these choices is that the firm must ensure that the method chosen matches the needs and goals of the organization. It is critical for ultimate success that the firm adequately plans for the acquisition by understanding these issues thoroughly and acting accordingly.

MANAGERIAL GUIDELINES

In planning for acquisitions, managers need to remember that:

1. Compatible goals are essential if an alliance between two firms is to succeed.
2. There are many options for obtaining technology from external sources, but matching goals to the method is important.
3. External efforts to obtain technology require diplomacy and pragmatism during the early stages.
4. Planning for the merger/acquisition is much easier than the actual implementation of combining the different organizational systems and structures.
5. Each organization involved in the external acquisition of technology will bring its own set of systems to the table. These include governance, information, human resources management, financial systems, culture, and a myriad of others.
6. There is expertise spread throughout all of the organizations involved. This expertise must be preserved and tapped into if the acquisition is to be successful.

Once the firm has decided to acquire technology, it must plan and then implement the plan while continually monitoring outcomes to see if they meet expectations. Chapter 7 is about the implementation process, and Chapter 8 discusses the process of strategic diagnosis and adjustments needed during the acquisition.

Guiding Questions

Research has shown that there are several things that are crucial to successful planning of the acquisition of technology. By examining the following questions, a manager should address many of the issues:

1. What is the value of the intellectual assets of the firms involved? Be aware of the economic value of intellectual assets. One of the most common mistakes is to undervalue intellectual assets. This often results in missing potential synergies during implementation. Intellectual assets are not individual items that can be separated into distinct units. Instead, intellectual property is a holistic unit so that, when obtained or released, the influence on a broad range of potential impacts must be considered.

2. Do your plans link the strategy and goals for the technology acquisition activity with the integration process? It happens that firms enter into agreements for very good reasons but fail to articulate them clearly and often while implementing the alliance. The result is unfocused actions that do not lead to desired outcomes.

3. What is the "mood" or "tone" of the alliance or acquisition? Friendly external methods of technology acquisition have a better chance of success than unfriendly ones. This should seem obvious; however, it is amazing how many acquisitions are misguided by politics or power hunger. In addition, an alliance based on devious, hidden goals has less chance of success.

4. If there is a merger, acquisition, joint venture, or similar type of activity, plan to use best practices. Remember, the overall goal is to improve the firm. What system will be used to discover the "best" practices?

5. What lessons have we learned from previous attempts to acquire technology through externally oriented strategies? As is true in most cases, "success breeds success." For example, Honeywell and Dow use two different approaches for information system integration. However, each has been successful with its own method. Because of that success, each feels its way is best. The success of past experiences makes them confident that they will be successful with that method in the future.

6. How do external stakeholders perceive our firm? Potential partners? Recognize that the company's reputation, as represented by its brand, must also be considered in any deal, particularly in alliances.

CASE **6.1** THE REAL WORLD
United Technologies

United Technologies is a multinational firm whose revenues are more than $55 billion. The firm has six subsidiaries that include such well-known names as Carrier, Otis, Pratt & Whitney, and UTC Power. The firm already uses contractors extensively for technology-related activities, but around 2000, United Technologies decided it could save further by using more subcontractors and by moving some of its subcontracted work overseas. The firm determined that the means to select the activities that should remain internal, be subcontracted in the United States, or be subcontracted overseas were:

1. If the work is strategic and central to competitive advantage, it remains internal.
2. If it is not strategic, it will find the lowest cost provider. If UTC is not that provider, the activity will be outsourced.

(continues)

CASE **6.1** (*continued*)

3. If the lowest cost providers are international, then UTC should examine how its subsidiary is structured. In those subsidiaries where the IT departments are centrally controlled, such as Otis, subcontracting internationally made sense. However, in settings like Carrier, where there are approximately fifty different facilities, contracting out internationally did not make sense; it would be too difficult to control and integrate such widespread systems and processes.

4. If subcontracted, security must be maintained so no company secrets were lost.

The outcome was that the firm determined what activities would remain internal and which it would subcontract. Next, the firm explored when subcontracting activities should be kept in the United States and when they should go international. Ultimately, UTC decided to issue requests for proposals for subcontracting information systems work in three broad domains: mainframe, e-commerce, and ERP programming. The firm looked not only at price for specific subcontracting jobs in these areas but also in 500 discrete areas such as current service offerings, management practices, procedures, and current business profile.

The firm decided that rather than look at subcontractors around the world, it would look only at subcontractors in India and the United States. There are multiple potential subcontractors around the world, but focusing on India and the United States, where established subcontractors and support systems exist, led to reduced transaction costs.

The outcome was that all help desk, network, desktop, midrange mainframe, and web hosting for the corporation were subcontracted to one company in India. The goal was eventually to have all application development also moved to various firms in India. The overall impact has been savings of about $50 million annually.

To illustrate how one subsidiary made the transition, Otis had some subcontracting in India prior to this latest effort. But the latest effort moved thirty different applications to India. Otis was pleased in general with the activities and was saving an estimated $500,000 by subcontracting activities to India. But the model of having so many separate projects that were run individually was proving difficult to coordinate. Therefore, Otis sought to set up a dedicated center to be run by the Indian firm Wipro. This center was in Bangalore and worked only on Otis-related efforts. This allowed the Otis projects to be centered in a single location which, in turn, allowed greater expertise and focus to be developed on specific issues of concern to Otis.

The transition to the new model took four months. The planning for the process was critical. The cost of the transition was an estimated $420,000. These were start-up costs that would be Otis's and not recouped from the subcontractor. There is now a twenty-person shop focused on Otis subcontracting in Bangalore, and the company estimates that it saves $1.4 million in application, development, and maintenance.

CASE **6.1** *(continued)*

In 2008, Otis remained profitable because of its growth in the Asian market during a time of flat sales in North America and Europe.

1. Do you think United Technologies was wise to focus only on subcontractors in India and the United States?
2. The Otis unit of United Technologies illustrates differences that occur in a worldwide firm as it manages alliances. What are some of the other differences illustrated? What other issues should be considered?

References

Overby, S. Inside outsourcing in India. http://www.cio.com/archive/060103/outsourcing.html.
http://www.utc.com/utc/News

CRITICAL THINKING

Relating to Your World

1. In planning alliances, top management teams often overlook critical issues. This is especially true in the area of information technology. Many alliances fail because the computer systems or internal information reporting systems are not well matched. For each type of alliance structure described in the text, what issues do you believe would be critical in planning for the integration and sharing of information in an alliance partnership? How do these issues differ by type of alliance?
2. If an alliance is in the future of the organization you are working for, what are some of the issues each level of the organization should be concerned with? In other words, what are the issues that top management should address? Middle management? Supervisors? Non-management personnel? What are the potential advantages of an alliance for each of the levels?
3. You have formed a number of alliances in your personal life at work, at school, and with your family. How have these paralleled the types of alliances described in this chapter? For example, marriage is a very formal alliance that has the potential for great synergies and great pain. Think about some of the types of alliances you have formed. What have been the keys to success for these alliances? What did you learn from these experiences about future alliances?

WWW EXERCISES

1. Use your favorite Internet search engine and find an example of a successful alliance that has led to a technological breakthrough and an alliance that was not successful. What attributes were described as contributing to the success of the alliance? What were the reasons given for the failure? How are these related to each other? What does this tell you about keys to success in planning an alliance?
2. Find a firm that has been very active in growing through acquisition of technology. Did the company have a clear plan for how it wanted to approach alliance formation? Has the company been successful in achieving its goals?

3. Find an article or website that provides guidelines for the planning of acquisition of technology. What do you think of the advice given? Compare the advice you find to the advice your classmates find.

AUDIT EXERCISE

In developing plans for alliances, especially mergers and acquisitions, the top management of the organization is charged with the responsibility of practicing due diligence. Due diligence is the process of identifying and confirming or disconfirming the business reasons for undertaking the alliance. If the purpose of the alliance is to acquire technology, then due diligence would require identifying and confirming the presence of the technology and the ability of management to take advantage of it. The exercise of due diligence can address a number of areas. These could include:

- Business climate
- Cultural environment
- Marketing channels
- Financial processes
- Operations systems and structures
- Human resources policies
- Environmental conditions
- Management of technology and innovation

Choose two of the areas listed and delineate eight to ten concerns to address in the planning of an alliance. For example, one of the operations systems that should be considered is the information system. The concerns in information systems might include defining the various hardware capabilities and networks, listing what software is used for what applications, and listing what software is used on which networks, and so on. For each area, you should be able to list eight to ten concerns that would aid in the planning for the acquisition of technology. By doing this at this stage of the process, you will develop a list of criteria to measure success against during evaluation and control (Chapter 8).

DISCUSSION QUESTIONS

1. Find a recent merger or acquisition that has been announced in the popular press. Is it related or unrelated? What are the implications for the merger or acquisition and plans for implementing the blending of the firms?
2. For the merger or acquisition in question 1, what were the strategic rationales offered for the action? Do you think the merger/acquisition will ultimately be successful? Why or why not?
3. Why do you think an industry like the airline industry has pursued alliances so aggressively?
4. McDonald's is a franchise-based organization with which you are likely familiar. What similar technologies (product and process) exist in McDonald's everywhere? Are there any differences in McDonald's in different locations? Why do those differences exist?
5. How would you describe the strategic efforts of Acer in terms of the types of alliances and acquisitions they have made? What are the advantages and disadvantages of using a variety of methods and approaches to technology and innovation?

PART THREE OPENING CASE: ACER

1. After reviewing Acer's process for technology acquisition and development in the opening case, what would you say are the strengths of their process and what might be missing in the process? How could Acer amend the process to make it better?
2. If Acer were to develop a joint venture with a company you work for, what would you want to happen? Would you want to work for the joint venture or stay with your firm? Why or why not? What circumstances would change your perspective?

KEY TERMS

acquisition 183

agency theory 184

consortia 175

due diligence 190

first mover 184

franchise agreement 175

horizontal mergers and
   acquisitions 187

joint ventures 174

licensing agreement 176

market power 185

merger 183

second movers 185

strategic alliance 174

subcontracting 177

transaction costs 174

vertical merger or
   acquisition 188

NOTES

1. Vance, A. 2009. Cisco and EMC form venture to serve data centers. *New York Times*, November 3. http://www.nytimes.com/2009/11/04/technology/business-computing/04cisco.html?_r=1.

2. Shirouzu, N. 2009. Corporate news: GE agrees to supply China aircraft Industry. *Wall Street Journal*, November 16, B.3.

3. http://www.entrepreneur.com/franchises/fasttek sonsitecomputerservices/324963-0.html

4. http://webhandling.okstate.edu.

5. Anonymous. 2009. IBM Helps Tracker Networks Enhance Business Resilience Requirements; New Managed Continuity and Work Area Recovery Solutions optimize and protect business and IT assets. *M2 Presswire*. Coventry, November 18.

6. Anonymous. 2007. China, Unregulated. *New York Times*, August 15, A.20.

7. Nuese, C., J. Cornell, and S. Park. 1998. Facilitating high-tech international business alliances. *Engineering Management Journal*, 10(1): 25–33.

8. http://www.america.gov/st/scitech-english/2009/March/20090309174506adxcilerog0.9232294.html. Obama Lifts Ban on Government-Funded Stem Cell Research.

9. Carr, C. 1999. Globalisation, strategic alliances, acquisitions and technology transfer: Lessons from ICL/Fujitsu and Rover/Honda and BMW. *R&D Management*, 29(4): 405–421.

10. Porter, M. 1987. From competitive advantage to corporate strategy. *Harvard Business Review*, 65(3): 43–59.

11. Kogut, B. 1988. Joint ventures: Theoretical and empirical perspectives. *Strategic Management Journal*, 9: 319–332.

12. Overby, S. 2003. Inside outsourcing in India. *CIO Magazine* (Jun. 1). http://www.cio.com/archive/060103/outsourcing.html.

13. Supplier voice: Calling for network solutions. 2000. *Outsourcing Journal* (Mar. 2). http://www.outsourcing-journal.com/mar2000-supp.html.

14. Grief, J. 2003. Technology acquisition planning: 10 guidelines to a better way. *Association Management* (Jun.): 16–17.

15. Henry, D., with F. Jespersen. 2002. Mergers why most big deals don't pay off. *BusinessWeek* (Oct. 14): 60–70.

16. Rappaport, A., and M. L. Sirower. 1999. Stock or cash? *Harvard Business Review*, 77(6): 147–158.

17. Wright, P., M. Kroll, and D. Elenkov. 2002. Acquisition returns, increase in firm size and chief executive officer compensation: The moderating role of monitoring. *Academy of Management Journal*, 45(3): 599–608.

18. Jensen, M. C., and W. H. Meckling. 1976. Theory of the firm: Managerial behavior, agency costs and ownership structure. *Journal of Financial Economics*, 3: 305–360.

19. http://itstrategyblog.com/skype%E2%80%99s-growth-strategy/. http://itstrategyblog.com/ebay-and-skype-strategy-analysis-again.

20. Kovar, J. 1999. Storage Tek signs Dell, SGI. *Computer Reseller News* (September 20) (860): 83.

21. Hayes, J., C. Shapiro, and R. Town. 2007. Market definition in crude oil: Estimating the effects of the BP/ARCO merger. *Antitrust Bulletin. New York*: 52(2), 179–204.

22. Blodgett, H. 2007. Ebay's unrelated acquisition of Skype? http://seekingalpha.com/article/35118-stumble-upon-ebay-s-newest-unrelated-acquisition.

23. Hitt, M. A., J. S. Harrison, and R. D. Ireland. 2001. *Mergers and Acquisitions: A Guide to Creating Value for Shareholders.* New York: Oxford University Press.

24. Grikscheit, A., and M. Cag. 2002. Extracting value from solid alliances. *Mergers and Acquisitions,* 37(6): 28–35.

25. http://www2.nortel.com/go/news_detail .jsp?cat_id=-8055&oid=100244956&locale= en-us.

26. Mergers and acquisitions; $305M acquisition of ESI Lederle completed. 2003. *Drug Week* (Jan. 24): 50.

27. Fitzgerald, M. 2003. Software systems: The missing element in M&A planning. *The Journal of Corporate Accounting & Finance,* 14(2): 13–17.

CHAPTER **7**

Implementation in Obtaining Technology

OVERVIEW

This chapter builds on Chapter 6 to identify how a firm implements the acquisition of technology or innovation. Once a firm decides to obtain a technology externally and the plans for these activities are developed, the focus must turn to implementing the plan. The relevant issues central to this effort that we explore in this chapter include:

- The people and the central role they play in the integration effort
- The potential pitfalls that limit the effort to integrate and utilize the new technology
- The need for a clear vision and understanding in the implementation process
- The importance of executing the implementation process quickly
- The importance of communication in the alignment process

INTRODUCTION

As noted in Chapter 6, there are a number of ways to obtain technology from external sources. These efforts can range from an informal temporary alliance to the purchase of another firm. No matter the activity, the processes to integrate and blend the different aspects of the organizations and technologies involved are critical. These efforts may vary in the energy that is required for implementation because of the type of external technology pursued. The fundamental issues, however, are similar for the various methods of obtaining technology externally.

As was noted earlier, most alliances, mergers, and acquisitions fail to reach their expected potential. The primary reason for these failures is the difficulty in integrating two firms. As a result, instead of capturing economies of scale or complementary technologies, the acquired and acquiring firms mismatched structures, competing cultures, and incompatible information systems lead to failure. Once the organization decides to embark upon a strategy for the external acquisition of technology, the focus should become how to blend quickly and effectively the entities involved in order to capture the desired strategic advantages and avoid the negative possibilities (or at least minimize them). Top management must address several key questions as they begin the efforts to obtain technology externally. These are the same questions addressed at the beginning of internal innovation implementation and include:

1. What should we be doing now and what can we do later?
2. What activities require the most time, attention, and/or specialized skills?
3. What should be delegated and to whom?

After answering these questions, the organization needs to address the four critical elements for implementing a strategy of technology acquisition: integration (firm due diligence, sharing lessons between firms, and blending), leadership (integrating procedures, strategy, and speed), execution (training, blending teams, and developing synergies), and alignment (rewards, common policies, and building fit). These elements are interrelated. The three questions and the four subsequent implementation items are examined in this chapter. Figure 7.1 summarizes these issues.

We will initially examine the key questions to ask, and then look at the four basic elements in the implementation of an externally focused strategy.

INITIAL QUESTIONS

You need to answer the three critical implementation questions no matter which approach is pursued in obtaining innovation or technology externally. Just as when pursuing an internal innovation process, the firm must decide the answers to these three questions before actually beginning the implementation process. These questions are: what should we be doing now, what are the key activities we need to focus, and what/whom to delegate.

Integration	**Leadership**
Due diligence	Integrate policies and procedures
Develop and share lessons learned	Develop strategy for critical activities
Blend structures and cultures	Speed

What to do now versus later?

What requires the most time, attention, or specialized skills?

What should be delegated and to whom?

Alignment	**Execution**
Tie rewards to integration	Training and development
Establish common policies	Blend individuals and teams
Look to build "fit"	Develop synergies

FIGURE **7.1** Key Implementation Issues in Obtaining Technology Externally

What Should We Be Doing Now?

This question addresses the reality that there are both long-term and short-term concerns that face a firm. There will be actions necessary today to ensure both long-term and short-term success. However, not all actions required to meet short-term and long-term goals must happen immediately. Instead, some must occur now, others must occur soon, and still others are necessary later.

Here we focus on actions that must occur immediately. When two organizations make an agreement to merge, align, or any of the other externally focused activities, it is imperative that a manager address a number of issues to move toward integration. The issues that need to be addressed include:

1. How will the external effort to obtain innovation capability or technology affect customers? How does the organization keep short-term commitments to customers while the internal chaos of blending takes place?
2. How do organizational priorities change following the externally focused activity?
3. Where should the primary focus of the organization be during the external effort to obtain innovation or technology? Too often, firms and managers get caught up in the nitty-gritty of the acquisition process and forget to focus on what needs to be done, now and later, to make the acquisition of technology successful.

To illustrate the importance of understanding such issues, consider what commonly happens in banks following a merger or acquisition. Banks seek external methods of growth like mergers or acquisitions to obtain economies of scale. Central to these efforts to obtain economies of scale is combining the banks' operations, especially technical ones such as data systems. Too often, in the process of combining their technology, banks ignore their customers so that simple things like recognizing deposits in given accounts and generating monthly bank statements for customers are botched. As a result, as much as 15 percent of the customer base is commonly lost during early implementation of the merger or acquisition. The net result is consolidation in the banking industry has often resulted in declining return on equity. For bank mergers that span international borders, the results are even more negative.[1] Thus, the priority on integrating technology systems without thinking of other concerns, like customer service, can lead to poor results that lead to the failure of the external effort to obtain technology.

What Are the Requirements for Key Activities?

The second question calls attention to critical areas that require the focus of management. This question is important no matter what type of externally focused effort to obtain innovative capability or technology. The key areas of concern include the amount of knowledge transfer required, the degree of integration needed, and the speed of integration necessary. Issues that will impact these evaluations include whether there will be sharing and/or protecting of particular technologies and whether there are product platforms present or to be developed.

To illustrate, in 2008, Bank of America acquired Merrill Lynch. Bank of America wanted to expand its reach in the financial industry. However, several of the desired star individuals from Merrill Lynch left the firm because they did not want to work in a "Southern commercial banking" atmosphere – Bank of America headquarters were then in North Carolina. There were further problems as it proved very difficult to integrate the two firms because they had very different cultures. In addition, many at Merrill resented the acquisition since they felt superior to anyone at a commercial bank. The result was that by January 2009 the *Wall Street Journal* called the acquisition by Bank of America the "deal from hell" with serious questions regarding the ability of Bank of America to survive. Companies employing M&A need to be sure that the two organizations can mesh well.[2] If not, it can result in failure of the effort. The process of due diligence (discussed in Chapter 6 and explored in more detail in Chapter 8) should help identify what activities need time and attention early in the implementation process to limit such negative impacts.

7.1 REAL WORLD LENS

Hewlett-Packard–Compaq

The combination of HP–Compaq involved not only units in the United States but also other parts of the world. For example, both firms were active in Australia prior to the merger. The integration of the two firms in that nation provides an excellent example of the range of issues in integration efforts in a merger/acquisition. Here we focus on one major domain – the two firms in Australia.

The two firms knew that their parents were potentially going to merge; therefore, they began operations integration efforts in Australia. The firms

REAL WORLD LENS *(continued)*

had similar product platforms for messaging, file, and print serving based on Microsoft products. This simplified the process, but totally integrating the two firm's domains was a major task. The choices on which applications to use were based on whether HP's or Compaq's applications were the most widely distributed. The pricing of HP's products utilized a global system, whereas Compaq used a country-specific system. The combined firm moved to the HP system in pricing. However, many systems such as downstream order management and HR systems, like PeopleSoft integration (despite both firms using the same system), would be more difficult.

The HP acquisition of Compaq appeared to be a failure for the first year to 18 months. However, after 4 years, the firms are fully integrated. HP's ability to integrate operations went relatively well even from the start; but the integration at the strategic level was slow in emerging. Once the firms fully integrated, the positive results have followed.

1. What advice would you give HP–Compaq to facilitate the integration process? Be specific and explain the advice given.

References

Gracer, S. 2007. Deal Journal: HP Compaq union: From controversy to success. *WSJ Blogs,* August 16, http://blogs.wsj.com/deals/2007/08/16/the-HPcompaq-union-from-controversy-to-success/

Withers, S. 2003. Succeeding in integration: Part I. *Technology & Business Magazine* (Apr. 11): Zdnet. http://www.zdnet.com.au/insight/soa/Succeeding_in_integration_Part_one/0,39023731,20273537,00.htm

What and to Whom to Delegate?

The simple answer to the question on what and to whom to delegate is that critical tasks are delegated to the best person to do the job. The best managers/leaders need to be incorporated from both the purchased and purchasing organizations into the new combined organization. The resulting combined firm should use the best systems from either firm for those tasks that require the support of any systems. The reality is that organizations often fall into one of several potential traps rather than choosing best people or systems.

1. The trap of compromise: Choosing one technology, one person, or one operation from each side. The goal is to get the best from each organization—not compromise.
2. The trap of misplaced beliefs: Keeping focused on the true purpose of the external effort to obtain technology is critical. Often a firm seeks to gain control of a technology but then during the implementation process focuses on other goals such as gaining market share or obtaining economies of scale. The result is the firm loses focus on integration. If the purpose is gaining technological advantage, stay focused on that unless

other compelling information causes a need to look in a different direction.

3. The trap of system superiority: Objectivity in the evaluation of systems and how they mesh is another key. The support system kept determines who is in charge of a given activity rather than the best person in the integrated firm. This can lead to poor choices and loss of talented personnel. The choice should be who is best, not the seniority of the person. Likewise, the criteria for selecting the best system architectures should be objective and not based on where it was developed or who developed it.

4. The trap of not blending: Sometimes individuals in a merger/acquisition or alliance may believe that keeping things separate will work and help everyone feel comfortable. The result is that the hard task of integration is not delegated to anyone, and the new organization develops inefficient policies and procedures. At one time, WorldCom had more than fifty different accounting systems that had to be reconciled every month by a team of accountants. The process of reconciling the data produced by these various systems took two to four days every month. The firm never decided who would be responsible for blending the accounting technologies. Such incongruent systems and inefficiencies contributed to WorldCom's eventual failure.

While it seems improbable that a firm can make mistakes such as those of WorldCom, consider what Citigroup has been doing for a decade. Citigroup has a long history of internal turmoil from acquisitions. One example occurred in 1998 when Travelers Group merged with Citibank. The merger led to a complex multiple-headed management structure that has endured, and undermined the business ever since. Until recently, for example, the bank had three co-heads running its fixed income operation. In 2010 Citigroup attempted to restructure to address integration of units acquired over the last several decades. The goal at Citigroup is to have one member of the bank's upper echelon management in charge of each major client's business.[3] The placement of a single individual in charge seems simple. However, Citigroup has had problems in the past delegating responsibility for customers and services. Although it seems that making choices on what systems should be dominant and who should run them are obvious, many firms do not make those choices until forced to by the costs associated with decision avoidance.

KEY ELEMENTS

After answering the three questions, the firm needs to address the four critical implementation elements. There are a number of concerns for each of the areas of integration, leadership, execution, and alignment. Here we address some of the issues within these four implementation elements. Because each external method for obtaining innovation/technology is different, the concerns can be different, but there are still some common features in the four basic elements we address here. Often, we discuss these points in terms of mergers and acquisitions because they are the most formal of the various external methods to obtain innovation and technology. However, the issues presented here in the discussion that follows are also of concern for alliances.

Integration

Integration is at the top of Figure 7.1 because it is the most critical item that must occur for the successful acquisition of technology. Blending or creating fit within an organization, whether it is an alliance, joint venture, or merger/acquisition, requires developing shared norms, capturing competencies from both organizations, creating compatible systems and structures, and integrating teams and functions.

To illustrate, Cisco was founded in 1984. Between 1989 and 1999, the firm had a compounded growth rate of 89 percent per year. Today, the firm targets a more moderate but still impressive 12–17% annual growth. The firm in fiscal year 2008 had sales of $39.5 billion. Much of this growth came through acquisitions based on Cisco's clear strategy to gain competitive advantage through the acquisition of new technology. Cisco's continued success depends on those acquired firms being integrated into the parent firm. As a result, Cisco has a well-established system that it follows for the acquisition of technology and technology-focused firms. This system includes:

- Strong due diligence prior to acquisition identifies all potential issues and problems as well as critical differences in the acquired firm.
- One-third of top management from the acquired firm fills slots in the new unit.
- A "buddy" system matches key employees of the acquired firm with key employees in Cisco.
- Customization of each integration effort matches the unique needs of the firm.
- Conversion of existing systems, such as computers, to Cisco systems takes advantage of economies of scale.
- Ninety days are allocated to complete integration of the acquired firm into Cisco.

The consideration of the various elements necessary for integrating firms can be overwhelming. However, managing these elements becomes more difficult as the size of the firms involved expands.

For example, when the acquisition of technology involves two firms, the organization suddenly must deal with two different cultures, two sets of standard operating procedures, plus two of almost everything else. There are two distinct methods to approach the integration of such issues. One is simply to say that one firm's activities or culture will dominate in certain domains. In the other method, management selects the best practices of each organization to carry forward to the newly formed company. However, if the organizations vary widely in size, the cost to change the larger firm to the methods of the smaller firm may be high and the benefits limited. Therefore, this decision is more complex than simply saying that one firm or the other should dominate.

A common mistake is that the acquiring firm tries to impose its culture and systems without thinking through how this will affect reaching the desired goals. Often, such an imposition is not a conscious choice, but rather, the firm driving the acquisition will impose its culture and systems on those in the acquired firm without regard to which firm has the best to offer the newly combined firm.

To illustrate the danger in such impositions, one only needs to examine one of the many aspects of the AOL–Time Warner merger. Following the merger, Time-Warner employee e-mail accounts were automatically switched to AOL accounts. Unfortunately, the Time Warner employees, particularly those who worked in domains like *Time* magazine, attached large files to their e-mails. The AOL technology was primarily for home computer users who do not need to attach large files to e-mails. Thus, the automatic decision to switch e-mail to the AOL system led to serious problems for some areas of the newly integrated firm—e-mail attachment size became an issue. The integration of the two firms was so poor that by 2009 Time Warner was attempting to spin AOL off into an independent unit.

Three critical elements in integration include:

- Due diligence
- Shared lessons
- Blended structures and cultures

Each of these elements will be examined next.

Due Diligence

Cisco's model, and most models of successful technology integration, emphasize due diligence, or the full examination of the firm or technology prior to the consummation of the deal. A key part of due diligence is the evaluation of the leadership team at the firm that will be purchased including key individuals associated with the technology of interest. This same due diligence should also be used in other less formal external methods of obtaining technology such as an alliance. People are a key part of the technology. If you lose the person who designed the technology the firm can be hindered in creating new generations of the project and also be creating new competitors if another firm hires that individual. The case of a merger or acquisition can be the most difficult to actually integrate because the blended firms can have two CEOs, two chief financial officers, two directors of marketing after the combination. Thus, it is often inevitable that someone may leave but such impacts can be minimized.

Technology firms are unique in that human knowledge and capabilities are typically the most important competitive advantage for the firm. Thus, for a technology-focused firm, the decisions on the integration of individuals are often the most critical issue to be negotiated. Well-developed due diligence can demonstrate how best to pursue the integration effort so that key individuals do not leave the firm but instead feel part of the new strategic goals, activities, and efforts.

Often in a merger situation, the top leadership decisions are made as part of the merger agreement; however, this does not mean the individuals involved will blend easily and the duties of the leadership team will be accomplished in an exemplary and smooth fashion. A year after the AOL–Time Warner merger, Gerald Levin, who had led the integration effort for Time Warner prior to the merger, left the newly combined firm because of conflicts with the leadership of AOL.

Due diligence can also identify difficulties about integrating the two firms' sets of activities. At the time of the merger of AOL and Time Warner, AOL employees were somewhat disdainful of their old-economy partner. AOL saw great potential synergies in creating a full range of advertising packages that would combine the old mature economy with that of the Internet. Additionally, AOL believed that there were benefits in trying to distribute Time Warner products over the Internet. However, AOL perceived that the new value was not going to come from the current personnel at Time Warner; instead, it was coming from new distribution channels for Time Warner products.

Initially, the AOL culture was dominant as Time Warner employees were encouraged to adopt new technology to replace what AOL employees viewed as the out-of-date processes of a mature and declining firm Time Warner. However, with the burst of the Internet bubble, the roles were reversed between AOL and Time Warner; AOL became the firm that was widely seen as being in decline. The combined firms had never truly integrated, and hostilities between them grew worse; the Time Warner employees began to view the AOL employees as pulling down the entire firm. Between 2001 and 2003, the stock of the new firm declined by over 70 percent. By midyear 2003, all major players that had promoted the merger in the management of both companies had left the combined company. The critical issue in this decline was that a stronger combined firm did not result from the merger. Instead, there were two firms operating under one name that were at war with each other. In part, this outcome was due to the failure of the partners to conduct adequate due diligence. If they had, they would have been more likely to understand the potential problems they faced and could have made plans as part of due diligence efforts on how to integrate staff and operations.

Share Lessons Learned

Regardless of the nature of the external method used to obtain the innovation or technology, the knowledge of the two entities that are aligning must be leveraged if the organization is to reach its goals. The knowledge shared may be a process that is technologically new to one partner, or it may be a new product. There are several lessons about sharing knowledge that should be in the forefront of external efforts to obtain technology.

1. Analyze and understand who knows what. If the organization is buying new technology by acquiring a smaller company, then recognize the ownership of the knowledge the smaller company brings to the table. Too often, there is a grab for the new technology, but the full potential of the technology is lost because the integration into the large firm leads to the exit or loss of those that either know or develop the technology.
2. Develop a systematic way to share knowledge. The execution of this sharing process is how you blend individuals and teams. Integration requires that there is some method in place to encourage this. There are a

number of ways, including blended teams, regular meetings, formalized liaisons, cross-training opportunities, and establishment of formal communication channels.

3. Take time to be sure that learning and know-how are shared. This is part of the evaluation of how to share knowledge as well as part of the implementation. If know-how is not shared, corrective action needs to be taken. After all, the reasons for an acquisition strategy are all about sharing know-how.

4. Remember, knowledge and information exist at all levels and should be shared with all who can use it. With mergers, the potential communications breakdowns add a new factor. Not only is there a potential problem across levels and functions but also across old organization boundaries.

The due diligence process should include some analysis of decision-making processes. The differences and similarities in decision-making policies should be clear. The lessons learned from the due diligence process could be used to develop procedures before problems arise. Developing common decision-making criteria and sharing them can help eliminate possible conflict. One firm may have a policy that midlevel managers can make capital expenditure decisions on their own with projects of up to $50,000, hire employees with minimal consultation, and have a formal project funding process. The partnering firm (or merged firm) may have a lower dollar amount for similar decisions or a more formal process for such decisions. The blending of these two approaches would require careful consideration by managers from both organizations. It would also require training the employees in the new blended processes adopted by the firm.

To illustrate the role of sharing knowledge in a merger and acquisition consider the merger of Commonwealth Bank with Colonial Limited, the largest corporate merger in Australian history when it occurred. The merger involved more than 37,000 staff, more than 135,000 points of representation, and a global presence in 16 international markets. To prevent inconvenience to customers and ensure accessibility of necessary data and systems, the bulk of the IT systems blending occurred during a single weekend. This meant migrating retail and nonretail systems used by Colonial customers into the Commonwealth Bank's technology network in just sixty hours. This integration required that they move 1.2 million customers into the new environment with continued access to more than 650,000 demand deposit accounts, 875,000 debit cards, and 500,000 credit cards. To accomplish this required extensive communication and understanding by all parties. Both firms in the merger had to communicate about their prior experiences with these databases and the pitfalls that could affect the merger on both sides.[4] Without sharing this information and the lessons learned, the effort to merge the two IT systems would have been bound to fail due to unrecognized potential problems. The sharing of knowledge between the parties ensured that this did not occur and the migration was a success.

Blend Structures and Cultures
The blending of structures and cultures centers on two factors: the type of external activity used to obtain the innovation or technology and the existing

structures and cultures in each of the organizations. Consider what occurs in a merger or acquisition where the greatest level of integration is required. In this setting, one of the key issues is whether the firms involved have similar technology. For technology firms, we can broadly assume that if their key technology is the same, then the skills necessary to be successful are the same, and we can classify the firms as related. Firms with technology that is not similar are unrelated. Figure 7.2 summarizes some of the key concerns in this situation.

Figure 7.2 shows that in an unrelated acquisition, restructuring is often required. This could mean a new strategic business unit or a new division is formed to house that technology. The formation of a separate unit requires less integration. But, in related acquisitions, there is usually much more integration, and the integration is far-reaching—to the individual and team levels. The amount of cultural change needed or expected is usually more extensive and based on several other factors: size of the merging entities, strength of the culture, and criticality of the knowledge each brings to the blended firm.

For the blended firm, the culture usually takes on one of four forms:

1. Separate cultures: The resulting firm maintains the separate cultures of the original firms. This usually emerges with unrelated acquisitions and with the acquisition of a firm that is large enough to be its own division or SBU within the integrated firm.
2. Dominant culture: This occurs when two entities merge and one has a very strong culture or is much larger. Usually, the larger firm is the dominant culture; however, if the reason for acquiring external technology is to change the direction of the firm, a smaller firm with a strong culture

Description of Acquisition of Technology Relative to Dominant Technology in Firm Currently	Characteristics of the Technology Acquisition	Key Fit Elements	Examples
Unrelated	Strategic High uncertainty Wide firm impact	Top management support Multiskilled team External alignment	Major restructuring of firm
Related	Less uncertainty Low organizational impact	Strong integration team Supportive internal environment Accountability processes	Integration and innovation at the team level

FIGURE **7.2** Characteristics of Types of Acquisitions

<table>
<tr><td>

7.2

REAL

WORLD

LENS

</td><td>

Merck Serono

In 2006, Merck began negotiations to acquire Serono, a biotechnology company with its headquarters in Switzerland. As the deal closed in January 2007, the integration team at Merck Serono International S.A. had identified 211 goals to ensure the successful integration of the two firms. Among the priorities were:

</td></tr>
</table>

- Communicate often with employees—via webcasts, meetings, newsletters and other tools—to ensure acceptance, keeping in mind Europe's complex collective labor arrangements.
- Redefine stock incentives and retention plans. Following an audit of total rewards, HR professionals created an interim compensation and benefits philosophy. Ex-Serono staff members eligible for long-term incentives now receive a new incentive package.
- Harmonize performance management systems with different philosophies. In Serono, the performance management system focused on meeting goals, with links to bonuses, whereas Merck offered fewer short-term rewards for meeting goals. The integration team had to create a middle ground.
- Attend to the culture. A survey identified how to bring people together better. To that end, staff attended more than 100 team-building workshops. Although the cultures were different, surveys identified common ties, including a focus on quality, innovation, teamwork, customer service, and a common vocabulary. A change management workshop worked to help employees adjust to the merger and create personal action plans.

In addition to cultural issues, the integration team faced two distinct approaches to internal business processing: Serono was vertically oriented with a functional structure, while Merck had a geographical based structured with a country head managing all activities in that country. Merck Serono adapted both styles by developing a hybrid structure that included elements from both the functional and geographical structures.

Because of clearly stated goals, personnel in both companies understood and accepted the business of the merger. This helped to make it a success. The integration of the two firms led to adoption of the best of both in many cases. While such cooperation may take more time in the short run, the results for Merck Serono have been excellent with faster growth than the pharmaceutical market, lower turnover, and increased employee satisfaction.

1. There is a list in the chapter of CEO activities for speedy integration (page 216). Based on the information given about Merck Serono what integration activities were addressed and what activities were ignored? Explain your answer.

Reference

Davis, N. 2008. Merger kept 'the best of both'; Merck KgaA of Germany acquired Swiss Serono in 2007 and set managers to the task of creating

REAL WORLD LENS (continued)

a powerhouose in Merck Serono International S.A., *HR Magazine*, November. http://findarticles.com/p/articles/mi_m3495/is_11_53/ai_n31008284/

---

may influence the new entity much more. In this case, the CEO of the smaller company is usually the CEO of the blended firm.

3. New culture: This occurs when the merging entities find enough common ground that a new culture emerges that truly blends the best of both. This is obviously desirable. It most often occurs between firms that are similar in size and when each firm brings to the alliance a knowledge base that the other needs to sustain success.

4. Multiple cultures: Subcultures exist within all firms of any size. A firm with multiple cultures allows the merging entities to keep their cultural norms and blends them to a smaller degree than with the new culture approach. This can be successful, especially in firms with divisional or SBU structures. The danger here is that the systems and processes in each area can easily evolve to levels of incompatibility. Then the differences in culture can create more friction in trying to change and integrate the systems. This type of culture usually emerges when a large firm with multiple unrelated businesses acquires multiple entities—a holding firm.

To illustrate how such integration should occur, consider the merger of Nokia Networks and Siemens Communications. The goal of the merger was to create a strong global company with strengths in both wireless and land based telecommunications, plus leverage a massive international sales force, and achieve economies of scale unavailable to either company. But, NokiaSiemensNetwork (NSN) involved the merger of two distinct corporate cultures. A team composed of executives of both firms was tasked with the integration of the cultures. This committee found several things that the two companies had in common, but there were also major differences. The most striking differences involved the sense of formality and structure in Siemens' culture, as opposed to a looser set of relationships and emphasis on flexibility at Nokia. To sort out what the new culture would be, NSN first had to define the cultures of their parents, both of which were ingrained after more than 100 years of operations. NSN hired a British consulting firm to help decide what to emphasize as the two firms combined. The outcome was a design to move the firms to a common culture that shared the best of both cultures. Today, there is wide acceptance of this cultural mind-set in the combined firm.[5]

For international alliances, there are often more complications. Cultural differences occur at three different levels: individual differences in experiences (history or functional background), the cultures of the firms involved, and country culture. In a study of British and Chinese managers, the impact of national cultures was clearly illustrated. The managers in this study found each other dedicated; positive in attitude (humor); cautious in actions, planning, and making rules; and willing to involve employees in decision-making. However, the British perceived the Chinese managers to be less capable in developing structure, internal communication, and in developing incentive programs. In fact, the British were much more critical of the Chinese managers than were the Chinese of the British. These perceptions may not be accurate, but an understanding of these types of perceptions can alert the organization to training that may need to occur or to develop systems that help to overcome British managers' biases. Understanding is also critical if blending the two organizations based in different countries is to take place.[6] The need to blend such culture can even be present in two firms such as Nokia/Siemens that are geographically close.

Here we have discussed blending of structure and culture in a merger or acquisition. It is clear from the discussion that issues involved in this blending can be significant. The requirements in such integration activities can encourage a firm to pursue a less formal means to obtain innovation and technology, such as an alliance where less effort is required because there is less integration of the two entities. However, in all cases, the firm needs to have the ability to build linkages so that the knowledge and lessons learned can be shared to a degree consistent with the nature of the external effort to obtain technology.

Leadership

Leadership is the second major element in the implementation of an externally focused means of obtaining innovation capability or technology. Leadership is the art of getting things done through others. In externally focused methods of acquiring technology, leadership has the responsibility to:

- Integrate policies and procedures
- Develop a strategy for critical activities
- Ensure that these activities occur in a timely manner

We will examine these component parts in turn. However, it is important to note that although CEOs and COOs are involved in negotiations in externally focused efforts to obtain technology, too often, the CIO of a firm or the head of research and development is often excluded from such negotiations. Many of the leadership issues identified here could be resolved by involving technology professionals such as the CIO, R&D managers, and engineering managers at all stages of the planning and implementation processes.

Integrate Policies and Procedures

The integration of policies and procedures when an acquisition occurs is fraught with opportunities to excel or to fail. Often during the integration process, managers do not look for competencies to share across the organization. This happens because the policies and procedures evolve through compromise rather than through a planned integration of best practices. The leadership of the organization is critical in building the integration effort. To illustrate this point in 2001 following the AOL–Time Warner merger there was a meeting in which AOL and Time Warner managers met to talk about *Fortune* magazine providing articles about AOL's Netbusiness product.

Netbusiness provided a wide range of Internet-related businesses and services including news and information. However, the Time Warner managers wanted to be paid for the articles, whereas the AOL managers insisted on getting them free because both *Fortune* and AOL now belonged to the same corporation.

The amount of money, an estimated $1 million in potential fees for using *Fortune* magazine's articles, was not significant for a firm such as AOL–Time Warner, but no agreement was made and executives stormed from the meeting.[7] Leadership by top management at times such as these is critical to support and encourage the integration process. A clear statement by the leadership of the new AOL–Time Warner on what would be shared between units and what would be paid for would have prevented these difficulties.

There are several abilities that the organization should nurture among the newly formed groups to enhance the integration of policies and procedures.

1. Ability to learn from the other firm: Too often, instead of being open to learning, the individuals involved are more interested in protecting territory. One way to combat this is to try to balance the membership of teams that will be addressing the integration issues.
2. Ability to meet performance goals: Repeat often to the employees that after the merger is complete all members of the organization will benefit from the organization's success. The decision about which procedure to use should be driven by the question of how best to meet goals today and in the foreseeable future.
3. Ability to build trust: Often, trust declines in external efforts to obtain technology in one or both of the firms involved. The uncertainty generated by the activity results in individuals and groups within the organization being more concerned about what each move or decision means. Firms build trust through consistency and open communications. Consistency comes from analyzing what policy or procedure is best and implementing it and then clearly communicating why you made that choice.

Many analysts questioned the HP–Compaq merger initially. However, HP realized the importance of trust in the merger if it was to be successful. The combined firm planned to gain efficiencies by eliminating an estimated 10 percent of its employees. HP felt that to build trust a critical aspect of the integration was to be up front about this fact and quickly address it. Although the merger had some initial difficulties, the integration methods used and, in particular, the firm's effort to build trust have been applauded.[8] The result was that shares of HP initially dropped but they then tripled between 2004 and 2007 as the company has surpassed Dell Inc., to become the world's largest PC maker in terms of units sold.

Develop Strategy for Critical Activities

The firm needs to address some activities more forcefully than others when conducting an externally focused effort to obtain innovation capability or technology. The ability to identify the key expectations in the following areas should take priority: service/product continuity, cost structures, productivity, internal administrative services, and information systems alternatives.[9]

In addition, personal relationships are critical in all aspects of an externally focused effort to obtain innovation or technology whether it is an alliance, joint venture, or merger/acquisition. In large measure, personal relationships are critical because people are fundamental to technology-focused firms. Thus, in externally focused efforts to obtain technology, it is important that there be a focus on people and relationships so that key aspects of both are maintained. The important concerns that need attention from the leader at the start of the implementation process to develop and maintain key people and relationships are:[10]

1. People/relationship issues involving problems related to communications, culture, and roles
2. Operations issues involving problems related to technical details of implementation of the strategic plan
3. Strategic agenda issues of problems concerning the goals and objectives of the alliance

More than 50 percent of the problems in implementing an acquisition strategy are related to people/relationship issues.[11] The problems most often are communications barriers, differences in organizational culture, and uncertainty over roles and responsibilities. The leadership of the implementation team should develop strategies and goals to address these issues. For example, defining roles and responsibilities quickly will allow the firm to take advantage of potential synergies before they are lost in the maze of the new organization. To build a unifying culture and improve communications, the manager should set up opportunities for success. If the firm has a clear strategic agenda, then it is easier to set up operations that will allow people to succeed.

Some examples of operation issues that can emerge during the implementation of newly acquired technology are:

1. Lack of understanding of each other's products/processes
2. Underestimating the cost of integrating and supporting
3. Lack of supporting documentation—everyone in the old organization just knew how to do that
4. Incompatible technology
5. More engineering/debugging/development work needed than anticipated

A leader's role is to try to prevent these problems from occurring. Because the total prevention of problems is unlikely, leaders need to develop strategies to address the problems as they arise. The greatest focus should be on activities that are the most critical to the organization. The actions developed for these critical activities must be executed in a timely fashion and aligned with the objectives of the firm no matter what external method of obtaining technology is employed. Leaders have to recognize the interests of the two firms involved in the activity. According to research, the path toward successful implementation in a technological setting should include the following:[12]

1. A goal that specifies desired business results should be defined.
2. Potential alternatives for achieving the goal need to be identified and evaluated.

3. Once the method to achieve this goal or goals is selected, enabling technologies should be evaluated.
4. After identifying the appropriate technologies, a plan is developed to integrate the technologies into one system.
5. Technology specifications, tied to desired business results, are tested by users.
6. Employee participation is part of every phase of the implementation process.
7. The ability of employees to adapt to the new technologies should be assessed.
8. Training programs need to be developed to address any areas of weakness.
9. Communication programs regarding the changes needed are fully developed and used.
10. Behaviors supportive of the new process are rewarded, and negative behaviors result in negative consequences.
11. Every component of the organization is analyzed and restructured, if necessary, to support the new technologies.

Speed

The last element is the need to act in a timely manner—speed. The consulting firm A. T. Kearney estimates that 85 percent of the benefits from the synergy of joining two entities come in the first year.[13] Thus, if organizations cannot act quickly, few benefits of the combination of the two entities will occur. After that time, the momentum for change is lost, and obtaining greater synergy will take considerable effort.

To facilitate acting quickly, the people in the two entities involved in the acquisition of technology need to know what their status is and what to expect quickly. For example, in a merger of two firms, the turmoil can be quite high. Previously we noted that talented people are one of the key assets in external methods to obtain technology. However, the most talented individuals always have the greatest number of opportunities. If there is high tension or uncertainty, these highly sought-after individuals can leave the organization. Recall that in the Cisco example that the firm seeks to have integration completed within ninety days. The ability to do so minimizes the anxiety of the employees and focuses the firm on moving forward.

However, speed must be tempered with an understanding of what must occur and when it must occur. Figure 7.3 indicates some of the issues with timing and criticality. Something may not be critical, but if not done in a timely fashion, it can become a critical problem. Likewise, something may be critical, but it does not need immediate action. The firm must determine what must be addressed and in what order. This determination is the part of implementation that answers the questions: What do we do now? What do we do later?

For the leadership, there are key activities that can speed the process. McCreight and Company[14] developed a mergers success checklist for CEOs. This checklist can be adapted for all alliance types as well as mergers. If the

IMPORTANCE VERSUS URGENCY

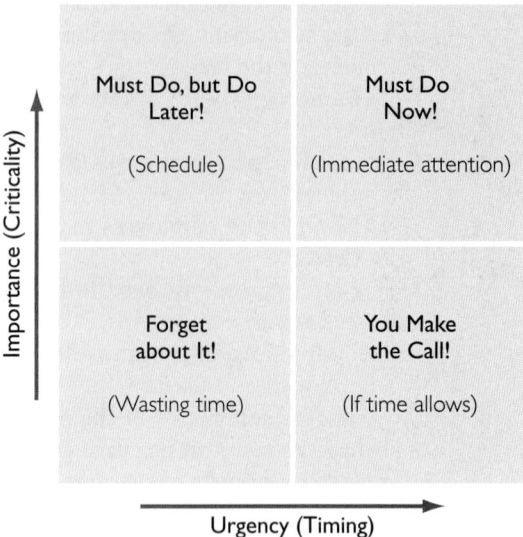

FIGURE **7.3** Determination of What to Do Now

CEO knows what to do before the implementation begins, then it should be a quicker process. The checklist includes the following:

1. Establish and communicate a "change vision" by putting together a team that looks at the nitty-gritty implementation issues. They should know why the deal was done, what is expected, what the expected timing is, and how to measure success. This team should include individuals from all partnering firms and should have well respected individuals with the power to make decisions and implement plans.
2. Meet and evaluate all the key executives in the partnering firms. These executives need to own the alliance success.
3. Define gaps in management talent. Too often, there is a missing management capability even in the merged or allied firms. Being proactive in filling these gaps will save time and energy later.
4. Set implementation milestones and measures of interim success. New competencies will emerge. Goals and focus should be presented early and measured often.
5. Install a sound organizational structure and management process. Determine how decisions will be made the morning after the deal is finalized. Be sure there is a list of how decisions will be handled, and that there is a timeline that the implementation team is working with.
6. Consider developing specialized reward systems to reinforce the alliance activities and to support the implementation team.

With these activities under control, the CEO and the implementation team are ready to execute the merger/acquisition or alliance.

Execution

Execution refers to the people-related issues that are addressed during the implementation process. Nothing positive happens without the individuals and teams within the organization working toward common goals, whether an alliance, joint venture, or merger/acquisition is involved. To obtain the complete effort of the employees, include:

- Training and development
- Blending individuals
- Developing synergies

Training and Development

During implementation of mergers or acquisitions, the employees can suffer from feeling a number of losses. These losses can include status, knowledge of the firm, network of informal relationships, control of the work environment, understanding of the future, and even understanding of what their individual job entails. Training and development should help employees understand what is happening in the organization and with their jobs. It is a lot easier for an employee to accept change if the employee knows how he or she fits into the change and where the organization is trying to go. Because there is an evolution of cultural norms, changes in policies and procedures are likely, as well as changes in reporting relationships and job tasks. The manager needs to educate employees about the changes, what they mean for the organization, plus what they mean for individuals.

One means of such education is through employee training and development. The firm should make this training and development formalized. For example, you would not simply ask employees to educate each other about how to operate machinery or the policies of the firm. Thus, as you pursue external methods of obtaining technology the firm also needs formal ways to train and educate employees. To illustrate, too often a software program that is new to many of the personnel in the newly combined firm is simply put in place and no training is made available to the employees. As a result, resentment grows because of unrealistic demands on the employees for productivity on a software program they do not understand. These employees are to learn this program on their own while learning a new job and developing new relationships. The result is poor performance and unhappy employees. The lack of training and development across integrated systems in newly combined firms is a key reason for many acquisition failures.

GE Capital Services uses a process called Pathfinder in which it identifies key issues in such mergers or acquisitions. In this process, the firm seeks to work with employees to educate them as to what is occurring and why. By applying the process, GE Capital has learned five lessons for blending individuals and implementing the deal.[15]

1. Start early.
2. Implement restructuring sooner rather than later.
3. Dedicate resources to the integration.
4. Integrate operations and cultures by focusing on outcomes.

5. Communicate strategically in a continuing manner with straightforward messages, modes, and regularity.

The firm continues this process beyond the immediate joining of the two firms to include long-term efforts, to educate the employees, and to make the blending work.

Blend Individuals and Teams

There is a need not only to integrate personnel from both organizations but also to create lines of communication that are relevant to the new organization. Just putting individuals together in a physical space is usually not enough. There should be clear objectives for the teams and individuals that are now working together as well as communication systems in place. The leadership of the combined firm should celebrate success loudly and work to learn from failures. It is important to celebrate success early and often because individuals and teams will blend more quickly if there are positive results coming from the venture.

However, conflict in the combination process often results. The differences and the amount of change involved can lead to the need to manage the conflict. Some of the most common sources of conflict and suggested solutions appear in Figure 7.4. Teams with members from different functional areas and both firms can often smooth these types of conflicts by recognizing and blending the strengths and experiences of the two firms that are forming the alliance.

Louisville Gas & Electric and Kentucky Utilities Energy were regional utilities. To achieve critical mass and fend off unwanted takeovers, the two

Source of Conflict	Suggested Solution
Blending Priorities	Develop a master plan compatible with long-term strategies.
Administration Procedures	Clarify roles, responsibilities, and reporting relationships at the beginning of the project.
Technical Opinions and Performance Trade-offs	Use peer review and steering committees to review specifications and design.
Human Resources	Develop a work breakdown and a corresponding responsibility matrix.
Cost and Budget	Develop overall budgets supported by detailed budget and cost estimates of subproject tasks and activities.
Schedules	Develop an integrative schedule that includes staffing and other resource constraints.
Personality Conflicts	Emphasize team building.

FIGURE **7.4** Sources of Conflict and Suggested Solutions

companies agreed to a merger in 2003. The goal was to have complete integration within six months. There were two concurrent work teams to make the transition a reality. The merger integration teams were to identify and manage information technology and human resources. The combined firm used teams both to smooth the integration and to identify ways to reduce the duplication of services and costs. The teams were so successful that the firm can now anticipate more than $760 million in gross nonfuel savings over a ten-year period.[16] The combined firm became part of the E.On group from Germany in 2005. The 2003 transition process was employed again in 2005. As a result E.On U.S. continued to be efficient in the utilities industry even as it became larger and more international.

Develop Synergies

To develop synergies, the leaders of the integration effort must find ways to gain efficiencies or to transform the firm. In settings where the acquisition of technology employs a merger or acquisition, the leadership of the firm should find synergies that reduce costs or improve the economics/competitive environment of the firm. These synergies can also help reduce the risks involved with the process of acquisition.

To execute such a plan for the acquisition of technology, the managers must keep in mind several issues.

1. Implementing an effort to acquire technology externally is a process that is challenging. The challenge and the task need to be respected.
2. There are no "conquering heroes" in implementing an acquisition strategy. The alliance process is about the two groups of people, selling the boats they have been sailing in and getting a new one that is better.
3. There are new roles to be learned and polished. If leadership conveys the reasons for the blending and the goals for the future, it will be easier for individuals and teams to find ways to move toward success.
4. Remember, the people issues are where most of the building of synergies needs to take place. Individuals need to understand the new processes, the new logistics, the new communications expectations, the new culture, and the new goals.

Alignment

The last critical element that needs to be considered in the implementation of the acquisition of technology is the fit between people, systems, operations, and strategic goals. This fit represents the alignment. To ensure the alignment is taking place, the leadership must ask several questions during the integration process. These questions give direction and provide a basis for evaluating how the blending of the two entities is proceeding.

1. Are you constantly looking for and finding synergies that would not have been possible without the acquisition?
2. Is knowledge being shared across groups regardless of which firm held the knowledge before the acquisition?

3. Is new knowledge emerging because of the interaction among people from different systems and structures? In other words, is there a capturing of the richness of differences in approaches?
4. Is trust increasing as individuals and groups are blending into the integrated organization?

Whether the technology acquired comes from an alliance of some type or an acquisition of a small firm with a new product by a dominant player in the environment, these questions must be addressed and three key actions must take place.

The three key concerns to help the organization to achieve alignment include:

- Reward systems of the blended firm must support integration.
- Common policies need to be established.
- Elements of each organization's dynamics must fit together. Task and technology, management, structure, and individuals and teams must fit with the goals and the processes to reach those goals.

These items will be reviewed next.

Tie Rewards to Integration

If the goal of the organization is to develop new technologies through acquisition, then the reward systems of the organization should reward individuals and groups that help the organization achieve that goal. The expected resource commitments should reinforce potential synergies from the alliance. If there are insufficient resources for the integrating processes, then the message to individuals is clear: This integration is not important to the organization.

Integration occurs because employees know that the company expects it and has dedicated the resources to make it happen. Not all efforts to integrate will be successful, but with a carefully thought-out strategy for integration and the resources and rewards to support the strategy, there should be more successes than failures. The willingness to reach out across the different groups and to help move the organization to one system need to be rewarded and encouraged. This calls for specific rewards that support such activities. If those rewards are not present, individuals may voice support for change, but their support may be superficial or short-lived.

Establish Common Policies

The alignment within the organization after a merger/acquisition is through common policies and procedures. For example, for human resources some of the key issues include information on pay, benefits, bonuses, employment regulations, third-party claims, employee relations, and safety. As much as possible, the common policies should be the best practices of either organization, not on what the largest firm is doing. This is especially true because one problem with large firms that we mentioned earlier is that they tend to be bureaucratic.

The policies and procedures of the new firm may require a change in structural form of the resulting firm after a merger or acquisition. The important point to remember is that in the end the organization must have systems and structures that work as efficiently and effectively as possible.

To illustrate the potential difficulties with structure, consider the combination of Chrysler Corporation and Daimler Benz in 1999. Chrysler was in deep financial trouble when the firm merged with Daimler Benz. However, the resulting integration of the very different firms with different technologies and cultures was problematic. These factors contributed to the failure to integrate the two firms structurally. Without sound structural integration, there was little ability to gain the efficiencies desired. The result has been that the Daimler Chrysler merger was unsuccessful—Chrysler ultimately became a separate company again. Daimler paid over $36 billion for the firm and ultimately sold it for less than $7.4 billion—a loss of approximately $29 billion.

Look to Build Fit

The implementation of the acquisition of technology strategy is all about integrating, blending, and building fit. The principal reasons alliances fail relate to the inability to integrate and build fit. These failures in building fit into three broad categories: people issues, operational problems, and strategic actions.

People issues in building fit include the following:

1. Talent is lost or mismanaged.
2. Power and politics become the driving forces.[17] The positives and negatives of politics are presented in Figure 7.5.
3. Defensive actions and motivations take over because of fear.
4. Culture clashes go unchecked.
5. There is a failure to send a consistent message.

The operational issues in the effort to build fit are also critical. The operational problems arise because of the following failures:

1. Transition costs are underestimated.
2. Impossible levels of synergy are expected.

Positive	Negative
• Can create social and financial resources • Can bring disparate groups together • Can help promote or leverage change • Can provide support for specific objectives • Political influence is power that can be brokered • Can help break impasses • Can help mobilize opinions and opinion leaders	• Can cast doubt on the credibility, trust, and ability of merger leaders • Helps to create schisms and divisiveness • Raises questions about the merger's purpose and expected outcomes • Externalizes a firm's problems • Encourages scapegoating • Creates the impression that anyone is dispensable • Can be coercive and abusive • Creates distrust and promotes favoritism

FIGURE **7.5** Positive and Negative Effects of Politics on Merger Outcomes

Strategic actions that can lead to failure include:

1. Expectations are unrealistic.
2. Strategy is hastily constructed or poorly implemented.
3. Focus of top management is distracted from the purpose of the organization.

The building of fit requires time and effort. However, a clear understanding of the need for fit and the wide variety of activities that need attention will aid in the implementation process. For example, many times systems are not capable of **autoadjudication**, or the integration of two systems into one. If an insurance company purchases another, their computerized claim systems may not work together. As a result, the firm needs to choose either to fit the systems together or to replace the systems. If the new combined firm chooses to do nothing then integration becomes problematic. The result is that until the firm finally begins to seek integration of these various systems, there will be gross inefficiency. Although it seems that making choices on what systems should be dominant and who should run them are obvious, many firms do not make those choices until forced to deal with the costs associated with decision avoidance.

Management should remember that obtaining technology externally leads to change in the organization. Whenever there is change, the level of comfort of those in the firm is affected. The level of comfort is increased by communication, trust and mutual assurance, building new relationships (chemistry), and recognition of differences as a strength (diversity).[18]

SUMMARY

This chapter has examined the implementation of an effort to obtain technology. It has highlighted the fact that most mergers and acquisitions fail as the result of the inability to successfully implement the acquisition. The rational calculations of buying another firm or forming an alliance with it may be good, but the actual implementation is a human activity that requires extensive skills and effort.

The analysis here turned on three questions that initially need to be asked for key elements of the integration effort. The questions include:

1. What should we be doing now and what can we do later?
2. What activities require the most time, attention, and/or specialized skills?
3. What should be delegated and to whom?

From these three questions, the firm then needs to address four key actions to be successful. These include:

- Integration
- Leadership
- Execution
- Alignment

If a firm will answer these questions and explore these key elements along the lines presented here, it will have far greater potential to be successful and avoid the poor performance often associated with attempting to acquire a technology or innovation through external means.

MANAGERIAL GUIDELINES

Leadership, execution, and alignment must take place to integrate the firms. For this to happen, managers must do the following:

1. Establish a clear strategic purpose for the alliance or merger. The joining of firms should be based on serving the needs of the organization, not management or any other one stakeholder.
2. Find a partner that has the potential to fit. There should be compatible goals and capabilities that complement and enhance.
3. Recognize the need for each partner to do what it does best. This requires specialization and trust.
4. Create incentives for cooperation among various groups that will interact. Blending does not just happen; it requires effort.

5. Share information, personnel, systems, and whatever else is needed to reach the goal of the alliance.
6. Treat the alliance partner as you want to be treated. Example is still the best way to set an agenda of behavior.
7. Exceed the expectations of the partnership. Give more if you can. Again, this builds trust in the relationship.
8. Be flexible. These types of partnerships are open-ended and dynamic. Sometimes the partnership happens easily; other times it is painful. Flexibility is important when trying to take advantage of new opportunities.

Guiding Questions

Figure 7.6 is a list of potential reasons for failure when technology is acquired. The focus of this chapter has been mostly on settings where technology is acquired through a merger and acquisition of two firms because these are the most

radical and the hardest to manage. However, the list has suggestions that also apply to alliances. If managers pay attention to the key questions and the four areas described in this chapter, they should be able to realize the synergies of the alliance and take advantage of the technological

• Unrealistic expectations • Hastily constructed strategy • Unskilled execution of strategy • Failure to send a consistent message • Talent is lost or mismanaged • Power and politics become the driving forces	• Impossible degree of synergy is expected • Culture clashes go unchecked • Transition management fails • Underestimation of transition costs • Defensive motivation and actions • Focus of top management is distracted from "Important" activities

FIGURE **7.6** Reasons for Failure of Technology Acquisition

SOURCE: "Debra Sparks in New York," Business Week, March 22, 1999: Iss. 3621, pg. 82; Gary Hamel, "When Dinosaurs Mate," Wall Street Journal (Eastern Edition). New York, N.Y.: Jan 22, 2004, p. A12; Africa Arino and Yves Doz, "Rescuing Troubled Alliances… Before It's Too Late," European Management Journal, London: Apr. 2000, Vol. 18, Iss. 2, pp. 173–182; Randall Schuler and Susan Jackson, "HR Issues and Activities in Mergers and Acquisitions," European Management Journal, London: Jun. 2001, Vol. 19, Iss. 3, pp. 239–253; Yves L. Doz and Gary Hamel, "Alliance Advantage: The Art of Creating Value Through Partnering," Harvard Business School: Boston Press, ©1998.

gains for the organization. This is, after all, what the major stakeholders of the firm want from any strategic action.

To build alliance competence is not easy. The three central questions and the four key areas all present complex challenges. The external effort is more likely to be successful if the firm examines the following questions:

1. What are the time and place to select an external means to obtain technology?
2. What are the alliance parameters that are keys for us?
3. What is our plan for implementing the integration of the new technology?
4. How clearly have we delineated the roles and responsibilities of the partners and the people in the blended firm?
5. Are we developing capable alliance managers?
6. Have we committed the appropriate resources to ongoing alliances?
7. And last, but certainly not least, have we established the competence to develop and maintain a continual review and modification system?[19]

CASE **7.1** THE REAL WORLD
IBM/PwC Acquisition

In 2002, IBM bought PwC Consulting for $3.5 billion in cash and stock. PwC was the consulting arm of PricewaterhouseCoopers Partners. The goal by IBM was to build up its consulting activities. IBM envisioned the acquisition as part of its IBM Global Systems unit, Business Integration Services (BIS) division.

IBM wanted to strengthen its expertise in industries such as pharmaceuticals, financial services, and retail consulting. In addition, IBM hoped to gain specific skills including ERP (enterprise resource planning), CRM (customer relationship management), and SCM (supply chain management) from PwC. The ability to obtain the new skill sets would allow IBM to become a vertically integrated business because the firm's existing expertise in technology would combine with PwC's ability to consult on how to use the technology.

PricewaterhouseCoopers needed to separate its consulting and auditing practices. The passage of the Sarbanes–Oxley Act encouraged this strategic choice. In addition, the ability to combine the unit with IBM would ensure the success of the spin-off. The resulting merger would combine the 50,000 employees in the IBM consulting unit and 30,000 in the PwC unit.

A large number of problems faced the potential merger. IBM operates as a typical corporation. In contrast, PwC was a partnership. In dealing with customers, PwC typically was more involved with consulting and solutions for clients, whereas IBM focused on infrastructure issues. Each of the firms also had a clear and distinct culture. Finally, PwC clients might be concerned that IBM products would be stressed over other products, which may actually serve their needs better.

To integrate the two businesses, IBM established a three-stage process. First, the firm was to close the deal, establish the unit's basic operating model, and name the unit's president. IBM realized that people were their key asset and, if that asset walked out the door, a significant loss in value would occur. Therefore, IBM offered retention incentive packages to

CASE **7.1** *(continued)*

encourage key employees to stay. The second phase created greater detail on operating models and developed teams to integrate the different key areas in the business. The first two phases took place in 2002. The third phase took place in 2003. During this phase, the IT systems of the two firms were integrated, and the actual integration of the business units occurred.

As can be seen from the timing, the integration process was relatively quick. The new head of the unit was an IBM person, but then various PwC leaders also had significant leadership roles in the new unit. To obtain efficiencies, approximately 5,000 of the total 80,000 employees in the two businesses were laid off. However, most of these layoffs occurred in IBM. This helped ease the integration of the PwC workers. Finally, the IBM Global Systems unit operated its services as a freestanding business. In fact, 45 percent of the servers in IBM's data centers are non-IBM.

The general result of the integration of PwC and IBM has been positive. IBM Global Systems is now the largest IT services provider in the world. The unit provides approximately 52 percent of the corporation's revenues. However, recently, the growth rate of this unit has slowed. Therefore, IBM has been making more acquisitions—Filenet and Internet Security Systems to name two.

1. What do you see as the primary integration issues facing the new acquired units? How did IBM address the issues? What should they do next?
2. This Global Systems unit is the largest in the company. How does the expectation for the unit affect the management of the unit? What possible problems emerge because of the expectations? What possible advantages?

References

Morrissey, B. 2002. IBM–PwC deal gets OK. *InternetNews.com* (Sept. 12): Article 1462471.

Olavsrud, T. 2002. IBM swallows PwC Consulting. *InternetNews.com* (Jul. 30): Article 1436271.

www.ibm.com

CRITICAL THINKING

Relating to Your World

1. We have organized this chapter by activities that need to take place to successfully integrate or blend the entities involved in an external acquisition of technology. There are many aspects to this blending process. What would you say are the ten most important outcomes for a manager to see from the implementation phase? Discuss what the manager can do to ensure that integration moves forward.

2. We have discussed related versus unrelated acquisition. What parts of the integration process are most affected by the type of acquisition? Why?

3. What advice would you give a manager who was chosen to lead a technology-based joint venture project between two organizations of relatively the same size? Remember, in joint ventures, two or more firms combine equity and form a new third entity. Joint ventures commonly have very detailed agreements covering what each party is to provide and expect and how they are to operate within the joint venture. What do you believe are the key integration issues that must be addressed? What would be easier in the joint venture than in a merger? What would be more difficult? Why?

WWW EXERCISES

1. Use your favorite search engine on the Internet and look for postings of technology-driven mergers and acquisitions. What are the reasons given for the mergers and acquisitions? What is described as coming from each of the organizations? Are there sound reasons given for the merger and acquisition that are likely to lead to success? What might be missing?

2. Not all mergers and acquisitions are successful. Find an example of a merger or acquisition where technology contributed to its failure to produce desired outcomes. What reasons were given for the failure? What actions might have helped ensure success? Hint: Many mergers and acquisitions fail because needed information is not processed. Integrating information systems and accounting systems is often difficult.

3. Find an article or website that provides guidelines for integrating and blending people and/or systems in an alliance where technology is a motivating factor. What do you think of the advice given? Compare the advice you find to the advice your classmates find.

AUDIT EXERCISE

The implementation task can be large, but it is possible to make the changes needed. Keeping track of what has been done and what needs to be done to integrate the various systems and processes of two entities can be daunting. One method often suggested is a value analysis process. The purpose of most alliances is to add value, and all organizations want to add value to themselves and to the marketplace. Value analysis answers the following questions:

• What are the parts, processes, or pieces that can add value to the alliance or merger/acquisition?
• What does the alliance do or what is it supposed to do?
• How much will the alliance cost (money, time, people, and other resources)?
• Is there any other way to get the technology in a more effective or efficient way?

• How much will the alternative cost (money, time, people, and other resources)?

The value analysis itself is a four-phase process.

1. Information phase: Define the problem.
2. Speculation phase: Generate ideas that could work.
3. Evaluation and analysis phase: Analyze the ideas based on cost, feasibility, and goal attainment potential.
4. Implementation phase: Propose how to implement the integration or idea that is best for the organization and the conditions.

If your organization were to acquire new technology by buying a smaller entrepreneurial firm, what actions would you suggest in each of the four phases to ensure that your firm would get the desired synergies?

DISCUSSION QUESTIONS

1. Why would a firm use an alliance, joint venture, and merger/ acquisition?
2. How will the implementation effort differ in each?
3. Why do so many firms still pursue mergers and acquisitions if the integration of such combined firms is so hard and the results are so unsatisfying?
4. We have discussed a number of issues related to alliance/merger implementation. For each of the four areas (integration, leadership, execution, and alignment), how would implementing a domestic alliance differ from implementing an international alliance?
5. Find a merger/acquisition that has produced less than satisfactory results. What would you have done differently?

PART THREE OPENING CASE: ACER

1. In what area (integration, leadership, execution, and alignment) do you think Acer excels? Explain why you believe that.
2. Acer has a new market development team. In this chapter, we have suggested that a strategic implementation team is important. Why do you think Acer has emphasized the market team approach? What are the advantages and disadvantages of this type of approach?

KEY TERM

autoadjudication 222

NOTES

1. Beccalli, E., & P. Frantz, (2009). M&A Operations and Performance in Banking. *Journal of Financial Services Research*, 36(2–3), 203–226.
2. http://www.forbes.com/2009/06/16/mergers-acquisitions-advice-leadership-ceonetwork-recession.html
3. SOUTHPAW: Crunch time for Citigroup in Europe. (2008, November). Euroweek,*, Retrieved, November 28, 2009.
4. Massive coordinated effort integrates IT systems in Commonwealth Bank merger. 2000. EDS case study. http://www.eds.com/services/casestudies/cba.aspx
5. Fitchard, K. 2009. Creating CULTURE. *Telephony*, July/August, 23.
6. Cui, C., D. Ball, and J. Coyne. 2002. Working effectively in strategic alliances through managerial fit between partners: Some evidence from Sino-British joint ventures and the implications for R&D professionals. *R&D Management*, 32 (4): 343–357.
7. Klein, A. 2002. A merger taken AO-ill. *The Washington Post* (Oct. 21): E01.
8. Fried, I. 2002. HP–Compaq merger: Worth the wait. *cnetnews.com* (Sept. 2).
9. Mergers and acquisitions—The neglected HRIS challenge. 2002. *HR Focus*, 79 (9): 6–10.
10. Kelly, M., J. Schaan, and H. Jonacas. 2002. Managing alliance relationships: Key challenges in the early stages of collaboration. *R&D Management*, 32 (1): 11–22.
11. Ibid.
12. Richter, M. 1996. Technology acquisition and implementation: Learning the hard way. *Government Finance Review*, 12 (1).
13. Bert, A., T. MacDonald, and T. Herd. 2003. Two merger integration imperatives: Urgency and execution. *Strategy & Leadership*, 31 (3): 42–49.
14. McCreight and Company. 1995. CEO's merger success checklist. http://www.implementstrategy.com/publications/chapter_3.htm
15. Stopper, W. 1999. Mergers and acquisitions: Fulfilling the promise. *HR, Human Resources Planning*, 22 (3): 6–7.
16. Accenture. 2003. Case study: LG&E Energy Corp. *White Papers zdnet.com* (Jan.): Document 47487.

17. Bruhn, J. 2001. Learning from the politics of a merger: When being merged is not a choice. *The Health Care Manager,* 19 (3): 29–41.

18. Duysters, G., G. Kok, and M. Vaandrager. 1999. Crafting successful strategic technology partnerships. *R&D Management,* 29 (4): 343–351.

19. Lambe, C., and R. Spekman. 1997. Alliances, external technology acquisition, and discontinuous technological change. *Journal of Product Innovation Management,* 14: 102–116.

CHAPTER 8

Evaluation and Control in Obtaining Technology

OVERVIEW

This chapter details the evaluation and control processes used in conjunction with external means to obtain technology. In addition, the chapter discusses the interconnection between evaluation and control processes and the planning and implementation processes. The topics examined include the five areas in which evaluation and control should occur. These areas and the other topics examined in the chapter are:

- Examining alliance/acquisition capabilities

- Performing due diligence prior to obtaining the technology

- Negotiating the deal

- Integrating the new technology into the existing systems and structures

- Ongoing evaluation and control of the process of obtaining and blending external technology

- Developing metrics

INTRODUCTION

An organization first establishes where it hopes to go (planning), it then takes actions to get there (implementation), and finally it must evaluate if it is making progress toward its goals (evaluation and control). This text has stressed that most mergers/acquisitions and alliances fail to meet their desired results. However, it has also stressed that the effort to obtain technology through these external methods will only increase in the future as the importance of technology continues to grow in society. A critical part of not falling prey to the failure that permeates most efforts to obtain technology externally is an appropriate evaluation and control system. The evaluation and control system is not an isolated activity, but instead is an activity that should occur throughout the organization on an ongoing basis. Thus, you should view the obtaining of technology as a process in planning and implementation connected to evaluation and control.

Evaluation and control are the most pervasive of the functions in an organization because they are ongoing activities that are typically constant in technology-focused firms. To illustrate, large technology-focused firms such as Microsoft actually conduct their evaluation and control almost daily as they analyze their performance and the actions needed to ensure the meeting of strategic goals. A firm like Microsoft fears that its environmental changes are so rapid that without such constant evaluation and control the firm can quickly find itself off course and missing its strategic goals or missing new, emerging opportunities. Figure 8.1 summarizes how evaluation and control fit with the other strategy processes of planning and implementation.

Evaluation and control are pervasive in organizations today, and their importance will increase in the future. This is because of the growing

Strategic Elements	Analyze	Act
Planning	What are the key issues for selecting a partner? What are our competitors doing?	Set alliance goals Set measurable objectives for alliance type and partner-seeking activity
Implementation	What do we do now? What can we do later? What should be delegated to whom?	Set up structure and process for blending Build fit
Evaluation and Control	Are we reaching the goals of the partnership? How do we make adjustments?	Develop means of analysis and information sharing Make adjustments

FIGURE **8.1** Questions to Address for Evaluation and Control significance of technology for information processing, new product development, and systems management and the increasing use of external means to obtain capabilities in these areas. Already, an estimated 70 percent of mergers and acquisitions are driven by the desire to acquire some specific technology or technological capability.[1] The need to obtain technology has led two-thirds of the companies surveyed by *The Economist* to expect their dependence on external relationships (alliances, joint ventures, consortia, and strategic partnerships) to significantly increase in the future.[2]

Thus, we have a situation where external methods to obtain technology will continue to increase. Nevertheless as noted in Chapter 6, these external methods including both alliances and acquisitions often do not produce the desired results. In fact, more than 60 percent of alliances/acquisitions fail to meet expected goals. As a result, the need for evaluation and control appears high and can be expected to go higher, as efforts to obtain technology externally continue, but face a high potential of failure. Perhaps, improving the evaluation and control processes will ultimately improve the success of external acquisition of technology.

WHERE EVALUATION AND CONTROL OCCUR

A key part of the better methods needed in the external acquisition of technology is to recognize that technology and innovation are not isolated activities in a firm but instead impact and are impacted by multiple functions within the organization. As a result, when firms seek technology externally they need to recognize that the evaluation and control process is complex and requires the coordination and integration of multiple functions within the organization. If these multiple functions are not coordinated and integrated, then barriers between operational areas will continue or grow, the use of politics rather than logical decision-making will expand within the firm, and key human resources may be lost.

The evaluation and control effort occurs at five different places in the external focused processes to obtain technology by an organization. The evaluation and control efforts in each of these five places should include the analysis of multiple dimensions through multiple methods. The five places that evaluation and control happen in an externally focused process are:

1. Examining alliance/acquisition capabilities of the firm
2. Performing due diligence prior to obtaining the technology
3. Negotiating the deal
4. Integrating the new technology into the existing systems and structures
5. Ongoing evaluation and control of the processes to obtain and blend external technology

Clearly, evaluation and control should be ongoing. The ongoing evaluation and control efforts of the firm should focus on the examination of the following questions, that we have seen previously (Chapter 5):

a. Where are we compared with where we wanted to be?
b. What lies ahead that can affect us either positively or negatively?

c. Where are we going in the future if we continue on the current path? Is it where we thought we were going when we decided to obtain the technology externally?

Each of the five places that evaluation and control occur and the three questions will be examined in detail in this chapter to ensure that the multi-dimensional aspects of evaluation and control are understood.

Alliance/Acquisition Capabilities

Before undertaking efforts to build an alliance or to begin an acquisition, a firm needs to evaluate its capabilities for success in such activities. The manager should do this evaluation in light of whether the firm will employ alliances or acquisitions in its effort to obtain the technology. Most firms will focus principally on either alliances or acquisitions to achieve expansion goals.[3] The capabilities needed in each method are slightly different. Therefore, before beginning the due diligence process, the firm must determine whether it will pursue an alliance or an acquisition and then see if it has the capabilities necessary to be successful. Figure 8.2 delineates the characteristics of the environment and the firm's goals for determining which of the two alternatives is more appropriate.

If an alliance is the more appropriate method for obtaining technology, then the type of alliance must also be determined. Chapter 6 discussed the different types of alliances and their relevant characteristics.

Issues	Conditions Encouraging	
	Alliances	Acquisitions
Resources	Less proprietary resources are available among the potential partners	More proprietary resources are available and needed
Synergies Sought	Are more sequential or clearly separated	Are networked or reciprocal
Competitive Environment	Less intense and cooperation can flourish	Very intense; Ownership is needed to guarantee control
Market Conditions	Market uncertainty is either very low or very risky	Market uncertainty is offset by size
Competition for Critical Resources	Resources are available	Resources are scarce
Firm's Past Performance	Success with alliances leads to tendency to build alliances; Failure leads to aversion	Success with acquisitions leads to more; Failure makes firm shy even if appropriate

FIGURE **8.2** Differences to Look for to Obtain Technology Externally

The process of evaluating the type of collaboration method most appropriate for meeting the goals determined during the planning phase should lay the groundwork for the due diligence effort. For example, DuPont has developed a Business Initiative Process that it uses when it is thinking of obtaining technology from external sources. In this process, the firm fully explores a domain before trying to expand into it. There are five separate but intersecting steps with clear yes–no decisions in each of these steps. These steps are:[4]

1. The business case is explored. This is an examination of the strategic ways of pursuing the technology/innovation under review. If there is not a strong business case the project is abandoned.
2. Evaluating the environment and all relevant parties and planning the potential change takes place. Due diligence is an integral part of this process. If there are not matches for the firm's strategic goals and if assurances of success cannot be established, then the project is terminated or redirected.
3. There is detailed development of the introduction plan and preliminary negotiations.
4. If the project makes it to this step, there is a scale-up of activities and definitive agreements are negotiated. This can be a time-consuming process, but it is vital for successful implementation.
5. Implementation and commercialization are the activities in the last step. The new business is brought into DuPont and implementation begins.

Due Diligence

Chapters 6 and 7 discussed that the technology-focused firm needs to conduct a thorough evaluation of many elements in the initial stages of the effort to obtain and integrate technology from external sources. This investigation is referred to as due diligence. Because evaluation and control are not isolated events but happen throughout the organization on a continual basis, evaluation and control are part of the very early stages of the effort to obtain technology externally.

The due diligence process should gather data in an orderly manner about potential candidates for an alliance or acquisition. The information gathered will form the basis for implementation if the alliance or acquisition is pursued. Ensuring that the right information is with the right people is part of the evaluation and control process in due diligence.

To ensure that the due diligence of a potential candidate for an alliance or acquisition is thorough, a checklist is often used. For example, a technology-focused firm such as Cisco, which conducts a large number of acquisitions, typically relies on such checklists. The value of the checklists is that they not only form the means to evaluate potential partners but also form the basis of implementation plans and progress checkpoints during integration. Because of the importance of effective due diligence, we need to examine it more closely. Therefore, this section of the chapter will look at the characteristics that organizations should examine in their due diligence of a potential alliance partner or takeover candidate. The chapter will then examine the characteristics that the organization should use in developing its own checklists. Finally, we discuss how an organization uses such checklists in its evaluation and control effort.

Evaluating Future Partners or Takeover Candidates

Firms should base their evaluation of the potential alliance or acquisition candidates on a detailed assessment of a wide range of issues, including the alliance/acquisition prospect's history, financial position, human resources, technology, systems and structures, processes, markets, and competitive positioning. The firm should consider more than financial and technological performance of specific products in this process. To create value from obtaining technology externally an alliance or merger/acquisition requires a meshing of the systems of the different firms. Thus, due diligence needs to focus on a wide range of items rather than just the financial aspects of the firm.

There are five nonfinancial concerns that need special attention in the due diligence evaluation of the potential partner or merger/acquisition target. They are:[5]

- The value creation potential: Examines the value of technologies in the potential partners. This value is relative to positioning, profitability, and growth activities as well as the creation of shareholder value.
- Assessment of the portfolio of technologies: Looks at a multitude of dimensions relative to the new entity's ability to continue an effective technology program (including innovation through R&D). The issues to examine include timing, level of risk, core competency development, and exploitation as well as the portfolio of technologies across the organization. Thus, the evaluation should look not at a single product but rather at the range of products that the future partner or takeover candidate possesses.
- Business integration: Indicates the commitment of the firms to use teamwork for technology exploitation within the new entity (no matter who developed it), and the processes and programs for developing new technology.
- Value of the technology assets: Examines the strength and viability of each firm's technology and the potential partnership's technology. It includes proprietary assets, knowledge systems, and experience of employees, and it indicates the ability of the organization to create future value.
- Support of innovation practices: Demonstrates the ability of the potential partnership/alliance to produce newness. It includes management practices such as project management, idea generation, knowledge sharing, cross-functional linkages, and other practices that enhance the ability of the firm to reach new ways of doing things and new product development.

A firm known for conducting such forward-looking due diligence is Alcoa. One outcome of Alcoa's approach to due diligence is that the firm has consistently outperformed its competitors. The strength of the firm has in turn allowed Alcoa to be recognized as one of the leading environmentally sustainable corporations in the world by the World Economic Forum in Davos, Switzerland because of its environmental performance. Central to the efforts of Alcoa to obtain technology externally is that the firm includes strong elements of nonfinancial concerns in its due diligence efforts.[6]

To be successful, due diligence must also be well planned and executed. To help do that, several rules have emerged to conduct due diligence successfully on potential alliance or merger/acquisition targets:[7]

1. Objectivity should be maintained. Too often, especially where new technology is involved, the acquirers "fall in love" with the deal. As a result, the evaluation is flawed since the evaluator has on "rose-colored" glasses. Many failed mergers occurred because emotion and ego overruled sound judgment.
2. Suspicion about the analyses provided by others as well as your own is healthy. Economic forecasts and environmental risk analyses are based on numbers supplied by the prospect and on judgment calls. A healthy cynicism accompanied by a critical eye will help alleviate many potential problems in the future.
3. Both the upside and the downside of an alliance or a merger/acquisition should be reviewed. If the firm desires process efficiencies from the blending of the two systems, then due diligence requires that you know the systems are compatible or linkable. Too often, firms lose all potential benefits and even increase costs because the technologies for systems such as IT are not compatible. Thus, a manager needs to develop a best case and worst case set of scenarios.
4. Keep the process quiet as long as possible, but do not rush because you are worried about potential leaks. Starting a process of alliance/acquisition has implications for both parties involved. These implications include a wide range of issues such as the effect on each firm's stock price. However, rushing through the process of due diligence to protect against leaks does not enhance the process. The manager needs to balance between protecting against leaks and rushing the process. This balance is different for each deal, but it is one that managers must actively seek to maintain.

Due Diligence Checklist To obtain the information discussed above in a systematic manner, the manager should use a checklist. Such a checklist helps systematically consider the widest possible set of issues in the firm's due diligence process. There are five key characteristics for such checklists.

1. Clarity of objectives
2. Comparison
3. Competitive understanding
4. Customization
5. Continuity

These characteristics of due diligence checklists are discussed next.

Clarity of Objectives The objectives for developing a due diligence checklist are not merely listed so that a box can be checked if the item is acceptable. Rather, the manager should use them to discover important, often hidden, potential expenses, costs, and liabilities (as well as benefits) that might affect the alliance/acquisition if it is completed. The checklist cannot cover every aspect, but it should be thorough and aimed at the reasons for considering the alliance/acquisition in the first place. In fact, the answer to the question of "why" should be the first consideration. If the individuals working on forming an alliance cannot articulate a clear answer with identifiable objectives, the alliance/merger will probably fail.

One area that has received attention because it is not considered by many organizations in their due diligence checklist is information technology. A sample due diligence checklist for information technology is in Figure 8.3. Notice the checklist deals with the physical assets of IT, the systems for protection and continuation, maintenance policies, management procedures, people issues, other contractual obligations, and potential problem areas.

The manager should develop a similar checklist for each of the key areas of the takeover prospect. By having a clear checklist, the management of the takeover candidate understands something about the culture of the acquiring firm, and the manager can communicate the objectives for the takeover. The danger with poor checklists is a merger like the one between Quaker Oats and Snapple. In this merger the food products company acquired the drink

Area of Concern	Issues
Hardware, software, IT service vendors	Viability, flexibility, financial position, market share, qualifications of major suppliers
Disaster recovery and security	Procedures, security surrounding contract management, procurement, accounts systems
Capacity	Time between failures, capacity/scalability, percentage of utilization
Maintenance	Service contracts for critical systems and components
Change management processes	Look for tight controls of system changes during the conversion period
Outsourcing and updating processes	Look for systems that are out-of-date, contracts that specify huge termination penalties
Personnel issues	Look at attrition rates, number of consulting contracts; Look for poor performance and/or chronic trouble spots

FIGURE **8.3** IT Areas to Examine

company in 1994, but then divested it by 1997. A principal reason for the failed effort to obtain technology externally was that Quaker Oats never considered the culture clash that emerged from trying to combine a mass marketer like Quaker with a quirky, distributor-oriented firm like Snapple. Similarly, Houston-based Lyondell Chemical and its United States affiliates filed for bankruptcy in January 2009, a little more than a year after the December 2007 merger with Basell to form LyondellBasell Industries. The merger, which came at the height of the buyout boom, was funded entirely with debt financing, and saddled Lyondell Chemical and its affiliates with a huge debt. Lyondell obviously expected the continued growth patterns that had led to the boom in acquisitions in the chemical industry during the early part of the decade.[8] However, as the market returned to a more conservative position, the firm's effort to obtain technology and economies of scale proved to be its undoing.

Comparison The leading partner in an alliance or acquisition cannot assume that its own systems or methods are the best. In developing the checklist, the leading firm should **benchmark**, or compare, its systems and the systems of the partnering firm with the best in the industry. The integration of two firms or technologies is very difficult. One danger is that firms do not deal with tough issues such as choosing which systems to use. In fact, rather than simply relying on one system or the other, the best choice may be to change both systems.

The benchmarking effort allows the acquiring firm to ensure that in making changes they eventually employ the best system possible. Such benchmarking can be useful in making personnel decisions and taking advantage of superior technology throughout the combined organization if the alliance/acquisition is completed. In addition, such comparison leads to the how of integration planning. In fact, if this part of due diligence is done correctly, the plan for integration should emerge with the best processes and systems being integrated into the new entity. Thus, the benchmarking effort provides a set of key measures in the evaluation and control process to ensure that the accomplishment of the firm's goals.

Competitive Understanding The due diligence checklist also needs to include an understanding of how the partnering prospect or acquisition fits into the competitive environment. If the firm desires the partner or acquisition to enhance the product line with new products and/or new distribution technology then there should be an understanding of the competitive environment of each firm. Often, organizations may view a potential partner or acquisition target as attractive, but the ultimate impact on each firm's strategic posture may not be as great as desired from the activity. The technology may be outdated or may not enhance the competitive posture of the acquiring firm. In fact, research shows that about two-thirds of mergers and acquisitions do not deliver the expected synergies. The odds are that about half of the firms will divest the acquired assets or dissolve the partnership in five to seven years.[9] Thus, there is a critical need to understand the potential partner or acquisition target thoroughly, including what the true strategic impact of joining the two firms will be.

Legal and environmental issues are critical in this area. Legal issues such as patent protection and concentration of economic power are important but often overlooked. The legal issues of concentration of economic power are important not only in the United States but also in other parts of the global marketplace. For example, for the HP–Compaq merger in 2002, it was not concerns in the United States but those in Europe that delayed the merger.

Environmental issues are also a critical part of the competitive understanding of a firm. The law in the United States is that if a factory is closed, the factory site needs to return to a natural state. This is because many manufacturing processes involve highly toxic materials. For example, the silicon chip manufacturing process involves harmful chemicals. If a firm acquires a silicon chip manufacturer and the acquired firm has not handled these materials correctly, then the acquiring firm takes on the liability of returning that factory site to an environmentally desired state. The costs of doing that may be far greater than any benefits obtained by having a constant silicon source or any emerging technology the silicon chipmaker may have been developing.

Customization   Some active firms in the alliance building or acquisition arena do not have formal checklists but rather have an informal due diligence checklist that emerges in a patterned way. Whether the checklist is formal or informal, it is important to remember that each deal is different. This means the due diligence checklist cannot be rigid and inflexible but rather must be customized to the potential deal. Identifying potential deal breakers or major hurdles early will help focus the process on the most effective approach to due diligence. If the acquiring firm is looking for certain outcomes, then the manager should examine the potential target first to see if the target has the organizational characteristics to provide those outcomes. If a firm wants to enhance its manufacturing systems and the potential target does not have a competitive advantage in that area, then other targets may be better suited.

Customization of the due diligence process is aimed at answering three basic questions.

1.   Are the goals for this potential alliance or acquisition clear and strategically significant?
2.   What assets and processes does the target firm have that will help reach those goals? What might hinder reaching those goals?
3.   What assets and processes does the acquiring firm have that will help reach those goals? What might hinder reaching those goals?

Continuity   If the due diligence checklist is well developed and applied, then it can become the foundation of the post-acquisition/post-merger operating plan. The work done during due diligence should tell the management of the blended organization where the strengths and weaknesses are as well as where the potential synergies for improvement exist. If the acquired organization personnel understand the nature of the analysis performed for both companies and if there is an effort to truly incorporate the best practices, they should feel less like they are being taken over and more like they are joining a bigger, better organization. This can help the implementation of the acquisition a great deal.

Figure 8.4 shows the principal reviews needed for obtaining technology from external sources. The six phases of reviews can all be predicated on a well-done due diligence process. If due diligence is done correctly and the questions asked are pertinent and thoughtfully answered, the integration process will be greatly enhanced.

Employing the Checklist in Evaluation and Control The preceding checklist for the potential alliance or acquisition of a technology-related firm generates information that is useful in evaluating those activities before they occur. However, the information generated can also be used by the firm to determine whether the alliance or acquisition is performing as intended and, if not, why not. This is why the due diligence process forms the foundation for the firm's evaluation and control process. The information generated while considering

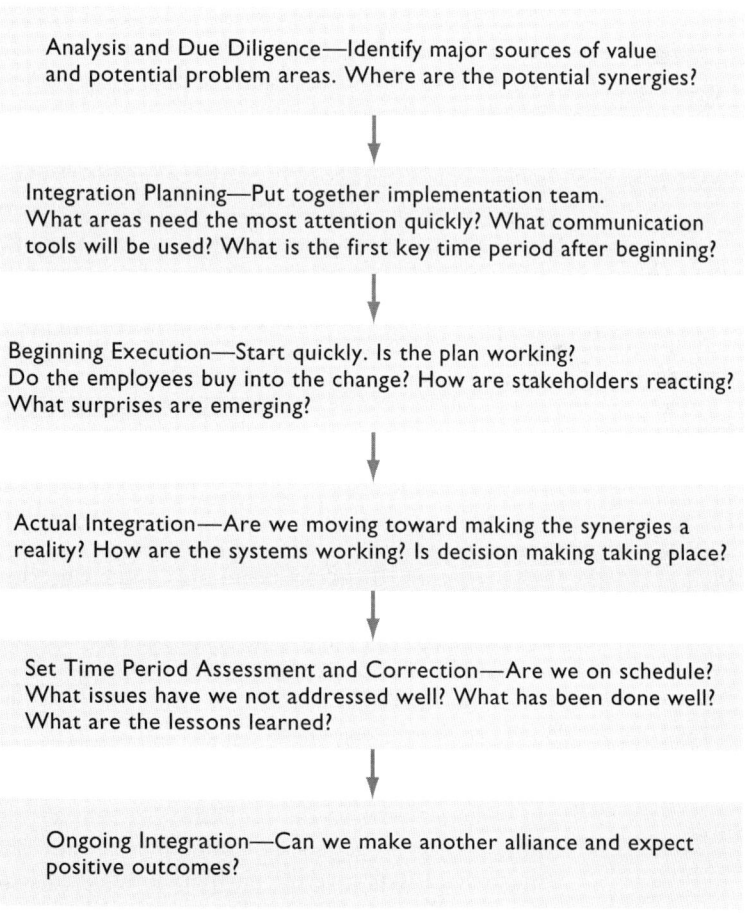

FIGURE **8.4** Needed Process Reviews

the deal is also useful if an alliance or acquisition does not produce the desired results as it can help generate learning on how to avoid such situations in the future. A follow-up to examine the due diligence process should ask questions such as:

- In the analysis of the alliance or acquisition, were there key questions and data that were not gathered?
- If the information was generated but somehow overlooked, why was it not given the importance that it should have been given?

The data generated will help in the actual alliance or acquisition activity. The data also provide a critical paper trail that the organization can evaluate as it goes forward to ensure those things that worked well are repeated and those that did not are avoided in the future.

Negotiation of the Deal

If after due diligence the decision is to pursue a deal, the evaluation and control continue through the negotiation process. During negotiations, the specific goals of the parties should lead toward a mutual understanding. It is important in multiparty negotiations to understand where the various parties stand as they approach the negotiations. Tera Allas and Nikos Georgiades, writing in the *McKinsey Quarterly*, argue that three key dimensions need to be understood in such multiparty negotiations.[10] These are: 1) their position or preferred outcome, 2) their salience or how much importance they place on any given issue, 3) and each party's clout or ability to influence a given decision about some issue. By examining each of the key items in the negotiations along these three dimensions, it is possible to determine which items can be compromised, which cannot, and the best approach to negotiation.

Once the potential partners have defined what is essential, what is negotiable, and what is nonnegotiable, then the process of analyzing where to build the bridges among the firms begins. This evaluation process will determine the organizations' blending and integration efforts and can help improve the potential for success. There are certain questions that need answering during the negotiation phase. These include:

- Where is the value creation for each organization and for the combined organization or alliance activity?
- What are the short-term and long-term objectives for each partner, and will the joining of forces help each reach those objectives?
- Who knows what in each organization?
- How will the joint venture, alliance, or merger be governed?
- How will the alliance or joint venture be terminated if either party becomes dissatisfied? This is not an issue with acquisitions; divestment is the choice of the acquiring firm.

The answers to these questions can then become the basis for evaluating the alliance or acquisition outcomes if they become a reality. Furthermore, the answers will give the partners a target for controlling the actions and continuation of the alliance. Figure 8.5 gives examples of some information to consider as negotiations begin and continue.

Strategic Factors	Knowledge Needed for Alliance	Examples
Type of Business	Operational requirements; Industry focus	Alliances in banking involve services; In airlines, gate sharing is key; etc.; Oil exploration sharing risk and expertise
Proprietary Knowledge	What needs to be protected; Shared; Level of trust	Manufacturing alliances overseas may involve only older technology
Partner's Experience	In alliances; In areas that firm needs help in	More experience leads to better anticipation of risks, better fit, better relationship building
Criticality of the Alliance to Each Partner	The financial and operational activities of each partner	Some alliances are for market accessibility, not operations; Some are for exposure, not necessarily growth
Individuals Involved	What level is the alliance for each firm?	Skill requirements and trustworthiness affect operations and management of the alliance

FIGURE **8.5** Evaluation Factors during Deal Negotiations

Integration

Once due diligence is completed and if the deal is consummated, then the technology-focused firm will either form an alliance or make the acquisition. The integration process (the fourth area of evaluation and control) is critical to the ultimate success of the external effort to obtain the technology. As mentioned in Chapter 7, this is the place where most alliances/acquisitions fail. There are a number of issues for evaluating the integration process. These include:[11]

- Clear, common objectives and definition of success
- Appropriate governance model with clear decision-making criteria determined
- Clear plan for integration and evolution of the plan if needed
- Clear metrics to track and measure success and areas that need attention

Each of these integration issues typically involves written documentation. Organizations should not only judge choices they make but also create measures for evaluating outcomes in these domains over time and take actions to change things if necessary. The evaluation for integration should focus on several items. These items are domains that pose the greatest risk to the success of the integration effort. Studies indicate that these factors include:[12]

- Financial systems
- Core business applications

<table>
<tr><td>

8.1

REAL

WORLD

LENS

</td><td>

Satyam Group

In 1987, The Satyam Group was a motley group of companies. There was Sree Satyam Spinning & Weaving Mills Satyam Constructions (renamed Maytas in 1998); Satyam Impex (exporter of shoe uppers); Satyam Homes (building residential apartments); and Oceanic Farms (aquaculture). The Satyam Group was controlled by the Ranu family (three brothers). However, Ramalinga Raju chose to leave the family firm and start a different business. Charmed more by cyberspace than spindles, he once told a reporter that he had set up Satyam Computer Services as a hobby. This hobby saw a 122-fold rise in net profit in a business that grew over 20-fold since incorporation within 10 years of its founding in 1987. By 14 years after its founding the Satyam Computer Services was listed on the New York Stock Exchange and subsequently joined the billion-dollar club by 2006.

For most of these years in Satyam, Raju was a thorough planner who thought long term and looked at the big picture. He was willing to take risks but most of them were calculated risks. However, in the 2008, Raju changed his actions and his business directions. First, he became more and more enamored with the "star" lifestyle and was seen with politically powerful individuals and leading business people. Second, he became obsessed with land acquisition and neglected his vision for IT services. His business decisions became less transparent and less analytical. He then destroyed shareholder value through two unrelated acquisitions. Ultimately Raju was arrested for fraud by Indian authorities and his business empire came to an end.

1. List at least five lessons Satyam Group shareholders and board members should have learned from this experience.
2. If you were a consultant to Satyam Group, what would you suggest they do in the future to rejuvenate the firm?

References

Sharma, E. 2009. The tiger rider; B. Ramalinga Raju was a meticulous strategist who was fuelled by a burning desire to make it big. He didn't know where and when to stop. *Business Today*. February 8.

</td></tr>
</table>

- Networked operating environments
- Systems compatibility

These items require an in-depth analysis of the processes in both firms. Benchmarking helps the firm to identify best practices in these various domains. To understand how the systems operate, the firms need to recognize the underlying technology and factors to align. Recall from Chapter 7 that the key alignment issues include the need to develop reward systems to match goals, establish common policies, and build fit. During the integration process, the blending for an alliance requires the development of (1) reward systems to support the goals of the alliance, (2) policies that each organization can support, and (3) fit in the structures and culture. Previously we noted that DuPont employs a *Business Initiative Process Guideline Manual* when it acquires technology externally. In this process for each new business development project, there is an assigned team whose job is to represent DuPont's interest throughout the five steps to ensure such an in-depth analysis. These steps include:[13]

1. A seminar to educate top managers and the project team on the issues and demands of alliance development
2. Partner evaluation and selection checklists and assessment worksheets
3. Guidelines for organizing and managing the agreement-negotiation process
4. Due diligence checklists with details covering all aspects of the effort
5. Detailed guidance on how to structure and integrate the new alliance into ongoing operations

Thus, a considerable part of DuPont's success in using external methods for obtaining technology is its thorough and systematic methods. These methods help DuPont understand how to fit and support a new partner, whether through alliance or acquisition. These systematic methods include both an established process and manuals that help managers generate success in their external acquisition efforts.

Ongoing Evaluation and Control

Until now, this chapter has focused on evaluation and control of the efforts that take place before the alliance/acquisition. Once the firms have begun the integration activities, evaluation and control become ongoing processes, which is the fifth item on the list of where and when evaluation and control take place discussed at the beginning of the chapter. The information from the due diligence process and the negotiations about what, who, why, when, and where provide considerable information on which to base the implementation effort. After due diligence and integration have been accomplished, the firms should evaluate whether they have been done well or whether there is a need to make changes in what the firm is doing. As stated earlier, during the efforts to obtain technology externally, there is also a need to conduct evaluation and control efforts as part of an ongoing strategic process of the organization.

The questions examined in Chapter 5 concerning the ongoing evaluation and control of innovation internal to the organization are also relevant for the ongoing evaluation and control process in the acquisition of technology as well. These questions are:

a. Where are we compared with where we wanted to be?
b. What lies ahead that can affect us either positively or negatively?
c. Where are we going in the future if we continue on the current path? Is it where we thought we were going when we decided to obtain the technology externally?

These questions will be examined next.

Evaluation of Current Status

The first question to determine if the acquisition strategy is working is: Where are we compared to where we wanted to be? This is an evaluation of how we are doing right now. In planning for the acquisition of new technologies, there should be specifically defined goals. These goals should be both short term and long term. The firm can judge its current status against the short-term goals and objectives and the progress toward future goals and objectives. It is important that the evaluation timing match the timing of the goals.

Understanding the timing and goals for an alliance can be particularly difficult because of the differences between the firms involved. Obtaining technology through external processes involves arrangements between firms that may have their own reporting processes and systems as well as their own motives and goals. Agreeing on how to measure performance can be difficult in alliances. One complication that often occurs in the evaluation process is tracking the costs and benefits. For example, to be consistent in cost/benefit analysis both firms need to agree on how to implement transfer pricing. In addition, depending on the type of alliance the controlling firm may not have in place the evaluation system needed to get the information required for the evaluation and control processes. This may result in the alliance not receiving the necessary management scrutiny from all the parties.[14]

For a merger or acquisition, the timing and goals can include the desire to merge with a customer or supplier. If the firm merges with a customer, the purpose of the merger is to extend value to upstream operations. The analysis for evaluation then becomes cost/benefit oriented. Does the benefit for the firm outweigh the costs of blending operations with that customer? If the firm merges with a supplier, the firm should be evaluating the quality of inputs and the improvements that the increased control of inputs provides the organization. For example, in 2009 Boeing wanted to acquire Vought, the supplier of Boeing's 787 fuselages. In large measure, Boeing pursued this acquisition in an effort to tighten quality control standards over a key input to its planes and to save money. By owning the fuselage manufacturing process, Boeing hopes to have stricter oversight and control.[15] Boeing believes the resulting benefits of the merger will outweigh the costs.

One goal that is part of the evaluation of the firm's current status that is similar for either an alliance or a merger/acquisition is the desire to improve a firm's processes through the combination. The evaluation and control then focuses on whether a firm's efficiency and/or quality is improved. Thus, there is a need to examine the immediate situation in the resulting alliance or merger and see whether the outcome is consistent with the goals established and the time desired.

Evaluation of the Future

The second question that must be addressed in the ongoing use of evaluation is: What lies ahead that can affect the firm either positively or negatively? This involves scanning the environment for opportunities and threats. All of the environmental issues (see Chapter 2) are relevant when addressing this question. However, more than that, the blended firm must use its new technologies and competitive advantages to look for and find future opportunities. Because of the costs of mergers and acquisitions, the firm may also need to be particularly aware of potential threats that were not present earlier. When British Petroleum (BP) acquired AMOCO, the world's premier drilling research facility closed. This drilling research facility was known throughout the industry for developing creative solutions to drilling problems. The facility had developed many of the cutting-edge drilling bits for the industry. BP decided to move the personnel to its research facilities in other states and nations. Many of the research engineers decided they did not want to move. The culture and facilities were lost. As a result, BP did not acquire the full research prowess it had hoped for when it purchased AMOCO.

Another potential threat to a firm derives from the fact that once a major player in an industry makes an acquisition, many other firms will look for ways to make a similar acquisition to "level the playing field." This can cause a great deal of uncertainty in an industry as products, processes, research facilities, and other technology-based assets change ownership. For example, today the cable television industry and the telephone industry are beginning to merge into a single industry. Firms such as AT&T, Cox Communications, and Direct TV offer a package of services that include some or all of the following—internet connection, cable television, cell telephone services, landline telephones. As little as 5 years ago, this was not the case. However, legal changes in regulation by the United States government have resulted in a series of acquisitions that have led to a largely integrated industry today. In the future, the level of integration for all telecommunications services should become stronger.

Evaluating the future involves monitoring how well the blending process is going for the organization after the initial integration effort. If employees are still identifying with their original companies two years after the acquisition/merger, then the building of future synergies may be hampered. The evaluation of the future should also consider unexpected internal opportunities. It may be that an unexpected blending has led to positive outcomes. If this works in one part of the combined firm, it may work in other parts. This is clearly a potential benefit for the future.

Evaluation of Where You Are Heading

The last question in the ongoing evaluation and control process is a fundamental evaluation of whether a radically different direction for the newly formed firm is needed. The firm should ask: Where are we likely to end up if we continue on this path, and is it where we thought when we developed the plan? Such radical evaluation does not occur as often as the comparisons between the goals and outcomes of the acquisition. If the firm is not heading in the direction that the managers believe will lead to long-term success, this more radical type of evaluation may lead a firm to look for acquisition opportunities. The evaluation process needs to periodically examine potential opportunities or paths that are different from those being sought by the organization. Breakthroughs in thinking and action can lead to new initiatives for the firm.

This questioning of direction is particularly relevant for industry leaders or those who have had long-term success. The industry leaders are targets for imitation and for strategic attack—frontal or guerrilla. The firms that have had long-term success may get complacent. This is commonly known as the **inertia of success** since firms assume because of past successes future success will occur.

The key areas to address when looking for future direction are:

1. Creation of value
2. Integration of systems, processes, and technologies
3. Opportunities and threats

Figure 8.6 illustrates the importance of each of these and the key questions for the organization's managers and other stakeholders as they look to the future.

Creation of value when evaluating the acquisition of technology depends on the emergence of processes and/or products that improve the competitive positioning of the organization. This can include the emergence of improved processes, the development of new products and other innovations, and the alignment of best practices. For the acquisition of technology to reflect positive outcomes, it is important that some type of technology improvement has emerged. The expertise of the two organizations should merge into a system that has improved returns and provides synergies that support the company's goals and desired positioning in the industry.

One other question should be examined if the firm is not on the planned path—if we continue on this path, even though unintended, what do we think will happen? The unintended activities in turn can be formalized into an emergent strategy that seeks to continue the activity. Emergent strategies can lead the firm to outcomes the management did not envision initially. The integration of systems, processes, and technologies requires consideration of the how, why, where, and when of process. When a firm creates any type of alliance, the integration will determine the eventual success. This process is difficult, at best, because of the newness factors and the identifications that exist.

Key Areas to Consider	Definition and Importance	Key Questions
Creation of Value	Emergence of processes, new ideas, and to put the best into practice	Have we created synergies that improve our ability to compete? What is the return?
Integration of Technologies	Focus is on process, culture, teamwork, and structure	Are we gaining? Are we stronger? Are there warnings of potential decline?
Opportunities and Threats	Competitors change or emergence of new competitive opportunities	How strong are others? Is a paradigm shift coming? What are others doing?

FIGURE **8.6** Key Areas and Their Measurement

For example, the culture of one firm may be very family oriented. However, if one of the firms in the alliance or an acquiring firm is not as family oriented, then integration can be more difficult. These issues may or may not be directly related to work outcomes, but they still affect the work environment and need to be taken into consideration. The team that evaluates this area should include members from both organizations and from multiple levels and areas of the organizations. If the firm does not integrate after an alliance formation, then inefficiencies appear, and other potential opportunities may be missed.

METRICS

A key aspect of evaluation and control that is part of the implementation process of evaluation and control is the ability to measure many of the issues raised in the preceding discussion. The generation of information during due diligence analysis, negotiation, and implementation impacts and facilitates some of this evaluation. But there still remains a need to develop **metrics**, or measures, that the organization can use in its evaluation and control.

Generally speaking, the development of metrics for use in the external acquisition of technology is more difficult than evaluation of internal innovation. This is because the acquiring firm does not control all aspects of the process and as a result may not have all of the information needed to do "hard" evaluation. The most common metrics and the ones that managers tend to feel most comfortable with are "hard" metrics or numbers on some aspect of the technology that can be compared to some standard. Financial information from the income statement and balance sheet are such measures. However, as indicated in Chapter 2, even these types of numbers are subject to interpretation, and it is important to take notice of what accounting practices the different firms may have. For example, different methods for valuing inventory may result in very different outcomes. There are areas beyond financial and similar domains that managers should also examine and develop metrics. Callahan and MacKenzie argue that domains that need to be measured include:[16]

- Partner motives—Clarity of partner's motives for the alliance
- Partner capabilities—Partner's skills to deliver desired result
- Partner resources—Partner has deep managerial resources
- Development processes—Development processes of two firms fit well together
- Organizational cultures—Key players in each organization accept the alliance

These types of metrics can be very detailed; just as in the due diligence process. When the firm is evaluating the partnership, whether it is an alliance or an acquisition/merger, it is important to be thorough. However, the firm needs a balance since evaluation can be costly to the firm. The balance is to gather the information needed to help the organization realize its goals in a timely fashion while being cost effective.

GAP ANALYSIS

A key evaluation and control tool that the organization can use in its implementation of evaluation and control is a **gap analysis.** Chapter 5 noted that the difference between goals and outcomes is a gap. Gap analysis seeks to identify the gaps before major problems arise. If a gap is recognized, then managers can take corrective action. These actions take two forms: change the desired outcomes or change activities that produce the outcomes. In strategic alliances, the gap analysis identifies the fitness of the alliance. The focus in this part of the chapter is on alliances and not mergers/acquisitions because the alliance can be changed or abandoned more easily than the merger/acquisition, which involves a permanent movement of assets.

In the gap analysis, there are four critical types of fitness examined: financial, strategic, operational, and relationship.[17] With the analysis of these four types of fitness the firm should be able to identify gaps in performance.

Financial Fitness

Financial fitness refers to the difference between the desired financial outcomes and those actually produced. In a merger, acquisition, or alliance, financial measures show the outcome of strategic actions. These measures are often readily available. The standard types of financial measures used in this area are sales revenues, cash flow, ROI, ROA, and net present value. (Refer to Chapter 2 for financial items and their calculation.)

Technology-focused firms should also consider other issues as part of their financial fitness gap analysis. Particularly, there needs to be an analysis of the organization's progress toward increasing efficiency. Examples include the measurement of the reduction of overlapping costs, transfer-pricing revenues, other increases in revenue attributed to the alliance, and cash outlays against expected returns. When entering an alliance, a firm should include specific financial objectives and then measure the outcomes against those objectives.

Strategic Fitness

If the purpose of the alliance is the acquisition of technology then there should be an evaluation of **strategic fitness,** or the ability of the organizations to align their strategic goals. Therefore, although strategic fitness is more fluid in its measurement, it is still important that the organization consider it and make adjustments as required. Just as with financial fitness, strategic fitness measures can indicate success and failure but are unlikely to help in specific diagnosis of potential solutions.

When acquiring technology or forming some type of alliance for technology development, the goals should reflect issues such as development of new technology, increased knowledge base, improved processes, and better service to customers through bundling of related products. The goals of the external technology obtaining activity should define the types of measurements used. For example, if the goal is to develop new products then the measure could be the number of new products brought to market, an increase in market share, or even the number of new patents obtained. This type of assessment often requires creative thinking to develop a set of measurements that truly reflect the outcomes the firm is seeking.

Operational Fitness

Operational fitness refers to the difference between the desired and actual operational performance. Managers examine a firm's standing in this domain by measuring the efficiencies that emerge from the combined activities in areas such as sales and manufacturing costs. These metrics can help reveal the underlying cause of poor financial performance as well as uncover potential future problems. When seeking technology externally for improving technology, there should be key operating goals such as economies of scale, increased input reliability, and increased quality of outputs. These areas need ongoing evaluation. For example, operational fitness goals could be to reduce the cost of goods sold by 5 percent. One way managers can accomplish this goal is by acquiring a competitor that has this advantage because of just-in-time inventory management. Operational fitness can also be indicated by measures such as cost of goods sold and lower inventory costs as well as the cost of the just-in-time inventory system.

Another area of operational fitness that is often overlooked is optimization of coordination efforts. Often, with an alliance or acquisition, the systems (especially operating systems like IT) do not get integrated. This can lead to human resources within the organization spending many hours reconciling the differences between systems. The decision for a firm then becomes whether to continue, rectify, or divest. In this case, the failure to obtain operational fitness may dictate strategic decision making rather than the preferred strategy dictating operational decisions.

Relationship Fitness

Relationship fitness is the difference between the desired and actual relationships within the firm. This fitness concerns a number of issues in the firm including: Are decisions made in a timely fashion? Is the proper information getting to the proper place within the organization? Are managers roles clearly defined? Is senior management involved? Are the cultures at least compatible? Are projects properly monitored? Are evaluation and control based on unbiased measures and processes and not on the source of the object being measured? Has the new organization truly adopted best practices?

For relationship fitness one problem is how to measure such items. Despite the difficulty in measuring such relationships the firm needs to establish clear expectations in this domain. The building blocks for relationship fitness are:

1. There must be integration and trust among the human resources at all levels of the organization.
2. There must be concern for other things besides "the numbers." While financials are important, they do not tell the whole story.
3. Oversight of technology must be flexible to promote idea generation. If there is integration and trust, this is easier.

8.2 REAL WORLD LENS

DuPont

Although DuPont has been very active in internal development of technology, it has also relied extensively on external means to obtain innovation and technology. The firm made its first acquisition in 1859. The coal industry at this time used a great deal of gunpowder to break up coal seams in the mine. The miners would then haul out the dislodged pieces of coal after such a blast. The movement of the powder from the Delaware factory to the coalfields of Pennsylvania was expensive. Therefore, in 1859, DuPont bought an explosives mill that was close to the coalmines of the time from Parrish, Silver & Company for $35,000. After purchasing the mill, DuPont upgraded the facilities to make it a state of the art mill.

Later DuPont did not want to be limited to business in the United States. Therefore, in 1910, it purchased a Chilean mine and formed the DuPont Nitrate Company. The firm produced nitric acid for the manufacture of smokeless gunpowder. The firm continued its expansion in South America in the early 1920s by joining with other firms to form the Compania Sud-Americana de Explosivos at Rio Loa, Chile. Alliances were also a critical means for expansion into Europe. In 1929, DuPont and ICI from Great Britain formed an alliance to share information about patents and research. The firms also agreed not to compete in certain geographical territories and established successful joint ventures in Canada, Argentina, and Brazil.

DuPont also continued its external efforts to obtain technology in the early part of the last century by purchasing specific products. Cellophane was invented in Switzerland and it was first produced commercially there in 1912. DuPont acquired the U.S. patent rights in 1923. The company continued to improve this product, and by 1938, various cellophane products accounted for more than 25 percent of the firm's revenues. In 1928, the firm bought Grasselli Chemical Company, one of the largest U.S. chemical companies of the time to help support it various operations. The firm then diversified further when it bought Remington Arms Company in 1933. In 1969, DuPont bought Endo Laboratories to enter the consumer pharmaceuticals market. Eight years later, DuPont Pharmaceuticals and Merck formed a joint venture known as the DuPont Merck Pharmaceuticals Company. In 1972, the firm bought Berg Electronics to enter consumer electronics, and in 1981, it bought Conoco Oil Company. Thus, throughout its history DuPont has used a wide variety of external means to obtain firms and products.

In recent years, the pattern of active external methods of increasing innovative capability through acquisition of technology has continued. In 1996, the firm formed a joint venture with Dow Chemical Company called DuPont Dow Elastomers. This joint venture offers a wide variety of products ranging from thermoset rubber polymers used by the general rubber industry to high-performance fluoroelastomers used by the chemical processing and automotive industries. At the end of the twentieth century, DuPont bought Pioneer Hi-Bred International to integrate agricultural biology into the company's science and technology base. The firm's last

REAL WORLD LENS *(continued)*

major purchase was in 2004 when it bought VERDIA, which specializes in genetically modified plants.

These external efforts to obtain technology, whether through alliances such as joint ventures or through mergers and acquisitions, are only a partial list of the wide range of activities conducted by DuPont. However, this partial list demonstrates a strong pattern by the firm to employ external methods to obtain innovation capability or technology. Today, Du Pont is focusing on building on the technology and skills it acquired in its external efforts as it seeks to address specific global trends it has predicted for business; the move from petroleum to bio-fuels and the need for home and commercial security products.

1. What benefits has DuPont gained from its pattern of acquisitions? Administratively? Technically? Product? Process?
2. Besides the potential benefits discussed, what else did DuPont probably gain? Are there potential losses it should guard against?

References

http://www.marketwatch.com/story/dupont-ceo-company-not-interested-in-major-merger

Karol, R., R. Loeser, and R. Tait. 2002. Better new business development at DuPont—I. *Research Technology Management* (Jan.–Feb.): 24–30.

Smith, John. DuPont: The enlightened organization. http://heritage.dupont.com

Stevens, T. 2001. R&D times two. *Industry Week* (May 7): 49–53.

4. Oversight of technology must be tight to prevent runaway projects and divergence along pre-alliance organizational lines.
5. Opportunity knocks. It is the responsibility of management to be ready.

SUMMARY

We established the dimensions of building an evaluation and control system for the acquisition of technology. Evaluation and control are more complex for blending two firms than for providing an innovative environment for the internal development of new technologies. Even though no two evaluation systems look alike, there are common elements that should be present.

1. Evaluation of readiness to create an alliance is critical for success. Numerous checklists can help to guide the process, and potential partners or acquiring firms should have one they are comfortable with. However, managers must remember that all involved with the acquisition or alliance should be evaluated. This will identify potential problems and synergies if the alliance is formed.

2. Top management should take the lead and be involved, of course. This is natural for this type of technology acquisition. The neglected areas for acquisition of technology are in operational areas of the organization. In other words, the alliance is made and then the problems emerge as the operations areas (i.e., human resources, manufacturing, IT) try to blend systems.

3. There should be a clear goal for the alliance, and the focus should be on reaching that goal. Too often, power struggles emerge as relationships are tested by the changes, and learning stops. Power and politics then emerge as the motivating energy rather than where the organization is trying to go.

4. Evaluation and control are ongoing processes, not just once a year phenomena. Monitoring systems that are appropriate for the level of detail and activity being demanded should be in place. The successful evaluation and control system gathers relevant information, verifies its reliability, is used for making good decisions, and spurs actions to improve the processes and products of the organization.

MANAGERIAL GUIDELINES

There are a number of points to remember when designing and implementing the evaluation and control system for obtaining technology from external sources. These include:

1. There will be a drop in productivity as energy and resources are used to accomplish the planning and blending of the alliance/acquisition. The evaluation and control process should recognize this.

2. There will be feelings of loss among the employees. These feelings of loss relative to the old way and the old systems and processes should at least be acknowledged. In implementing control, such a loss should be acknowledged while still seeking to move the organization in new directions.

3. Try to avoid the "conqueror" outcomes. Especially in acquisitions and controlling contractual agreements, people from the acquired company may feel like losers and may not feel welcome in the new way of doing things. People from the firm with the least power after the blending need the ability to express concerns. These concerns can provide valuable information to the evaluation and control process.

4. Remember, it is unlikely that the new entity will actually display the "best" of both firms. The goal is to display the "best possible" outcomes as the process unfolds. Control processes will allow the firm to make adjustments as the need to change becomes more evident.

5. Be sure that subtle aspects are not overlooked when blending functions and operations. For example, a sales force accustomed to retail distribution may have trouble with sales to manufacturing customers.

6. Do not assume everyone will understand the strategic value of obtaining external technology. Remember, there is no guarantee that everyone in top management will, so why should that be true throughout the organization? Communication is critical to the successful evaluation and control effort.

7. Acquisitions and new ways of doing things do not just blend in naturally. Just because the acquisition makes sense, it does not mean that sense of the acquisition will be made. Many acquisitions of technology that look good on paper have failed. Evaluation and control are critical to help avoid such problems.

Guiding Questions

There are a number of checklists that have been developed through the years to determine or enhance the potential success of an alliance or acquisition. Asking the following questions and evaluating the information indicated should move the management team toward success in mergers/acquisitions or alliances.[18]

1. Does the potential benefit/reward warrant the risk of failure or excessive cost as well as management distraction?

   - Examine the expected shareholder returns and compare them to the industry position.
   - Delineate the value creation opportunities in the deal and determine how risky they are.
   - Realistically assess how much management time will be absorbed by the blending activities.
   - Determine what opportunities will be missed or delayed by pursuing this particular alliance/acquisition.

2. Is the strategic rationale for obtaining the external technology well grounded?

   - Evaluate the basic business model for its potential for success.
   - Determine what the alliance/acquisition improves. This needs to be specific.

   - Delineate what the obtaining of the new technology will do for the firm's competitive advantage.
   - Realistically evaluate how the change will affect the firm's projected competitive positioning relative to competitors.

3. Is the integration plan well designed and realistic?

   - Clarify where the most value is to be obtained.
   - Specify what needs to be done to obtain the most value from the alliance/acquisition action.
   - Assign responsibility for the integration effort with appropriate authority given to the integration team.

4. Are top managers establishing a strategic plan and model for long-term success?

   - Determine the meaning of success for the firms involved.
   - Agree on how the success of the alliance/acquisition will be measured.
   - Set up a process for resolving conflicts and potential value destroying activities.
   - Agree on how the alliance will be monitored and determine how, when, where, and who will decide if the alliance will continue or expand.

CASE **8.1** THE REAL WORLD
Sport7

Sport7 was a consortium formed with the intention of creating a dedicated sports channel in the Netherlands. It was to be a European version of what people in the United States see when they view ESPN. At the time, this would have been a new innovation in the Netherlands. This consortium won the rights to broadcast the games of the Royal Dutch Football Association in 1996.

The investors in the consortium included Endermol (a TV production company), the Royal Dutch Football Association, ING (an insurance company), Nuon (a utility), and Philips (an electronics company). The

(continues)

CASE **8.1** *(continued)*

consortium included some of the most powerful firms in the Netherlands. Additionally, at the time of the formation of this consortium (1996), one of the local Dutch clubs had won the European Championship League title. This had built an environment where there seemed to be a very high demand for football (or soccer to Americans). Thus, the environment and the participants in the consortium thought the venture would be successful. However, ultimately, it failed.

There was a wide variety of issues that the consortium had not considered in its due diligence and implementation effort. These included three principal issues: (1) the opposition of the existing public broadcaster, (2) the opposition of some cable companies, and (3) the opposition of some of the leading football teams. Each of these issues will be examined in turn.

In the Netherlands, only the public broadcasting system may broadcast over the air. Other broadcasters must broadcast using cable technology. Historically, the public broadcasting system had televised football games in the Netherlands. Not everyone had access to cable television because of its extra expense, and the public broadcasting station was the most widely available in the country. The public broadcaster saw the arrival of Sport7 as a major competitor. This is especially true because the firm was taking one of the public broadcaster's most popular programs, football. As a result, the public broadcaster responded much more aggressively than Sport7 had anticipated.

Sport7 planned to broadcast football 25 percent of the time but needed other sports programming to complete the other 75 percent of the schedule. However, these rights were often held by the public broadcasting system. Additionally, both the public broadcasting system and Sport7 aggressively sought out minor sport associations to contract with. The public broadcasting system also retaliated by increasing the time spent broadcasting sports. Sport7 had dramatically underestimated the response of this competitor. As a result, the nature of the material it could broadcast and its expense were much higher.

The consortium also did not accurately evaluate how cable companies would respond to the development of the network. The consortium believed that the demand was so high that it could change the existing model for how networks interacted with cable stations. Historically, cable companies charged the network to broadcast their programs. The network would be expected to obtain their revenue by charging for advertising. However, Sport7 wanted each cable firm to pay 2 guilders ($1.12 U.S.) for each subscriber. As an inducement, the consortium reserved 15 percent ownership in Sport7 for cable companies. It also had the support of the leadership of the national association of cable companies. However, the existing cable companies refused to go along with the proposed changes in the model. The result was that, as the new network

CASE **8.1** *(continued)*

tried to get started, it faltered because it first offered the service free hoping to create demand for the product. But instead, offering the service for free created an expectation that it would be offered in the traditional manner of other networks.

Finally, the leading football clubs in the nation did not support the creation of the network. These clubs voted against the broadcasting contract for the new network. They wanted greater exposure to more fans and a greater part of the proceeds of the contract. These teams were critical to the perception of the value of the network. One of the teams then filed a lawsuit against the network that called the validity of the network into question.

Sport7 started with very strong backers in an environment that would have led one to initially believe that it would succeed. The absence of full due diligence prevented the participants from completely understanding the nature of the opposition they would face and why. This led ultimately to television channel Sport7 failure.

1. What were the key evaluation and control processes that Sport7 failed to use? Why did you pick the processes you chose?
2. Most consortia involve only two or three different types of organizations. Make a chart of the key members of this consortium. What were the goals of each key member? How did this hinder the consortium? What could have been done to overcome the potential problems?

Reference

Sminia, H. 2003. The Failure of the Sport7 TV-channel: Controversies in a business network. *Journal of Management Studies*, 40 (7): 1621–1649.

CRITICAL THINKING

Relating to Your World

1. We have discussed pre-acquisition evaluation and post-acquisition evaluation. We know that systems of evaluation and control should be connected to each other. Following is a chart of pre-acquisition goals. What should you know before and after the acquisition to determine if the goals can be met and if they have been met?

2. We talk a great deal about getting buy-in with the strategic direction of the organization. How should managers develop buy-in from the acquired company personnel? How would getting the support of those individuals aid the merging process? How would you evaluate if you have the support or not?

3. What advice would you give a manager who is charged with the responsibility of blending IT systems after the acquisition of a smaller competitor? What do you believe are the key integration issues that must be addressed? Would it be easier in a joint venture than in an acquisition? What would be more difficult? Why?

Pre-acquisition Goals	Pre-acquisition Data/Information	Post-acquisition Data/Information
Integrated product line by the end of the year		
Assimilate acquired firm's technology as a core competency		
Consolidate the vendor base to lower costs		
Ensure employees they will be treated fairly		
Develop a learning tool that could guide future acquisitions		

WWW EXERCISES

1. Identify a well-documented merger/acquisition that was motivated by technology acquisition. Find the goals for the merger/acquisition and then find how well the organization met those goals. What evaluation issues are identified in articles or comments about the merger?

2. Find a website that illustrates an evaluation process for merger/acquisition activities. How does this process compare to the issues identified in this chapter? How important is technology blending in the evaluation process? What are the strengths and weaknesses of the process that you find?

3. Find an article or website that provides guidelines for the evaluation and control of technology-based acquisitions and mergers. What do you think of the advice? Compare the advice you find to the advice your classmates find.

AUDIT EXERCISE

Goal	Metric	Goal/Results	−	Met	+
Financial Fitness					
Strategic Fitness					
Operational Fitness					
Relationship Fitness					
Readiness					

List four metrics for each of the fitness areas. What types of goals/results should the company look for? It is likely that not all of the goals will be met in any merger/acquisition. What should determine the company's readiness to acquire technology again?

DISCUSSION QUESTIONS

1. How would the evaluation and control effort differ in an alliance effort like a licensing effort compared to an acquisition of a firm?

2. Evaluation and control appear critical to the success of any effort to obtain technology externally. Do you think evaluation and control efforts contribute to the high failure rate in such external efforts? How and why? Or why not?

3. This chapter has a number of different evaluation and control frameworks represented. This is intentional; however, there are many more. What do you think are the five most critical issues in designing an evaluation and control system for obtaining technology externally? Give a brief description of why you believe these issues are critical.

4. Compare and contrast the purpose and process of due diligence and ongoing evaluation efforts. How should they complement each other?

5. Describe three types of measurement that can be used to determine the ability of a firm to be successful in strategic alliances.

PART THREE OPENING CASE: ACER

1. What are the special evaluation needs for a company such as Acer? What characteristics of Acer have the most influence on how well it evaluates progress toward stated innovation goals?

2. What steps in evaluation and control would you suggest Acer be most diligent about performing? How would its choices affect cost/benefit factors in alliances and acquisitions? Explain.

KEY TERMS

benchmark 237

financial fitness 248

gap analysis 248

inertia of success 246

metrics 247

operational fitness 249

relationship fitness 249

strategic fitness 248

NOTES

1. Gorman, M., and P. O'Grady. 2001. *Integrating Mergers and Acquisitions—A Perspective from the Technology Industry*. Dublin, Ireland: Prospectus.

2. Anslinger, P., and J. Jenk. 2004. Creating successful alliances. *The Journal of Business Strategy,* 25 (2): 18–22.

3. Dyer, J., P. Kale, and H. Singh. 2004. When to ally & when to acquire. *Harvard Business Review* (Jul.–Aug.): 109–115.

4. www.dupont.com

5. Tipping, J., and E. Zeffren. 1995. Assessing the value of your technology. *Research Technology Management,* 38 (5): 22–39.

6. May, M., P. Anslinger, and J. Jenk. 2002. Avoiding the perils of traditional due diligence. http://www.accenture.com/xd/xd.asp?it=enweb&xd= ideas%5Coutlook%5C7.2002%5Cstrategy_avoiding.xml

7. Milligan, J. 1990. The Ten Commandments of merger due diligence. *Institutional Investor,* 24 (7): 87–90.

8. McLaughlin, D. 2009. Corporate News: Lyondell Creditors Seek Probe of Merger. *Wall Street Journal,* March 11, B.2.

9. Hollander, D. S. 1998. Smooth post-merger integration hinges on detection of technology hurdles. *Bank Systems and Technology,* 35 (12): 56.

10. Allas, T., and N. Georgiades. 2001. New tools for negotiators. *McKinsey Quarterly,* 2: 86–97.

11. Anslinger, P., and J. Jenk. 2004. Creating successful alliances. *The Journal of Business Strategy,* 25 (2): 18–22.

12. Eckhouse, J. 1998. To navigate an M&A payoff. *Information Week* (Nov.9): 103.

13. www.dupont.com

14. Korman, R. 2001. Why acquisitions can turn bitter. *ENR:Engineering News-Record*, 247 (7): 15.

15. Sanders, P. 2009. Boeing Tightens Its Grip on Dreamliner Production—Company Is in Talks to Buy Fuselage Factory From *Supplier*; Supply-Chain Woes Have Dogged 787 Program. *Wall Street Journal*, July 2, B.1.

16. Callahan, J., and S. MacKenzie. 1999. Metrics for strategic alliance control. *R&D Management*, 29 (4): 365–377.

17. Bamford, J., and D. Ernst. 2002. Tracking the real pay-offs from alliances. *Mergers & Acquisitions*, 37 (12): 34–37.

18. Armour, E. 2002. How boards can improve the odds of M&A success. *Strategy and Leadership*, 30 (2): 13–20.

PART 4

Strategic Success

GOOGLE: A PATTERN OF SUCCESS

The last part of this text examines how to build sustainable success in MTI. Such success does not come easily and must be part of the strategic process of the firm. For a firm to build a sustainable competitive advantage two things are critical: (1) the capabilities necessary to generate success and (2) learning and knowledge management systems. We will examine both of these important issues in this part of the text.

Google is one of the most successful firms in the world with a technology and innovation focus. Therefore, initially, we will examine the history of Google to see how it has addressed these and other critical issues.

Google: The Firm's History and Strategy

Google's former CIO, Douglas Merrill, described Google's mission as "Gathering all of the world's information and make it universally accessible and useful." While this is very ambitious, Google's culture and history support taking on the impossible. Google has accomplished this through the use of MTI techniques and balancing the costs of internal and external means of acquiring technology are keys to its success.

Google, Inc., was founded in 1996 by two Stanford Graduate students who created Backrub, a search engine for the university that tracked web pages based on links to the page and not just word scans. By 1997, the usage of the site had overwhelmed the servers at Stanford and the company officially formed and changed the name of the service from Backrub to Google. After gaining a few investors, the company began its quest to gather all of the world's information and to make it useable. By 2000, Google was operating in over fifteen languages.

The company provides its service free to users but makes its money principally through advertising. The advertising arm of Google, Search, brings in over 97 percent of profits. The firm, however, does not stand still and dedicates 14 percent of its revenue to its Research and Development budget in an effort to develop new products. Google maintains a staff of engineers to improve and create new products as well as acquire other companies as illustrated next.

Google's Competitive Advantages

One way Google encourages innovation is by freeing 20 percent of the engineer's time for pursuing innovative ideas. Although this idea is not new,

3M has been implementing a similar program for over 50 years. Google has found this to be a successful means to generate new products. Marissa Mayer, Google's Vice President of Search Products and User Management, stated that, "fifty percent of what Google launched in the second half of 2005 actually got built out of 20 percent time." To illustrate, after the 9/11 attacks in New York, an engineer at Google, Krishna Bharat, got tired of searching through all the major news networks to gather information about the attacks. So, Bharat created a program to "crawl" through the information on the major news networks, and then clustered the data so he could access all the articles about one topic without having to go to different web pages. He used this program for a few months, and then presented it to management. The idea was further developed and Google News was born. When Google News was new, Google engineers could not decide if it needed a search tool to group articles by date or by category. They did neither but shortly after release customers requested a date organizer.

In addition to internally creating new projects, Google has relied heavily on external sources of obtaining technology. They have done this through the acquisition of companies or through joint venture agreements. The most publicized acquisition was in 2006 when Google bought YouTube for $1.65 billion. Although YouTube was only valued at around $600 million by some authorities, the Google CEO stated that he bought it at a billion dollar premium because it was far more popular than Google Video and already had a large number of users. Paying such a premium also allowed Google to skip the process of having to compete with other video sites in a bidding war or spending the money to build up market for Google Video.

Google managers followed a similar process when they were looking at making mapping software. In 2004, they discovered that a company in Australia already had the technology available to make interactive online maps. Google decided that they would purchase this company, Where2 LLC, and in the process found another company that had the technology to map the entire world through satellite imagery, Keyhole. By purchasing these two companies, Google was able to create both Google Maps and Google World, which have been used extensively in academia as well as military relief efforts after natural disasters such as Hurricane Katrina and the 2004 Asian Tsunami.

Google also uses acquisition of external technology to make incremental changes to existing products. In 2003, Google purchased Applied Semantics and Sprinks to improve the company's biggest revenue producers AdSense and AdWords. By doing so, Google was able to improve its advertising processes and introduce new video forms of advertising. Google has also signed numerous joint ventures with companies to provide them with Google's search database. These include Yahoo, America Online, and the Latin America company, Universo Online. In all three instances, Google made long-term agreements with these joint venture partners. Google is sharing its ability to build search engines for cash and partners who will expand Google's advertising footprint.

Google is unique in that the firm has a very balanced company in terms of innovation development. Google has robust internal R&D programs but also relies heavily on outside partnerships and acquisitions to maintain the company's position atop the online technology world. Google has developed a unique capability in the online search industry. However, to maintain its position, it is constantly seeking new ways of doing things—either by internal R&D or by external acquisition and alliance. They are constantly learning and developing new ways to manage information and knowledge so they can stay ahead of the competition.

Overview of Part Four

This part of the text will examine how a technology-focused firm can develop processes to sustain its competitive advantage over the long term. Chapter 9 will focus on how the firm can ensure that it has the capabilities to be successful. Chapter 10 will examine how knowledge management and learning processes influence future success of the technology-oriented firm. For successful MTI over a long period, a firm must be able to learn and change. Without such ability to learn, the organization will not be able to adapt and change in a world that does not stand still.

SOURCES

Google Inc. Sept 2009. http://www.google.com/corporate/tenthings.html

Mayer, Marissa 2006. Nine Lessons Learned about Creativity at Google. Stanford University Entrepreneurship Corner. Stanford University, California (May 17). http://ecorner.stanford.edu/authorMaterialInfo.html?mid=1554.

Mediratta, Bharrat 2007. The Google Way: Give Engineers Room. *NY Times* (Oct. 21). http://www.nytimes.com/2007/10/21/jobs/21pre.html?_r=1.

Merrill, Douglas 2007. Innovation at Google. Google Inc. (Aug. 1). http://www.youtube.com/watch?v=2GtgSkmDnbQ.

Sandoval, Greg 2009. Schmidt: We paid a $1 Billion Dollar Premium for YouTube. *CNet News*. (Oct. 6) http://news.cnet.com/8301-31001_3-10360384-261.html.

Walters, Helen 2009. Google: How Does Your Innovation Garden Grow? *Business Week* (Apr. 23). http://www.businessweek.com/ainnovate/next/archives/2009/04/google_how_does.html.

CHAPTER 9

Building Capabilities

OVERVIEW

The ultimate goal of a technology-focused firm, whether it seeks to obtain technology internally or externally, is to create value for the firm and the firm's stakeholders. The best way for a firm to create value that will continue over time is through a sustainable competitive advantage, a fact we noted earlier in Chapter 2. The foundations for this sustainable competitive advantage come from the capabilities of the firm. This chapter will integrate ideas developed earlier in the text to discuss these issues. The topics examined in this chapter include:

- Capabilities and how they are developed

- How a firm develops a sustainable competitive advantage

- Value creation for the firm through competitive advantage

- The role of the creation of industry standards

- The fundamentals of venture capital

- How to turn around a troubled firm

———————

INTRODUCTION

Parts Two and Three examined the two major ways to obtain technology: internally through innovation and externally through alliances or mergers/acquisitions. Regardless of the method employed to obtain technology, the organization must have the necessary capabilities associated with that technology if the firm is to gain or maintain a competitive advantage. Capabilities, you will recall, are those internal resources like leadership, culture, and training that allow a firm to implement a given strategy. The capabilities necessary are not only for today's competitive advantage but also to build and ensure that there is a competitive advantage in the future.

The development of the capabilities necessary for such success begins with the initial steps taken by the organization. It was noted in Part One of this text that the successful management of technology and innovation requires that an organization maintain a clear view and understanding of where it gains its competitive advantage and creates value for important stakeholders. From this understanding, the organization targets resources and skills that it needs to employ a given strategy. If the necessary capabilities are not present, regardless of whether the organization uses internal or external efforts to obtain the technology, it will not be able to build a competitive advantage. The goal is to have a sustainable competitive advantage that is hard to imitate and as a result, it remains a strength over a period of time. Although developing a competitive advantage is a conscious process for the firm, the competitive advantage should be something that occurs because of the firm's capabilities and their management. As a company strives to develop its capabilities, there are several ground rules.

1. Success depends on a clear strategic logic for processing information and sharing knowledge.
2. The appropriate structures and processes must be in place for both technical and nontechnical activities.
3. Employees must be motivated to develop and take advantage of capabilities.
4. Organizational fit must allow resources to be captured by the right people at the right place at the right time to make a competitive difference.

Figure 9.1 illustrates the successful building blocks for developing capabilities. Reviewing the model, the external environmental factors shape the resulting capabilities of the firm. The elements of the external environment discussed in Chapter 2 included: economic, social, political, and technological as well as the competitive environment itself. For example, as the economy slows, demand for many products declines. This decline in demand intensifies competition initially and may cause some firms to look for other opportunities or to change their competitive mix. In Chapter 2, we also discussed how a firm's industry influences the competitive actions of that firm. The specific model examined was that of Michael Porter and was referred to as the five-forces model.

Therefore, the concern for the venture (as discussed in Chapter 2) is to develop the appropriate strategic capabilities to help the organization successfully

FIGURE **9.1** Process for Building Capabilities

meet its goals and objectives as well as maintain competitive viability. The strategic capabilities the firm needs include the following:

1. *Leadership* to provide a clear strategic focus with visible top management commitment.
2. A *culture of support for innovation* to ensure a willingness to share knowledge, to invest in resources, and to recognize the abilities of all individuals and groups within the organization. In other words, decisions should not always come from the top.
3. A *structure* that fits with the goals and activities of the organization. The more innovative the organization is, the flatter and more networked the structure should be.
4. The *skills* necessary to implement a given strategy. These skills include not only those necessary for the given technology but also time,

knowledge, and space management skills. For example, the ability to work within the given space of the firm is critical. The construction or rental of space is expensive, and the ability to manage space can lead to savings that become a competitive advantage. Thus, a wide range of skills is necessary.

As illustrated in Figure 9.1, strategic capabilities enable the organization to make appropriate strategic decisions. We have discussed the firm's necessary strategic decisions throughout this book. To illustrate, we discussed the issues of corporate strategy and business-level strategy in Chapter 2. These are the basic choices by the firm such as which industries to compete in and how to compete in those industries. Other strategic decisions include where the firm will enter the life cycle of a product (Chapter 2) and whether the firm wishes to innovate internally (Chapters 3, 4, and 5) or acquire technology from external sources (Chapters 6, 7, and 8).

We have discussed the planning, implementation, and evaluation/control of internal innovation and external acquisition of technology separately. In reality, most firms will mix these methods to various degrees. No matter which strategic approach a firm uses, the firm's strategic decisions should be based on the capabilities of the organization. Therefore, the goal of the firm creating a sustainable competitive advantage and creating value for key stakeholders is intertwined with the capabilities of the firm.

This chapter discusses the building of capabilities, and building a sustainable competitive advantage. To accomplish this, the chapter initially examines the concept of competitive advantage, and next when that advantage is sustainable. The chapter then looks at the two principal strategies that a business-level firm can pursue in building a competitive advantage: low cost and differentiation. We then contrast the concepts of value creation and sustainable value creation to competitive advantage. This discussion includes the building of capabilities to accomplish a sustainable advantage and create value through technology and innovation. Finally, the chapter discusses how the organization can turn itself around to rebuild its capabilities and value if it loses its competitive advantage.

COMPETITIVE ADVANTAGE

After a business evaluates its environment and its internal capabilities, it develops a strategy based on how its capabilities can add value within the environment. It then implements that strategy. As noted at the beginning of this book, the goal of the strategy should be to create a competitive advantage for the firm. A competitive advantage is something that the firm does better than any other firm. This difference should be something that motivates a customer to buy the firm's product or service. The difference can be a more efficient production process that lowers cost and in turn, provides a lower price to consumers. Alternatively, a competitive advantage may be specialized services the customer finds attractive.

To illustrate the need for a competitive advantage, a firm may develop a new accounting software program for small businesses. This firm needs to ask itself: Why would a business switch from the program it currently uses to the new software program? The new software program may be wonderful, but what will it do that existing programs do not? Even if the new software does something that other software packages do not or if it does the existing activities better, will the difference be enough to motivate a purchase? There may be a fear that a new firm will not be in business over the long term to service its software. This makes switching to the new software even harder. The result is that there must be a clear performance or cost advantage that a customer will actually value enough to incur switching costs. If it is something for which a consumer is willing to switch, then the firm has a competitive advantage.

For technology- and innovation-focused firms, their competitive advantage is a function of the technological complexity of the environment in which they compete, the capabilities of the organization such as the skills and knowledge of employees, and the ability of the organization to learn from its activities. Figure 9.2 illustrates the linkage of technological complexity and human skill and knowledge capabilities of the firm. The narrowing of the funnel in the illustration indicates the pressure on the firm to have sufficient human and knowledge capabilities as technological complexity increases. In other words, human and knowledge capabilities must increase if the technological complexity increases. For example, purchasing new computer software does not mean that processes will improve. If the people in the firm do not understand how to use the software, then the software becomes a cost item with little benefit or value. The difficulty for a firm is to be sure that whatever level of technology it chooses, it has the full range of human skill and knowledge capabilities present to implement successfully the given level of technology.

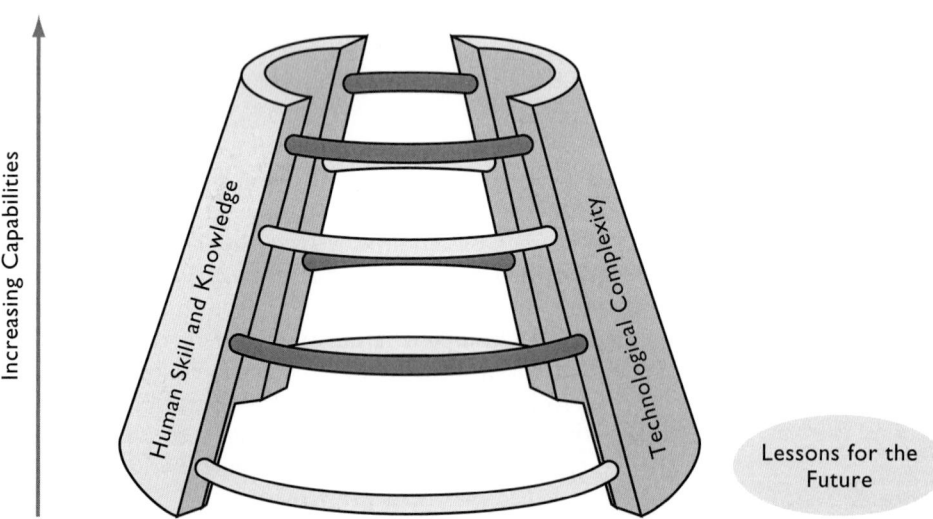

FIGURE **9.2** Human Skills and Technology Complexity

An illustration of the impact of not having the proper human skill capabilities is the difficulty in getting TiVo adopted. Initially, TiVo sold a product that allowed individuals to tape any program on television they desired. The customer could then treat the taped program like a video disk and fast forward through the commercials. But, TiVo did not grow as quickly as was predicted. The firm could not explain to potential customers the added value of the product. Most customers viewed TiVo's product as more costly than beneficial. Today the problem is also that there are cheaper ways to do the same function. For example, today, many individuals can do essentially the same activity within their standard cable package. Thus, the firms developing the TiVo technology had human skills in technology-related domains, but they did not have the human skills in marketing necessary for success. Today, TiVo partners with various media outlets, including some cable companies, as well as markets through retail stores to consumers. In addition, TiVo has won several major lawsuits against firms who infringed on their technology and "sold" it better. TiVo still leads in technology development but has changed its marketing to develop sustainability.[1] A firm needs to ensure that it has all of the necessary skills and capabilities if it wants to succeed in a given domain.

If people walk down the discount aisles of large electronics retailers, they may find a wide variety of products that are there because there is no market demand for them. These products typically have a technology enhancement but no competitive advantage that motivates their purchase. Each of these products illustrates an absence of some skills or capabilities that the firm that produced them needed to better align with its environment. If there had been better alignment, the firm would have been able to better formulate a product that met consumers' needs.

SUSTAINABLE ADVANTAGE

The firm must also consider whether its competitive advantage can be maintained over time or whether another firm can match that competitive advantage. The hope is for a competitive advantage to last at least two to three years. One difficulty that most firms face is that the period for which a competitive advantage can be maintained has gotten shorter as technological skills have increased. In addition, there is now effectively a worldwide market. Richard D'Aveni refers to this as hypercompetition.[2] **Hypercompetition** occurs in a rapidly changing environment where competitive advantage is hard to maintain because competitors imitate the strategies of successful organizations or leapfrog their technology rapidly. The characteristics of industries experiencing hypercompetition are:

- Flexible, aggressive, innovative competitors moving into established markets
- Constant disequilibrium and change
- Increased uncertainty, dynamism, and heterogeneity of players
- Hostility

Today, many technology-based industries are hypercompetitive.

A review of the history of Internet search engines illustrates the speed of competitive change and shows the difficulty in developing a sustainable advantage. In 1994, EINet Galaxy was one of the first search engines. Its technology became the foundation of Yahoo!. After the initial dominance of Yahoo! Google became the dominant force in this industry. Now other competitors are aggressively attacking Google, offering more backlinks and the ability to scan even more web pages than Google. For example, in 2009, Microsoft introduced Bing to challenge Google's dominance as a search engine. Bing promotes itself as having a better interface with the searcher and the search algorithm. In spite of heavy marketing by Microsoft, Google remains the dominant player. But this has not prevented Microsoft from continuing to try to displace Google in this hyperactive marketplace. Cell phone companies have faced a similar pattern of rapid changes.[3] In 1998, Nokia was the leading cell phone manufacturer emphasizing small phones. By 2002, Nokia was number four in the market, and Samsung was the leader with color screens. In 2003, market leadership was again shifting with camera phones being the dominant product. Today, cell phones are moving to become full cellular devices with the Iphone and Google phone providing a wide range of Internet opportunities through a mobile phone. For example, today's mobile phones can take pictures, text messages, store music, allow you to respond to your e-mail, play games, and find maps. Apple's application store has grown from less than 25,000 apps in 2007 to more than 100,000 in 2009. It is expected that this rate of growth will continue over the next three years. The mobile phone industry started a little more than 20 years ago, but manufacturers have created many new technologies that keep cell phone users coming back for more. They continue to increase the number of capabilities and services to accommodate the growing needs of today's "on the go" culture. Thus, the goal is to develop a competitive advantage that is sustainable for a number of years, but the ability to do so is more difficult in hypercompetitive environments.

The goal of the firm is not only to have a competitive advantage but for that advantage to be sustainable or hard to imitate for as long as possible. Firms do not face a uniform environment in trying to maintain their competitive advantage. Some environments are relatively tranquil; as a result, there is little pressure on the firm to change once it has built capabilities that provide a competitive advantage. However, other environments are very turbulent; the firm's competitive advantage, and the capabilities that provide it, face constant threats and the firm needs to adapt rapidly.

In the management of technology and innovation, how rapidly the environment changes is a key concern. The ability of the firm to build a sustainable advantage is somewhat dependent upon the type of environment in which it competes. There are three broad divisions used to describe the cycle of change in industries. They are:[4]

- Slow cycle: The change in this environment is less strenuous, with fewer radical innovations occurring. The result is that once a firm develops capabilities, these capabilities can be more durable and enduring than in some other environments. Typically, a slow change environment is less technology intensive than environments where change is rapid. The management of technology and innovation in a slow change environment seeks to maintain its competitive advantage by monitoring the environment and continuously tweaking existing technology for improvement. Binney and Smith, manufacturers of Crayola, are a prime example of a firm that has continuously improved its product, production machinery, and processes to maintain a competitive advantage. However, the changes faced by the makers of Crayola are relatively slow, so while the firm has maintained its advantage, it has not been under pressure to embrace major changes very often.

- Standard cycle: The change in a standard cycle industry is faster than than in a slow cycle. Often, such environments have industries that possess relatively high levels of technology, but they are mature industries. The rate of change is not the fast pace it once was, but there is still a degree of change. In these industries, change efforts focus on process improvements and efficiency improvements designed to lower costs of the firms in the industry. Companies with standard cycle resources (e.g., automobiles, fast food, credit card processing) face the possibility of losing the uniqueness of their processes through competitive imitation. Thus, in automobiles, American firms such as General Motors and Chrysler did not lose being the dominant innovative force in their industry overnight. Instead, over a number of years, other firms built better capabilities in a variety of domains that General Motors and Chrysler failed to monitor and respond to in a timely manner.

- Fast cycle: Organizations in a fast cycle environment face the highest rate of change and typically are technology intensive. There is little effective shielding for these firms' competitive advantage, and there is a need for rapid adaptation in the capabilities of the organization. As a result, continuous improvement is the key for the firm. Products such as mobile stereo systems like Sony's Walkman are examples of products that have changed a great deal over the last decade as competitors tried to gain a competitive advantage if only for a relatively short period of time (two years or less). Today the Ipod is the dominant player in this industry but its position is under constant attack. It is in these environments that firms are most likely to face hypercompetition.

STRATEGIC DECISIONS AND COMPETITIVE ADVANTAGE

As noted in Figure 9.1, the organization can create a competitive advantage based on its capabilities at the levels of both corporate and business strategy. Corporate-level strategy includes choices made by the organization about the portfolio of businesses in which it will be active and compete. A small business typically competes in only one business area. Larger organizations, however, can choose to compete in a wide variety of businesses. In fact, many large corporations have so many different types of businesses that they group them into strategic business units (SBUs). Thus, firms with numerous units, like General Electric, will try to group the units together in some rational manner. For GE, the various radio and television businesses are grouped in one SBU, and the different appliance businesses would be in another SBU. The SBU has the benefit of allowing like businesses to be in the same group. The head of each SBU would then report to the CEO. However, one cost of this structure and size is that it creates another layer of bureaucracy in the organization.

Chapter 2 discussed the key aspects of corporate strategy—whether the firm would pursue related or unrelated diversification. This diversification can occur by either internal development or mergers and acquisitions. The chapter noted that related diversification is widely recognized as providing better performance than unrelated diversification. A key aspect of this success is that the capabilities developed are applicable across the various units with related diversification. The capabilities of the organization in such a situation can be applied in multiple units rather than having to develop distinct capabilities that are only relevant to one unit of the organization.

Each type of business in the portfolio of a corporation will have its own business-level strategy. If the firm competes in only one business, it has a business-level strategy only. Michael Porter argues that these strategies can be broadly viewed as focused on either differentiation or low cost.[5] If the firm chooses to compete based on cost, it will strive to be the lowest cost competitor in the market. The alternative to compete on differentiation requires that the firm's product or service has unique features for which the consumer is willing to pay a premium. Thus, a firm such as Acer has a low-cost strategy for its principal business. It seeks to build its products for a lower cost than its competitors. In contrast, Cray Computers clearly focuses on selling a differentiated product—high-powered supercomputers.

The key to building a competitive advantage with either a low-cost or a differentiation strategy is that the various strategic capabilities of the firm are consistent with the strategy. A low-cost strategy focuses on delivering a commodity-like product to the marketplace at the lowest cost. Thus, key capabilities for Acer are in manufacturing efficiency and just-in-time supply management. Research and development are not part of the key capabilities for the business unit that manufactures laptops. The organizational structure associated with a low-cost strategy like Acer's is one where there is tight evaluation and control.

In contrast, Cray is all about creativity to produce cutting-edge products. Thus, while efficiency in production and delivery is desirable, they would not be key capabilities for success. Instead, the firm would focus on cutting-edge R&D. The firm's organizational structure would also be different because Cray is more concerned with creativity and knowledge sharing than with efficiency. Thus, a decentralized system that encourages risk taking would be appropriate. No matter what strategy a firm pursues, it needs to ensure that it has the capabilities that support the selected strategies.

If a firm has multiple SBUs, it is possible that it has a mixture of strategies among them. For example, one SBU can be oriented toward being a low-cost producer, while another seeks differentiation. The low-cost producer may focus new technology initiatives on production improvement, while the differentiator seeks innovation in products and markets. In this case, the overall corporation does not have an MTI strategy, but rather, each SBU does. These different approaches need to tie to overall corporate goals. For example, GE's goal is to be number one or number two in every market. That overall corporate strategy is implemented through decisions and actions at the SBU level.

Once a business-level strategy is established, the impact of that strategy will cascade to the functional level of the business unit. The functional level of an organization includes areas like accounting, finance, marketing, engineering, and manufacturing. These areas do not have a strategy that stands alone. Sometimes managers refer to the planned activities of the functional areas as **tactics**. Such tactics should be consistent with and supportive of the business-level strategy. It is at this level that the capabilities of the organization become the clearest and most quantifiable. Thus, the organization can determine what skills are in each functional area and how the individual-, group-, and firm-level skill sets compare to those of competitors.

Connecting It All

Creating a competitive advantage requires that the organization connect its various activities (i.e., manufacturing, marketing, developing, hiring, etc.) into a coherent whole. As noted earlier, most large firms will have multiple levels of activities, which make such interconnections difficult. For example, Procter & Gamble is very active in obtaining technology through both internal innovation efforts and various external methods.[6] The method that the firm uses in these internal and external methods is referred to as the *Connect and Develop* program. In this program, the firm recognizes that not only does P&G want the technology but seeks to connect the people associated with the new technology to others in the firm. The program is a recognition by P&G that creative ideas develop when people from different backgrounds meet. P&G has taken the idea of open innovation to heart and states that *Connect* and *Develop* is not an experiment; it is the strategy. The program has been very successful. Already, more than a third of the company's new products come from ideas from such interactions. The goal for P&G is to make inspiration for innovation a routine part of its culture.[7]

Each approach to obtaining technology, internal and external, has its own level of complexity. Figure 9.3 summarizes the complexity of implementing each activity. The firm will need to coordinate these complex activities with the firm's corporate, business, and functional strategies. The greater the complexity of the organization as it tries to connect these activities, the more difficult it is to make such connections successfully. This situation in turn requires greater effort to develop knowledge skills, managerial skills, physical skills, processes, culture, and structure. Too much complexity can drain the firm of scarce resources without providing the value needed to sustain the firm.

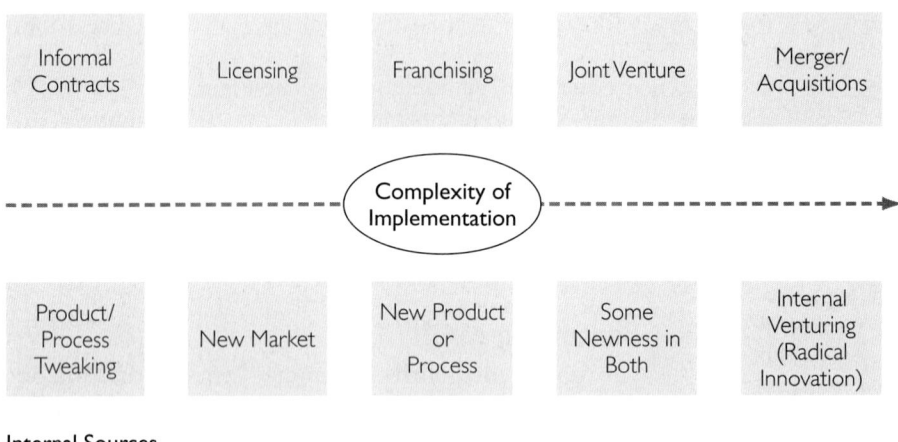

FIGURE **9.3** Implementation Complexity

In dealing with the complexity of the organization, the firm can determine where to focus its efforts by asking the following questions:

- Where is the value-adding potential, and how should the potential be weighted?
- What is most critical for long-term and short-term success? These must be balanced.
- What methods will we use to make the determination of the value and criticality?

Creation of Value

The three questions that help a firm connect its efforts all concern value creation. The result of a competitive advantage is the creation of value for the firm, not profits. The profits are an outcome of having a competitive advantage. The difference is more than semantics. If the firm focuses all its evaluation efforts on profits, it will make short-term decisions that can lead to profits today but problems for the firm tomorrow. Therefore, the firm needs a richer set of criteria when evaluating its strategy; one that includes more than profit. The creation of value at the firm level is the development of innovative capacity and the exploitation of that capacity to introduce new products, services, and methods that alter the competitive landscape.[8]

To evaluate the creation of value by the firm's strategy, the resource view of the firm is very helpful. This view maintains competitive success comes from the internal resources of the firm; recall that capabilities consist of internal resources that are well-managed. This view of success is in contrast to Michael Porter's industrial organization view of strategy, which was summarized in his five-forces model examined in Chapter 2. The industrial organization view of success argues that industry is the principal determinant of profitability.

However, the resource view of the firm argues that profitability comes from the internal resources the firm obtains, and how the firm uses those resources. These resources can be either tangible—such as equipment, buildings, and financing—or intangible—including brand recognition, perceptions of quality, culture of the firm, and innovation. To create value, resources need to be:

- Nonreplicable
- Unique
- Rare

For most organizations, it is not a choice of whether industry or resource utilization determines the firm's profitability. Instead, both have an impact, although the relative impact for a given firm may differ depending on the industry and the competitive environment. Thus, most firms consider both the external environment and their internal resources to be critical to their ultimate success.

When considering resources one of the most critical is knowledge. Many firms produce a competitive advantage through intangible resources—resources that are not imitable (e.g., a patent) or are person specific. If you can buy the equipment, your competitors can also buy the equipment; hence, tangible resources usually do not provide a sustainable competitive advantage. Instead, most innovative firms rely on creating a new product before their competitors develop a similar product. Thus, knowledge and innovation generate the most common and best sources of competitive advantage.

Value creation is not something that occurs automatically or quickly. Instead, value creation is a process that develops over time. Such a process requires resources and commitment. The stages of the value creation process include:

1. Develop capabilities/investments that increase innovation: these may be either internal innovation efforts or through the acquisition of such innovations.
2. Recognize and deploy innovation by increasing product activity or the number of new products introduced.
3. Exploit innovation through sales growth.

4. Continue improvement through investment in support capabilities and/or improvements in efficiency and productivity.

The first step recognizes that capabilities are the foundation on which value creation occurs. It also indicates that creating value requires commitment by the organization not only to develop such capabilities but also to maintain and build them. The development of the capabilities to produce value requires the organization to invest resources.

For technology-oriented firms, it is important for managers to understand that value must be customer driven. After all, if the customer is not willing to pay for the increased value then the profitability of the firm will decline. Too often, the focus is on technological capabilities to the exclusion of the ability to address customer needs. Earlier in this chapter, we discussed innovations for which consumers were unwilling to pay—while these products were nice from a technical standpoint, their value was not apparent.

Thus, for internal innovation, the organization needs to ensure that the right people are in the organization to create the knowledge necessary to be successful. This means that to promote critical capabilities firms should hire key individuals with the desired skills, plus sponsor the training and education of the individuals who will support that knowledge creation. Individuals need to know for whom they are creating value since it helps to focus their activity. Similarly, if the firm is to acquire technology, it must understand what types of innovations are needed, properly identify the sources of that innovation, and make sure appropriate plans are in place and implemented to ensure that the acquired technology is integrated effectively. Thus, value creation does not occur by itself; instead, value creation is a process that requires focus and the devotion of resources by the organization. It also requires that the context of the value creation and the relationship of the processes involved are compatible with the current processes and systems within the organization.

The second item in the value creation process is the recognition that new products and new applications of existing products need to flow from the innovation and technology management efforts of the firm. The innovation process is not an academic process. Instead, value creation requires the necessary investment in people. Value is a trade-off between costs to develop and benefits for which customers are willing to pay. When the opportunities arrive to sell products developed through the innovation process, the firm must act on those opportunities—the third point in the list of how to create value. As noted earlier, the value creation process relies on the creation of new knowledge. However, the ultimate success of that innovation process requires that the firm produces products and then sells them. The strategic management process for technology is not simply encouraging creativity, but an outcome of actions that create value and success for the firm.

The last point in the value creation process is that the firm must ensure that once it starts the process, it acts to continue the process. Value-based management is a mindset. In the management of technology and innovation, good ideas may abound, but it is the ideas that create value that help the organization sustain itself. There must be new innovations and new technologies added and old technologies eliminated to ensure that the firm stays successful. For example, Philips is Europe's biggest electronics company and the world leader in lighting. The firm has developed a new type of light bulb that will ultimately replace the traditional bulb. Incandescent light bulbs have not changed much since the early 1900s until this development. Today, consumers are interested in more energy efficient bulbs but unwilling to pay for costly long-life bulbs or fluorescent tubes. As part of an ambitious re-branding, Philips has introduced a light bulb that is completely different and much "greener" than incandescent or fluorescent lights. The new LED light generates the same amount of illumination but uses far less energy.[9]

9.1

REAL WORLD LENS

UPS

United Parcel Service daily serves 1.8 million shipping customers from its worldwide hub in Louisville, Kentucky. The firm must process these packages between 11:30 p.m. and 3:30 a.m. each day. The firm has to develop both technology and human resource capabilities to meet this strenuous schedule.

Sorting packages could not all be done manually simply because of the volume of packages. The firm found that there was also a need to increase the use of technology due to human resource pressures. In the 1990s, UPS realized that the heavy reliance on manual efforts was resulting in high employee turnover. Over the last 10 years UPS has spent $100 million on new software to upgrade its package flow technology. The results are:

- Increased productivity (as much as 50 percent) by package loaders
- Increased route efficiency for drivers—saving over 25 million miles of driving and 3 million gallons of fuel
- Increased internal connectivity and better communication
- Faster loading of trucks

The firm's efforts to develop its capabilities have not stopped there. Their IT staff has grown to over 5,000 people, as UPS has become the leader in package delivery services. The competitive battle between UPS and FedEx has been described as a war, with both firms in a technological arms race. The result is that both firms have made extensive evaluations of each firm's human skills and knowledge capabilities.

1. What sustainable competitive advantages has UPS tried to acquire?
2. What happens if a new competitor appears in the environment? How should UPS change its strategic focus?

References

_____ 2009. United Parcel Service, Inc.—Financial and Strategic Analysis Review. *M2Presswire*, April 7.

Mitchell, R. 2008. Special delivery. *ComputerWorld*, April 21, 38.

BUILDING CAPABILITIES FROM THE START

The preceding discussion demonstrates that a sustainable competitive advantage and the creation of value for the firm cannot occur without attention to the development of capabilities. This focus on capabilities should be present in the firm from the start. This start may be either entrepreneurial or intrapreneurial. The term **intrapreneurial** is used to describe entrepreneurial activities that occur within organizations. It is also referred to as *corporate entrepreneurship* or *corporate venturing*. In evaluating the capability of a firm to enact these types of activities, four key resources need to be developed.[10]

- Entrepreneurial resources: individual knowledge and experience in being innovative and managing new products and processes
- Human resources: the makeup of the workforce for the firm, including the technical knowledge, level of training and development, and type of reward system
- External network resources: the existing linkages and potential linkages to outside resources for potential collaborative efforts
- Economic resources: profitability and marketability of the technology being developed

These resources are of particular importance to the start-up technology firm because these firms have fewer slack, or excess, resources than do large firms. In the United States, entrepreneurial start-up ventures are at the heart of the economy. In 1980, Microsoft had only $8 million in revenue and thirty-eight employees. Today, it has more $60 billion in revenue and employs over 93,000 people in over 100 countries. It is estimated that 75 percent of new job growth in the United States comes from entrepreneurial ventures.[11] Most firms in the *Fortune* 500 have declining employment. The growth in employment has come from new entrepreneurial ventures. In the last generation the major entrepreneurial successes were firms such as Microsoft, Federal Express, Cisco, and Intel. However, much of the growth has occurred in other technology entrepreneurial ventures whose names may be less familiar to you but whose impact is significant, such as SST (see Appendix 3 at the end of the text).

As a result, firms need to focus on the development of necessary capabilities from the beginning. If firms develop such capabilities early, they will increase their ability to prosper and grow in the future. However, if the firms do not develop such capabilities at the beginning and have to develop them when they are already facing fierce competition, they can find they are out of business while trying to recover from the absence of capabilities.

There are two topics important to organizations and their development of relevant capabilities from the beginning—establishment of industry standards and the funding of firms through venture capital. Each of these topics are examined next.

Establishment of Industry Standards

A particularly important issue facing firms as they seek to establish their capabilities from the beginning is the establishment of industry standards. The standards for technology can be set several different ways. One way in which standards can be set is by industry-related professional or official groups. Thus, for 3G, or the third generation of wireless networks, standards were established by the International Telecommunications Union. However, such standards are typically not set until there have already been technological advances in the domain. Thus, 4G system of wireless is being released in 2010 but the standards will not be established until after the release. If a firm does not accurately monitor and participate in standard setting, it can find that its technology may be innovative but not relevant in the marketplace that exists because it does not meet the industry standards that everyone else is following. Therefore, the new technology-focused firms in cutting-edge domains should have individuals who have the capability to work with industry groups to set standards.

A second means for setting industry standards is in the marketplace. If there is not a standard set by an industry group, then the marketplace will ultimately choose which standard to follow. When the market sets the standard there are several factors that organizations seek to manage in their development of capabilities. The first is that the best technology does not always win. Entrepreneurs cannot assume that even if they have the best technology, it will prevail in the marketplace. The reputation of the key parties to the technology can play as critical a role as the technology itself. Previously radio frequency identification (RFID) was identified as one of the growing technologies in retail. There were several major competitors with different technologies in the domain, but the one selected by Walmart was the winner. It may or may not be the best technology, but the technology used by the largest retailer directs the choice of others. If suppliers are going to sell to Walmart, they need to use RFID technology. Thus, Walmart has effectively chosen the dominant technology.

The interconnections among the firms in the industry and others in the community also affect the emergence of a technology standard. The emergence of VHS instead of Betamax technology was the result of JVC making alliances with a large number of other firms in the industry, not the technology. Similarly, Apple may have had better, more usable technology, but when the firm did not make it easy for software writers to produce software for their computers, the DOS operating system won the technology battle. It was accessible to others who were developing software, while Apple's technology was not.

Therefore, in examining entrepreneurial ventures and the establishment of key capabilities, the business needs to focus not only on the technology but also on the strategic choices. The success of the technology and the firm itself depends on far more than whether the firm has the best technology or whether the firm can create value for its customers.

Venture Capital

A key domain in which building capabilities early affects a new technology venture is in the ability to obtain financing for the venture. The United States is unique in the world in its level of entrepreneurial activity. A critical factor in the ability of new entrepreneurial businesses to develop is venture capital, particularly if the business involves an innovation with high capital needs. New firms often get started with financing from friends and family, but these loans and investments are typically limited in scope. Banks typically lend only to established businesses, not to new ventures. However, bank lending usually happens long before a firm could hope to obtain any public equity financing through the stock market. Thus, a firm that has high capital needs will often experience a gap in its financing between what friends and family can provide and what banks or public markets will provide later. This gap has limited the potential growth of entrepreneurial ventures in many parts of the world.

In the United States, a type of financing has arisen that fills the gap between friends and family when the firm is beginning and the bank or public equity financing emerges later. This type of financing is venture capital. We first review what venture capital is and then examine the critical role it plays in new technology-based firms.

Venture Capital Basics

Venture capital is a type of private equity. A firm may have stock even though it is not publicly traded. For this situation to exist, the investors must be professional investors or very high net worth individuals, and there must be a limited number of such investors. Thus, a professional investor such as a venture capitalist can invest in the company and obtain part ownership in the firm. The presence of venture capital in the United States has allowed firms such as Microsoft and Intel to develop because there was financing available throughout the early growth of these firms. The source of the venture capitalist funds are typically pension funds and insurance companies. These institutions have a large amount of capital to put into a full range of investments to create a broad-based (diversified) portfolio. The pension funds and insurance firms have the capability to understand and invest in low-risk/low-return investments such as real estate and the ability to invest in public equities. Where they have had difficulties is in understanding and making successful, direct investments is high-risk/high-return ventures such as technology start-ups. Therefore, they invest with venture capitalists as an alternative investment class in their total portfolios.

One principle in venture capital financing is that all parties face risk in the investment and also enjoy the benefits. Thus, the venture capitalist is also expected to invest in the limited partnership in which the funds are organized. Consistent with this view, the venture capitalist will then invest in entrepreneurial ventures where the entrepreneurs also have a high percentage of their net worth invested in the venture. Typically, venture capitalists are more concerned about the entrepreneurs' percentage of their total wealth invested rather than some absolute dollar amount because they want the entrepreneurs to be highly motivated.

Venture capitalists' returns are based on how well the venture does because they typically have what is referred to as *interest carried*. This means that venture capitalists will obtain a certain percentage of the final return on venture capital funds. Thus, while venture capitalists get a small annual fee to operate the fund, their profit typically comes from the return the fund generates at the end of its life. One outcome of this arrangement is that venture capitalists are typically not passive investors. They are trying to maximize the success of the venture. Clearly, the result is a highly motivated venture capitalist as well as a highly motivated entrepreneur, each seeking to make the venture a success.

The venture capital limited partnership typically lasts no more than ten years because venture capitalists do not want to be partners forever. Instead, they want to exit the investment either through selling the firm to a strategic buyer or through a public listing of the firm. The annualized return on venture capital averages 27 percent, three times the average public equity returns over time. The success rate of venture capital supported new ventures is approximately 80 percent. This is in contrast to the 80 percent failure rate that is often reported for new ventures in general.

Venture Capital and Capabilities

The most critical element for a venture capitalist funding a new business is the leadership of the proposed new venture. The classic comment by venture capitalists is that they do not invest in the wrong venture; they invest in the wrong people. The venture capitalist typically will review more than 100 business plans before investing in one of the ideas. There is no lack of good ideas, but the success of the venture hinges on the leadership of that venture. Thus, human skills and knowledge capabilities determine the success of the venture.

For the new venture, it is critical to focus on obtaining individuals with skills and knowledge before seeking the venture capitalist investment. It is also important that the new venture be able to clearly articulate that these individuals are present and vested in the success of the venture before seeking investment.

OTHER TYPES OF VALUE PROCESSES

Until now, we have focused on the creation of value by the organization. This is the most common type of value associated with the management of technology and innovation. However, there are other types of value that must be managed by technology-focused organizations. Figure 9.4 delineates the four types of value processes along with the driving force, operating measures, and potential financial measures for each. We discussed value creation earlier, so we will discuss the other three processes here—valuation appropriation, value protection, and value destruction.

Value Appropriation

Appropriation is a process by which the firm seeks to imitate others or leverage its technology to go into a new domain. Earlier, we discussed several methods to aid in imitation of other firms as you enter a new domain such as **benchmarking** (Chapter 5) for best practices. In benchmarking, the firm seeks out the best products or processes in other product units or against other firms and seeks to find what they are doing that it can imitate either to

Value Process	Driving Force	Operating Measures	Financial Measures
Creation	Innovation; New venture; Risk taking	R&D; Investment; New markets	Sales and profit growth; Moderate returns
Appropriation	Imitate; Reassign/acquire resources	Product and geographical extensions	Growth; Increasing/high returns
Protection	Barriers; Market power	Market share; Efficiency	Stable (high) returns
Destruction	Competition; Divestiture; Gaming	Cut investment; Lost share; Dramatic action	Decline; Various profit outcomes

FIGURE **9.4** Value Processes

produce a better product or do an activity better. Thus, from benchmarking, the firm learns what activities it should imitate.

Another common way to capture value is through building platforms of products based on technology or process. A **product platform** (see Appendix 3 at the end of the text) is an integrated set of subsystems that allows a variety of related products to be produced. As a result, a firm is able to leverage its existing technology to move into new domains. When Hewlett-Packard developed the technology for inkjet printing, it did not develop a single inkjet printer; instead, it used that technology and the systems around it to produce a wide variety of products, each of which met a set of customer needs. For start-up firms, the production of a product platform can be prohibitively expensive. However, the potential for the entrepreneurial venture is greater if a full range of products is produced. The presence of a product platform also helps create entry barriers to new entrants because it does not allow a new entrant to find an opening to connect to customers. Therefore, building a full range of products can help protect the venture. For example, Google has done this through its development of various specialized software tools such as Google News. In this product, as you recall from earlier in this chapter, the firm brings news stories together that are related and found through Google search processes. Thus, another product for customers using the company's search engine has emerged. These various products fit together and form a product platform for Google.

Value Protection

The process of value protection seeks to build barriers to others entering a market, or developing market power. One strategy for pursuing protection is **horizontal acquisition**, or acquiring a firm similar to yours. For example, the

2009 acquisition by Verizon of Alltel—both of which are wireless phone operators. A horizontal acquisition would be a firm more similar than a related acquisition. A horizontal acquisition contrasts to a **vertical acquisition**, which would be the acquisition of a firm that is either a supplier or customer in the value chain. Thus, if a firm like Amazon actually acquired a book manufacturer it would be a vertical acquisition because it would be acquiring a firm that produces one of Amazon's inputs. The economies of scale from a horizontal acquisition can help a firm become dominant in a technology or product line.

A firm can undertake other actions to create value protection. The firm may develop a sustainable source of competitive advantage through a better set of technical capabilities. These technical capabilities must be identifiable, well developed, and exploitable. They should allow the firm to build barriers and/or competitive weapons that will allow the protection of its differentiated technological capabilities. For example, a firm can develop an expertise in a given technological domain such as miniaturization, which allows it be the best at this activity.

The operating measures for the value-related processes are market share and efficiency. If the technology is viable and continuing, this type of value process should lead to high, stable returns. The problem with a value protection process is that the firm may become too comfortable with the success of the technology that is in place and miss changes in the environment.

Value Destruction

Obsolescence may require the firm to destroy value in one area to gain resources to undertake value creation and/or appropriation in another. The firm needs to have this destruction to focus its attention and resources in a given area that offers the greatest opportunity. In the 1990s, IBM had to use value destruction to survive. The firm had to undergo radical changes that involved restructuring units and laying off employees for the first time in its history. The actions were necessary so that IBM could promote new products that would allow it to prosper in the future. IBM has continued this process of constantly changing itself to stay on the cutting edge. For example, in 2004 the firm sold its laptop unit to the Chinese firm Lenovo. Today, IBM is largely a computer services company rather than a computer manufacturer.

The types of actions relevant to value destruction also are part of a wider effort to create organizational revival. Organizational revival is a turnaround, and is discussed next.

Turnaround

A firm may build a value adding business that has a sustainable advantage based on the appropriate capabilities, but over time, that firm may find it is in trouble. A firm may find that its performance and strategic position are such that it needs a dramatic change to survive. The reversal of such a firm's situation is a **turnaround**. It is difficult to achieve a turnaround. There are a number of dimensions of the firm that typically have to be addressed to successfully achieve a turnaround, including retrenchment, speed of response, leadership of the organization, and operating versus strategic turnaround strategies. Each of these dimensions is reviewed next.

Retrenchment A firm in decline may be in need of a dramatic change to survive. The first step in this situation is that the firm needs to recognize it is in decline and then seek to gain control of its cash flow. The common actions that occur in such a process are that the firm will seek to delay payments to suppliers and creditors and will quickly reduce any unnecessary expenses, such as a corporate jet. The firm will also cut operating costs, such as personnel costs. These initial dramatic actions are referred to as **retrenchment**. Occasionally, practitioners may also use the term "stopping the bleeding" to describe the retrenchment effort. The analogy is to an emergency room where the staff must take immediate steps to save an individual's life. Following those initial steps, more detailed and long-term steps can be taken, but some actions must occur immediately. For example, Starbucks retrenched in 2009. The model of growth that propelled it to such success had run its course. As a result, Starbucks began rethinking its store models and has even changed its product mix by introducing instant coffee. The chain has also changed how it promotes itself. Now, it is not promoted as a high-priced, status symbol coffee, but rather as a luxury that is still affordable. The firm is also reshaping how it deals with real estate by selling off much of its real estate holdings. Thus, Starbucks is changing its product line, its available services, and adding music in some locations in an effort to regain its positive momentum.[12]

Speed One issue for a firm needing a dramatic turnaround is that the faster the firm can move, the better. Firms in decline have been described as experiencing a spiral of decline.[13] One thing leads to another in the declining firm, and it may risk further losses such as losing its most productive employees if it does not reverse direction quickly. The employment market is always seeking employees who have the greatest capability. Thus, in an environment where employees are under severe pressure and the future of the firm is not clear, the best employees may leave. The organization needs to act quickly, or it will complicate its situation by losing its best capability—the employees it can least afford to lose. In 1996, IBM had lost 80 percent of its stock value and was losing ground to a number of competitors including Microsoft and Intel. Lou Gerstner became CEO at that time. Gerstner quickly cut costs and began its recovery.

For IBM, the result of this focus on speed resulted in Gerstner responding once to a question about the "vision of IBM" that the last thing IBM needed was a vision. This response indicated that the firm needed to act immediately and move where it could retrench; if it did not do that, it would fail. Thus, spending time developing a vision for the ultimate business was not what he focused on; instead, he focused on action. Later, when the crisis was easing, Gerstner spent time developing a new vision for IBM.

Leadership The individuals who lead the firm into a need for dramatic change typically are not capable of leading that change. These individuals' view of the world is what led to the problems in the first place. If they knew how to turn the firm around, they would do it. Thus, the top management of the firm is often changed if a turnaround effort is needed. Without a change in view and leadership, the cause of the difficulties may never be recognized.

Lou Gerstner's background was not computers. He had headed the Travel Related Services Group for American Express (Amex). Then he left Amex to lead RJR Nabisco. He was the first CEO of IBM who did not have computer-related degrees and experience. The result was that he brought to the firm a new perspective on what needed to occur. He brought a new paradigm that helped solve the firm's problems in new ways. In particular, he saw operational and marketing problems were hurting the firm. Previous managers had focused on technical solutions that principally involved building larger and faster mainframe computers.

Strategy Once retrenchment occurs, a firm needs to pursue longer-term turnaround efforts to complete the change in the organization. These longer-term strategies can be at a strategic or operating level, however, they must respond to changing market conditions.[14] The strategic type of turnaround seeks to move the firm in new strategic directions. Thus, a firm may decide to diversify into new areas or to reduce the diversification that it already pursues. Alternatively, an operating turnaround strategy seeks to continue the same strategic activities that the firm was already pursuing while increasing the efficiency of those activities.

For IBM, Gerstner primarily pursued operating turnaround strategies. Initially, when he took over IBM, he had key employees survey the market and find out what was missing. They quickly realized that while there were many firms supplying one piece or another of the IT needs of potential customers, no one firm supplied all of the various IT needs. The result was the birth of the concept "The IBM Solution." This operating solution built on one of IBM's core competencies: customer service. It required that new methods of operation and coordination be developed. The old system of each unit of the corporation acting independently was no longer tolerated. The outcome of these operating changes was the firm recovered, as did its stock. Gerstner provided IBM with the grounding it needed to realize its abilities in the marketplace. For the last several years, IBM has had a clear strategic logic (The IBM Solution), appropriate structures and processes (customer service orientation and bundling of products), motivated employees, and fit within the organization and in its environment.[15] The result was a successful turnaround effort. The flexibility and creativity created in that turnaround effort continues today as the firm has evolved and become a computer services firm.

Nortel Networks

Nortel Networks is a telecom equipment maker that was hit hard by the decline of the telecommunications industry in the late 1990s. However, its CEO, Frank Dunn, pursued an aggressive turnaround agenda. He continued to expand the firm's traditional market for carrier equipment in new geographical areas such as Asia. He also continued to focus R&D efforts on very attractive market niches such as Voice over Internet protocol. The firm then also pruned nonessential staff and sold noncore businesses, which lowered costs. The result is that the firm appeared is a survivor in an industry that has seen some hard times.

Unfortunately, it was later discovered that the firm had used questionable accounting practices. The impact was to overstate the positive results, which resulted in heavy bonuses for the top management of the firm. When this was discovered, the CEO, CFO, and controller were forced to resign and now face potential criminal charges. The difficult part for the firm is that the path to turnaround appeared solid but not as quick as the overstatement of the financials made it appear.

An interim CEO, William Owens, who was a member of the board of directors, was appointed CEO. Mr. Owens previously had been the president of a satellite firm, Teledesic. It is unclear how the loss of the CEO of the firm under an ethical cloud will impact the firm's ability to attract top innovative talent and its alliance partners. In 2005, Nortel appointed Michael Zafirovski as CEO. Nortel has undergone a restructuring of its core businesses since then. They have divested a number of businesses and have acquired others in an effort to align themselves with what they view to be the future of telecommunications applications—hyperconnectivity. Hyperconnected workers use seven or more devices for work and personal use in order to text message, instant message, web conferencing, social networking, and calling. The myriad of changes at Nortel has led to the creation of a Chief Restructuring Officer position. Today Nortel has emerged from the taint of the unethical activities. The firm is strong enough that it was the 2010 Winter Olympic Games Official Converged Network Equipment Supplier.

1. What do you think the impact on the firm was when CEO Dunn resigned under an ethical cloud?
2. What do you think Nortel learned in the early years of the 21st century? What else can they do to enhance their business model?

References

Austen, I. 2005. Nortel names Chief Executive from Motorola. *New York Times*, October 18, C.9.

Solomon, H. 2003. Nortel Networks set to rebound? *Computer Dealer News*, 19 (4): 1–2. www.nortel.com

SUMMARY

This chapter has integrated the various elements discussed throughout the text to develop an understanding of how a firm develops a sustainable competitive advantage. The key to a sustainable competitive advantage is the capabilities of the firm. The capabilities are skills and abilities the firm develops internally that ultimately allow it to do things its competitors cannot easily match. The sustainable competitive advantage should allow the firm to create value. Other aspects of value creation for a technology firm were also introduced in the chapter, including the establishment of industry standards, venture capital, and turnaround.

MANAGERIAL GUIDELINES

To develop a sustainable competitive advantage through the management of technology and innovation, there are several things that managers should do.

1. Use competitive intelligence to provide an early warning process for opportunities and threats that could influence the type of value processes needed.
2. Make the organization better able to respond quickly and appropriately by monitoring the industry and competitive environment.
3. Use relevant and timely information to make decisions.
4. Provide a systematic audit of the organization's competitiveness and the firm's relative position.
5. Risk being distinctive in how you view and react to the environment. This will change the environment.

6. Be sure members of the firm and key stakeholders can describe competitive advantages in easy-to-understand words. It is hard to develop, maintain, or protect what cannot be described.
7. Make the competitive advantage as unique a resource as possible so that it is hard for competitors to copy.
8. Be persistent and consistent in pursuing sustainable competitive advantages through processes and procedures in the organization.
9. Design and implement a proactive competitive intelligence process to determine what competitors are doing now and what might happen in the industry and in the environment in the future.

Guiding Questions

Examining the following issues and evaluating the resulting information should move the management team toward success in building the capabilities necessary for success.

1. What are the capabilities of the firm today, and how do they mesh with the predictions for the future?
2. Do the compensation systems and other reward systems support the development of the necessary future human skill and knowledge capabilities of the firm?
3. If a gap exists in the necessary capabilities to create value for the firm, are there ways to fill those gaps quickly?
4. What unexpected changes in the environment could shift the capabilities that are required by the firm?

CASE **9.1** THE REAL WORLD
Sony

Sony Corporation is one of the leading firms in Japan and in the world. However, the firm has undergone a major restructuring in recent years in an attempt to rebuild its capabilities. The key to this shift is the insight of CEO Howard Stringer—the first non-Japanese CEO for the firm. Mr. Stringer recognizes that there is a significant need to continue to change the firm to remain a world competitor. As part of this effort, Sony plans to halve the number of its suppliers to save 500 billion yen ($5.2 billion) in 2009 alone. In addition, he has continued some of the nontraditional activities his predecessor, Nobuyuki Idei pursued. For example, Sony Hawaii cut 44 percent of its workforce, through buyouts and layoffs in 2009. This type of action is culturally counter to the lifetime employment that typified many large Japanese businesses. As these firms have become more global, the cultural norms of the founding country have been strained.

The changes during Sony's restructuring effort have been widespread. Historically, the firm had many world-class analog engineers. However, when the industry standard moved to digital, the capability of those engineers to compete fell radically. The leadership of many business units had no framework to understand how to compete in the new competitive environment. The result was that management had to find new leaders in those areas of the firm, which brought a new focus on leadership and how to develop those leaders to the organization.

Several actions were outcomes of this need for new leadership. The recognition that there was a shortcoming in its technological standing due to changes in the environment allowed Sony to make changes in its very conservative culture that would not normally have been possible. The need for change and the need to take risks became increasingly clear to most of the firm's employees. Thus, when there was a suggestion to put game chips in the firm's televisions, Sony paid attention to the potential of the idea rather than being constrained by the idea of what televisions were in the past.

Another outcome was structural changes in the firm. The changes in the company resulted in units being combined in a new structure. This new structure allowed greater communication between the units. To illustrate, there were previously separate units for TV projection, picture tubes, and flat panels for the TV. These units were blended in the new structure. The new structure allowed cross-functional teams that help produce new products that are more innovative and come to market faster than the firm was able to do previously. As a result, Sony has seen its Bravia line of liquid-crystal-display television sets become a quiet hit. That has helped restore some luster to the company's electronics division.

Consistent with this structural change, Sony began to shift its focus from buying its semiconductor chips from others and now produces them internally. The concern for Sony is that it needs to produce proprietary

CASE 9.1 *(continued)*

products that are unique and for which it can charge a premium. Almost by definition, if the firm is not producing those semiconductor chips itself, it cannot claim that differentiation. The development of these chips requires a capability for creativity that was present in Sony at one time but had largely atrophied. Therefore, Sony has had to reinvigorate the creativity in the firm. To take full advantage of the semiconductors, the firm is also developing the software necessary to employ its proprietary semiconductors.

Thus, Sony has sought to develop or redevelop its capabilities in leadership, culture, structure, and technological know-how to make the firm competitive in a world environment that is rapidly changing. Those changes had previously left the firm less competitive than it desired. Sony is attempting to catch a moving target as it seeks to regain its leadership role in world consumer electronics.

1. What do you think the impact of Sony being a Japanese company has had on its effort to redevelop its capabilities?
2. Do you think Sony will be successful? What would you suggest they do next?

References

Anon. 2009. Business: Breaking free; Corporate restructuring in Japan. *The Economist*, 392(8636), 67–68.

Dvorak, P. 2004. Videogame whiz reprograms Sony after a 10 year funk. *The Wall Street Journal* (Sept. 2): A1, A4. www.sony.net/SonyInfo

CRITICAL THINKING

Relating to Your World

1. What would you say are the five most important things that a manager needs to understand to build organizational capabilities? How should the manager determine whether those things are happening? Discuss what the manager can do to ensure the organization is ready to take advantage of future technology-focused opportunities.

2. We have discussed building capabilities. Are there differences in the capabilities needed between an innovative firm and a firm that builds alliances and makes acquisitions for technical advancement? Explain.

3. What advice would you give a manager who was chosen to develop new capabilities in an entrepreneurial firm? In an established firm that is retrenching? Which do you think would be more difficult? Why?

WWW EXERCISES

1. Use your favorite search engine on the Internet and look for postings of technology-driven retrenchment. What are the reasons given for the mergers and acquisitions? What is described as coming from each of the organizations? Given what you know about turnaround strategy, are the reasons given for the

retrenchment reasonable and likely to lead to success? What might be missing?

2. Find an example of an organization with multiple strategic business units. How does the organization define its corporate goals? The strategies of the SBUs? Are the strategies seeking innovation and technology enhancement internally or externally? Why do you classify the SBUs as you do?

3. Find an article or website that provides guidelines for building and/or renewing technical capabilities. What do you think of the advice given? Compare the advice you find to the advice your classmates find.

AUDIT EXERCISES

1. If your organization were to enact new technology, what actions would you suggest to ensure your firm would get the results it hopes for? How would you determine whether your process for implementing the building of capabilities is fair, timely, and successful?

2. Earlier in the text (Chapter 2), we discussed measuring performance using financial data. Capabilities do not lend themselves well to such measurement. In Figure 9.1, there is a list of strategic capabilities. How would you measure each of the capabilities? Be specific.

DISCUSSION QUESTIONS

1. Why are capabilities and value creation so interconnected with the management of technology and innovation? Be sure your answer is in the context of its strategic importance.

2. Hypercompetition and the speed of change in turnaround are both related to quickness in the management of technology and innovation. Discuss the five most important reasons managers should be concerned with speed. In addition, what are the disadvantages of being too fast?

3. What are the roles of the different value processes in the strategic management process of technology and innovation?

4. What are the major decisions that impact the strategic management of technology and innovation?

5. Compare and contrast competitive advantage with sustainable competitive advantage. What does *sustainability* mean to the firm for future strategic planning?

6. What would be some of the strategic issues that a firm like Google would need to focus on as it seeks to improve its strategic capabilities in the management of technology and innovation?

PART FOUR OPENING CASE: GOOGLE

1. How has Google used its capabilities in technology and strategic management to be successful?

2. What strategic concerns would you have for Google in the future? What are its biggest threats? Opportunities?

KEY TERMS

benchmarking 281
horizontal acquisition 282
intrapreneurial 278

product platform 282
retrenchment 284

tactics 273
turnaround 283
vertical acquisition 283

NOTES

1. http://www.tivo.com/abouttivo/aboutushome/index.html.

2. D'Aveni, R. 1995. *Hypercompetitive Rivalries.* New York: Free Press.

3. Crocket, R. 2004. Wireless. *Business Week* (3888): 78–80.

4. Williams, J. 1992. How sustainable is your competitive advantage? *California Management Review,* 34: 29–51.

5. Porter, M. 1985. *Competitive Advantage.* New York: Free Press.

6. Sakkab, N. 2002. Connect & develop complements research & develop at P&G. *Research Technology Management,* 45 (2): 38–45.

7. Lafley, A., and R, Charan. 2008. Making inspiration routine. *Inc,* 30(6), 98–101.

8. Moran, P., and S. Ghoshal. 1999. Markets, firms, and the process of economic development. *The Academy of Management Review,* 24 (3): 390–412.

9. ____ 2005. Business: In a new light; Philips. *The Economist.* 377 (8446), 71.

10. Capaldo, G., L. Iandoli, M. Raffa, and G. Zollo. 2003. The evaluation of innovation capabilities in small software firms: A methodological approach. *Small Business Economics,* 21: 343–354.

11. U.S. Small Business Administration Office of Advocacy. Small Business FAQ. http://www.sba.gov

12. Caldwell, C. 2009. Reaching for the Starbucks. *Financial Times.* February 21, 7.

13. Kanter, R. 2003. Leadership and the psychology of turnarounds. *Harvard Business Review* (Jun.): 58–67.

14. Alsever, J. 2009. How to Innovate: A step-by-step guide to fostering business creativity. *FSB: Fortune Small Business,* 19(8), 68.

15. Gerstner, L. 2002. *Who Says Elephants Can't Dance?* New York: HarperCollins.

CHAPTER **10**

Knowledge Management and Organizational Learning

OVERVIEW

One of the keys to success for a technology-focused firm is the ability to learn and manage knowledge. Learning in an organization is more difficult than it may first appear. There are different types of learning, and the type employed needs to match the culture and structure of the organization. Thus, the firm needs to have the ability to know what information it actually has and how to use it to be successful. When managers can use information it turns that information into knowledge; however, the firm must be able to manage that knowledge to be successful. The ability of the organization to learn and the management of the resulting knowledge can be key success factors for a technology-focused firm. This chapter looks at both activities in depth, including:

- Activities that are part of learning
- Types of learning
- Structure's impact on learning
- Dimensions of knowledge management
- Creating fit between organizational components and knowledge management

INTRODUCTION

In Chapter 9 we saw that knowledge is a key resource. However, organizations must learn to build the resource of knowledge. As a result, a central concern in implementation is the learning that occurs among the members of the firm. Another central concern in implementation is the management of the knowledge that results from that learning.

Learning and knowledge management are related and interconnected concepts. As can be seen in Figure 10.1, **learning** in the organization involves the gathering of knowledge, from internal or external sources, and sharing that knowledge. In turn, **knowledge** involves the insights and experiences gained from gathering of data and converting that data into information. Knowledge can be either explicit or tacit. **Explicit knowledge** is knowledge that can be codified or written down and is written down as rules or guidelines. **Tacit knowledge** in contrast comes from experience and is internal to an individual.[1] Both types of knowledge have a role in learning, although they impact it differently.

This chapter reviews organizational learning as a process necessary in knowledge management. Through learning and sharing knowledge, the firm is able to bring the abilities and innovative thinking of individuals together to create competitive advantage. Then we examine knowledge management as a tool for enhancing the management of technology and innovation processes.

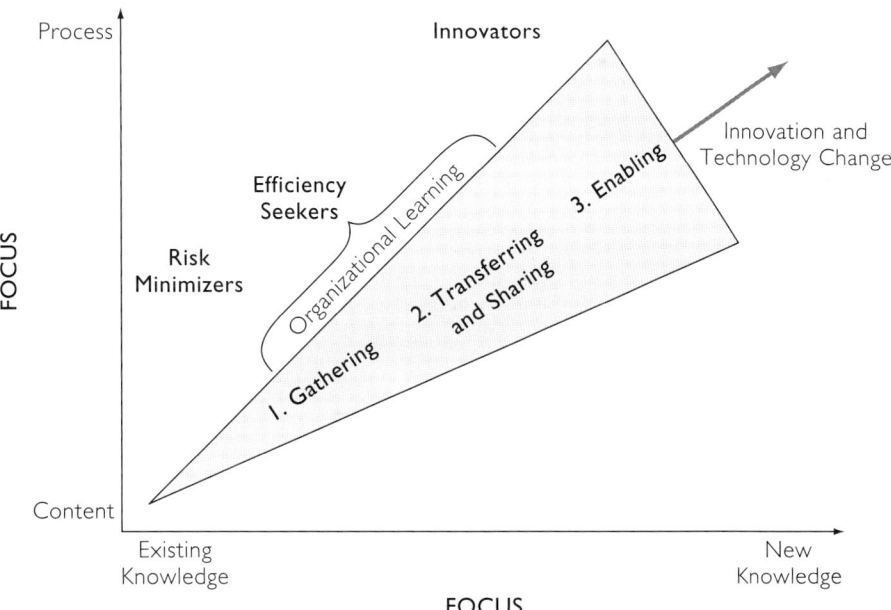

FIGURE **10.1** Firm Development in Knowledge Management

ORGANIZATIONAL LEARNING

For an organization to gain from its experiences (both positive and negative), there must be sharing of information, knowledge, and lessons about those experiences. This sharing of information requires communication and information processing, a process commonly referred to as *organizational learning* or *knowledge acquisition*. More formally, **organizational learning is** the acquisition of knowledge through the application and mastery of new information, tools, and methods. Learning should lead to "bettering" the organization. The type and the amount of learning that must take place for the firm to compete successfully depend upon the organizational and technical complexity of its internal and external environments.

Learning is essential if the organization is to adopt changes in technology and innovations. As noted earlier, resource theory argues that the root of a firm's competitive advantage is in the resources that it develops and that other firms do not have.[2] The resources can be tangible or intangible. Tangible assets are things that can be measured, such as financial resources or the location of the outlets. Intangible assets are things that cannot be easily measured, such as the culture of the organization. Intangible assets are the key to a firm's competitive advantage because they cannot be matched easily.[3]

The firm's ability to learn is a critical intangible resource. As shown in Figure 10.1, organizational learning ultimately depends on three stages of activity.

1. Gathering of data and information
2. Transferring and sharing information through communication
3. Enabling learning in the organization

The nature of the processes that occur in each of these stages impacts the type of knowledge sharing that takes place and the emergence of new knowledge within the organization. We will examine each of these stages of learning next.

Gathering Information

For an organization to learn it must have data. The greater the number and range of things to gather data on, the greater the potential information processing and organizational learning that can occur. However, if the complexity in managing a firm's knowledge processes is great then the potential for failure is great also.

A critical part in the gathering of information is scanning the environment. We have noted previously that scanning the environment is important to the management of technology and innovation. You will recall that scanning involves the examination of a wide range of issues in the environment that might impact the firm (see Chapter 2). For example, scanning the external environment tells the firm about issues such as competitors in the environment, their actions, and general trends in technology. The firm should also scan internally to understand the tasks and processes in the firm and the skills currently in the organization as well as skills and resources needed in the future.

The information gathered can help the organization meet its goals. Scanning will help the firm identify the critical issues, which the organization can then monitor to ascertain the impact of the issues and changes needed. The information gathered through scanning may also result in changes in goals and their priorities. The scanning process provides information that indicates that the goals of the organization are no longer appropriate if the environment has changed dramatically. Thus, the firm should change its goals if the environment has changed significantly. The firm may find that the environment has changed, but in response, it does not need to change its goals but simply change the priority of a particular goal or its timing. Through information gathering the organization also learns (knowledge acquisition).

The various information-gathering processes should answer the two questions that were introduced in Chapter 2 and that have been addressed in various ways throughout this text.

1. Where are we now? The answer to this question involves the internal factors mentioned earlier as well as the firm's place in its environment.
2. Where do we want to be? This includes gathering information about goals, potential future activities, and the types of skills and processes needed to meet the goals as well as what competitors might do.

We have stressed throughout the text that culture plays a critical role in the organization. Culture also plays a critical role in how the firm gathers and interprets information. To illustrate, if an organization has a very conservative culture, it may find limits placed on the type of information that is gathered. In addition, the sharing throughout the firm is restricted; that is, information through the firm travels by the paths shown in the organization chart, not by where the information needs to go. The result can be very negative with key information missed, or lost in the bureaucracy of the organization. Too often, the same people use the same sources to gather the same information multiple times. As a result, organizations are stuck in ruts of information gathering and miss important opportunities. Even worse, the management of the organization misses threats from the environment.

IBM, for example, completely missed the important role that personal computers would ultimately play in the computer industry. Mainframe engineers dominated the firm and they gathered information from sources that were familiar to them. The personal computer was seen as a play toy with no serious potential. The result was that IBM continued to focus on mainframe computers as consumers rushed into the market to buy personal computers. The nature of scanning and monitoring processes at the firm relied on narrow sources of information that confirmed what it wanted to believe, and significant market leadership opportunities missed. Today, IBM has exited from the laptop industry—the Chinese firm Levono bought IBM's laptop division in 2009. This decision was the correct interpretation of the industry that laptops were moving to become a commodity. Thus, today IBM is much more of a computer services firm than a manufacturer of computers.

Therefore, managers of an organization continually need to ask whether they are listening to all the relevant sources, not just the easy ones or the ones they know best. One of the reasons that new entrants in an industry are more likely to introduce radical innovations is because existing firms develop culture, structure, and information-gathering processes that are not open to the new technology, process, or product. More than twenty companies rejected Chester Carlson's new product—a photocopying machine—because making a plain paper copy seemed valueless. Because of these rejections, he founded Xerox instead of selling his invention to an existing company. The history of new technology has numerous examples of new ideas that were rejected by existing companies because the culture, structure, or information gathering processes failed to recognize the potential value of the technology. The information was there but was not communicated, or processed so that managers recognized potential value.

Transferring and Sharing Processes

Once the information is gathered, it needs dissemination in the organization. Dissemination of information is necessary before an organization can act on it. In other words, just because someone knows something does not mean that the organization "knows" the information. The information must be communicated to individuals for them to learn and apply the knowledge. **Communication** is the transfer of meaning from one source to another. The information must both be communicated and understood for it to have its full benefits. Combining information from different sources often leads to breakthroughs within the firm. Interconnectivity is critical for organizational learning to occur. Recall the example of Post-it Note development at 3M from Chapter 3. Many other examples of transferring and sharing processes exist. For example, at HP, an integral part of the culture is to leave projects out in plain sight so others walking by may examine them and make suggestions.

The more an organization embraces change through technology and innovation, the more frequent and pervasive the communication should be. Complexity adds to the amount and frequency of communication needed. In creative work environments, like an R&D laboratory, the communication will flow up, down, and across the organization much more freely than in a traditional organization. The tasks in such a creative work environment are more ambiguous and not strictly defined. Thus, communication channels need to be more fluid, and the types of communication need to be more personal (e.g., face-to-face rather than memos and reports).

The characteristics of communication processes in an organization where learning is taking place are:

1. Free exchange in, across, and between levels and functions within the organization
2. Recognition of the value of both the formal and informal networks where knowledge and information are exchanged
3. Encouragement of free exchange by all levels of organizational leadership
4. Open dialogue
5. Continual transformation of the organization systems for the processes of communication
6. Support of a culture that encourages meaningful interaction and exchange in, across, and between levels and functions

The firm's intranet (internal to the organization) typically aids this process. To illustrate, Cisco Systems originally designed its intranet to aid employees in their support of customers. However, today, Cisco's intranet has become a critical resource in providing communication in the worldwide company. The company's internal system now has over 15 million hits per year as communication speeds throughout the organization.[4] Not only does Cisco have a strong intranet system for itself but it produces Unified Communications Manager call-control systems for other firms. For example, WesBanco Bank invested almost $2 million in Cisco products and systems. The result is that on intrabank transactions and communications WesBanco is saving $1.2 million per year. WesBanco is developing other uses for the technology to make processes more efficient. Cisco is a firm that practices what it sells and is reaping the benefits.[5]

Enabling Learning

The organization has to ensure that it creates factors that enable the learning. Enabling factors involve two important aspects. First, the purpose of learning for the management of technology and innovation is to have the right information at the right place with the right person at the right time to make the right decision. To use the information effectively the information needs to be available when the individual needs it. Thus, if industry information is to have a significant impact, it needs to be part of strategic planning and the evaluation process.

The second factor that is important for learning is providing time for people to work on their ideas and their application of information to the tasks they are to accomplish. People need time to process information and make connections. Managers need to allocate time if they expect individuals to be innovative and to look for ways to improve the organization. In Chapter 3, we discussed 3M allowing time for employees to explore ideas for new products. Google has also followed this model successfully. This attitude in areas of the firm (besides R&D) can and often does lead to improvements in processes and products that would be missed otherwise.

Types of Learning

There are a number of ways to classify learning in organizations. For the management of technology and innovation, the type of learning relates to whether the firm obtains technology through internal innovation or external acquisition. Each of these strategies employs a different approach to learning. The learning approaches are not mutually exclusive, and firms may use either one or both. However, just as the firm typically relies on one strategy or the other to obtain technology, it also relies on one learning approach more than the other. Internal innovation of technology relies on interpretive learning, whereas the acquisition of technology relies on systematic learning. Figure 10.2 illustrates the differences between these two approaches.

Interpretive Learning

The interpretive approach to learning is more common in innovative firms that develop new technologies internally. This approach recognizes the ambiguity and trial-and-error enactment processes that are used in innovation

Characteristics	Innovation/Interpretive	Obtaining Technology/Systematic
Nature of Information Needs	Ambiguous—unknown, need more interaction	Uncertainty—more searching for information that exists
Primary Goal	Creation of new knowledge and meaning	Reduction in uncertainty
Actions to Take	Sense making, creative action, exploit successes, and learn from failures	Collect data, analyze data, share information, look for potential deviations
What the Organization Wants to Learn	Information that creates new knowledge for the organization	Information that indicates the reasons for acquisition are invalid

FIGURE **10.2** Approaches to Learning

processes.[6] In this view, action leads to understanding rather than understanding leading to action. True innovation requires that the organization accept some ambiguity and uncertainty. Early in new product and process development, there are multiple interpretations of what will and will not succeed. Through the testing and interpretation of results by individuals and groups in the organization, new understandings and interpretations emerge. These understandings lead to new activities, and an interpretation is made of the results of the new activity. In innovative organizations, much of the process leading to new products and new ways of doing things is iterative—done by trial and error.

The results of the interpretive approach to learning in the innovative organization can be summarized as follows:

1. To reduce ambiguity, undertake creative and innovative actions throughout the organization. This assumes that the firm allows time for such interaction to take place.
2. Distribute information in all directions—vertically, horizontally, and diagonally—in the organization. This is especially true of lessons learned from successes and failures.
3. Develop creative insights through an iterative process using multiple viewpoints. Learning occurs as different areas of the organization use their lenses to interpret the results from other functions, projects, and/or teams.
4. Look for new ways to apply known information. Learning about the basics of a successful innovation in one area of the organization should lead to exploration of other applications.

Systematic Learning

The systematic learning approach occurs typically when firms acquire technology externally through acquisitions or alliances. The understanding (learning) occurs through data collection and analysis processes. Managers employ the information gathered to make decisions and take actions.[7] When an organization is trying to improve its competitive position in areas of technology through an acquisition or alliance, it is important to take a rational view of potential partners. Too often, companies fail to realize expected benefits from these strategies because the processes used to analyze the takeover candidate or potential partner are not systematic. This approach to learning about potential partners will lead the organization to rigorous data collection and analyses that in turn should reduce uncertainty about potential outcomes. The underlying assumption is that the organization can "know" its environment and has the ability to gather and process the information. Google between 2001 and 2009 acquired 55 companies. In this process, Google focused extensively on ensuring that they fully understood the firms they were acquiring and their technology. One way that Google did this was by encouraging developers to create web applications using their proprietary databases and application programming interfaces first.

Thus, as firms begin to look to partner with or take over other firms to acquire technology key concerns included:

1. Develop a clear understanding of how potential partners can integrate with each function of the organization. Managers should distribute the resulting rigorous analysis within each functional area.
2. Have a group of specialists within the organization who understand the potential synergies and problems associated with blending people, processes, and resources from multiple companies. These specialists should be able to analyze and interpret data from both organizations to facilitate the adoption of best practices.
3. Enact a clear set of guidelines with which to interpret the potential for success if the acquisition of technology takes place.
4. Develop guidelines that will be reinforced and become more entrenched. These lessons may be useful in the near future, but the firm needs to periodically review them relative to the current and future environment.

Most organizations emphasize one type of learning over the other. Therefore, the issue becomes what type of learning will dominate. A variety of factors impact the exact mixture of interpretive or systematic learning the firm will employ. For example, we have seen that the impact of culture of the firm on sharing of knowledge, as a result we know firm culture will also influence the mixture of learning styles. Similarly, the size of a firm is also important since a smaller firm is subject to greater impact from its environment. For example, Pegasus Solutions is the dominant Internet hotel reservation firm. The firm has implemented internal and external efforts to obtain technology. However, its learning method is almost totally interpretive. The rapidly changing environment requires that the firm adapt quickly and constantly by looking for environmental changes. In contrast, a dominant firm in a domain may rely on a systematic approach. The RFID technology discussed earlier initially had a number of different technologies that were competing against each other. However, Walmart employed a systematic approach to analyze which technology it would employ. Because Walmart is such a dominant force in global retailing, once it chose a technology based on its analysis, that technology became the standard for the industry.

10.1 REAL WORLD LENS

Ericsson

Over the last decade, Ericsson, the Swedish telecom giant, has acquired new businesses and divested old businesses in an effort to keep up with the changes in the telecommunications industry. In this industry, companies gain a competitive edge through the learning capabilities and knowledge sharing abilities of their employees.

While pursuing Nortel's wideband code division multiple access (CDMA) business, Ericsson moved toward ending its joint venture with Sony and let Sony have the handset part of its business. Ericsson sees Nortel's CDMA unit as being cutting edge while the handset business is fiercely competitive and getting more so. Ericsson over the past few years has acquired Marconi, a leading telecommunications manufacturer, as well as the UIQ software business for smartphones from Symbian. These are just a few of the business adjustments that Ericsson has made to realign itself for future sustainability in telecommunications and data communications industries.

How does Ericsson keep learning ongoing with so many changes? Ericsson recognizes and practices three kinds of learning among employees: learning basic knowledge, co-creating new knowledge, and learning changing knowledge. In creating new technologies or innovative processes, learning basic knowledge is necessary. The outcome of this knowledge is a basic understanding that enables the person or team to begin work with a new product, service, protocol, or application. In co-creating new knowledge, Ericsson employees, especially engineers, often find as they interact with new products, systems, etc., they experience "eureka" moments—knowledge that did not exist anywhere before. Because Ericsson is trying to stay on the leading edge of its industries, its engineers find themselves where no one has explored. The learning of changing-knowledge is a process of continually reconstructing knowledge to solve today's issues and potentially tomorrow's.

In 2009, Ericsson made a deal with Sprint Nextel Corporation to take over day-to-day operations of Sprint's network while Sprint focuses on new products and customer retention. Sprint Nextel will pay $5 billion to Ericsson to provide these services, but will retain ownership of the network. This is the first time that a United States telecom operator has outsourced network operations. This means that both Sprint Nextel and Ericsson will need to share knowledge and learn together how to make this work.

REAL WORLD LENS (continued)

1. What kind of learning do you believe Sprint and Ericsson employees have to employ if this arrangement is to be successful? Be explicit.
2. How is the acquisition and divestment of business affecting Ericsson's knowledge base? Should they worry about losing knowledge?

References

Cheng, R. 2009. Corporate news: Sprint to outsource operations to Ericsson. *Wall Street Journal*, July 10, B4.

Doos, M, L. Wilhelmson, and T. Backlund. 2005. Functioning at the edge of knowledge: A study of learning processes in new product development. *Journal of Workplace Learning*, 17(8), 481–492.

Veverka, M. 2009. Ericsson tucks Nortel away. *Barron's*, 89(33), 21.

Structuring for Organizational Learning

One of the keys to success in implementation discussed throughout this text is having the right structure in place to encourage innovation and blending structures and cultures when obtaining technology from outside the firm. To facilitate organizational learning managers must address structural issues and cultural practices. There are specific characteristics of the firm to implement and monitor if the firm is to promote learning activities.

The organizational structure that enhances learning and knowledge sharing is flexible and decentralized. The flexibility is apparent in flatness or lack of hierarchical ladders, with decision-making lower in the organization—down to the operational level. In the learning organization, top management determines the vision for three to five years into the future, identifies where to develop value, and determines the skills needed. Operational leaders must find the skills, facilitate the learning processes, and push the organization forward. In addition, keeping up with innovation is part of everyone's job in this type of structure. Johnson and Johnson is a well-known example of a decentralized and learning organization. The core to the firm's success is creating leaders who can manage and create such a learning environment in the firm.[8] An organization that is structured for learning is characterized by the following:[9]

*   Decision-making processes based on shared and transferred knowledge with highly developed, integrated communication systems.
*   Many processes encourage boundary spanning within the firm. These include cross-functional teams and job rotation–based training programs.
*   Internal network development is encouraged within the organization.
*   Reward systems tied to team and system-wide goals. In addition, risk-taking behaviors, plus linking ideas from diverse areas of the firm are rewarded.

For the learning organization, the structure must support a corporate culture that encourages each person to expand and to share individual knowledge. This means the firm must be sure individuals:

- Have a respect for what each function does and contributes to the organization
- Are able to act within the clear vision of the firm
- Are oriented toward helping and sharing rather than protection and building silos within the firm

Keys to Successful Organizational Learning

The success of learning in the organization depends on a number of factors. These keys to success are not all-inclusive but rather they indicate what managers should look for to determine whether organizational learning levels are appropriate and happening in the firm.

1. Organizational learning is a function of a firm's competencies, culture, and comfort.
2. Organizational learning requires the communication of ideas and observations.
3. Just because one employee knows something does not mean the organization can act on it. Too often, organizations focus on position and power rather than knowledge. Learning for the organization is not the sum of what everyone in the organization knows but rather is the sum of what is shared.
4. Organizational learning needs to be stored in organizational memory.
5. Organizational learning processes need to be intentional in the organization, but the organization also needs to recognize learning by "surprise."

As noted earlier, a prime example of open source software is Linux. There is no fee for the system. Instead, it is freely available to all who wish to use it. Thus, it represents a prime example of learning that occurs through shared information and exchange of knowledge. Although such software is not firm specific, the enhancement and debugging of the software have taken place in a true learning environment with a unique culture, new ideas, and shared knowledge. Improvement of the software occurs through the active sharing of information by a wide variety of parties. Forward-looking organizations often seek employees familiar with such activities who increase not only their own knowledge base through participation in these types of activities but the organization's as well.

KNOWLEDGE MANAGEMENT

Successful management of technology and innovation requires that the firm effectively manage the knowledge gained from the learning. This management includes the tools to develop and store the information. Knowledge management involves not only maintaining existing knowledge but also developing new knowledge. If learning is capturing and sharing information, then knowledge management is maintaining and creating smart intelligence systems. The intelligence systems in turn gather data and provide excellent information for decision-making.

Just as with organizational learning, knowledge management depends on individuals within the organization interacting. The people serve as agents in the intelligence systems of the firm. Individuals participate in a number of knowledge-exchange relationships and create a number of perspectives. These perspectives help improve decision making in the organization by bringing fresh and different insights to the process. It is the abilities of individuals based on experience, education, information available, and "gut" instincts that provide the knowledge base that needs to be managed.

Knowledge-Management Definitions

Before discussing what is involved in knowledge and its creation and management, it is important we define *knowledge* and *knowledge management*. Knowledge is "a fluid mix of framed experience values, contextual information, and expert insight that provides a framework for evaluating and incorporating new experience and information."[10] There are several things that are true about knowledge. First, it exists in the minds of individuals. Those with knowledge hold the key to knowledge management in the firm. If individuals are not willing to share knowledge, then organizational learning does not take place. Second, managers can document knowledge, and knowledge can become embedded in the processes, practices, and culture of the firm. Both 3M and HP have cultures that encourage the sharing of knowledge. Third, while knowledge is developed and resides with individuals, everyone in the organization holds this knowledge. For the firm to utilize the collective knowledge of individuals, it must have a knowledge-management system in place.

Knowledge management (KM) processes combine data, information, and individual learning in a synergistic manner. This is accomplished by defining, developing, and processing the innovative and creative abilities of the firm's employees individually and collectively.[11] The organization needs to manage its knowledge in a way that leads to the acquisition, selection, organization, sharing and leveraging of business information and expertise. Knowledge management captures synergistic integration of information processing and combines it with the creativity of personnel to maximize the responsiveness and flexibility of organizations.[12] KM combines the infrastructure of the technology and organization in such a way that value is created from intellectual assets. The first step in developing such a system is to understand how knowledge is created within the organizational context.

Dimensions of Knowledge Creation

There are two types of knowledge: tacit and explicit. We defined these two types of knowledge earlier in this chapter. Explicit knowledge can be codified, whereas tacit knowledge occurs through the experience of the individual. The management of explicit knowledge is somewhat easier, although it requires communication between the concerned parties to be successful. Designing and maintaining effective communication systems are always a challenge.

Tacit knowledge also requires such communication systems, but its management is more complex than explicit knowledge because it is rooted in the actions, experiences, ideals, values, thinking, and emotions of an individual. Because of these factors, tacit knowledge is hard to formalize for transference and thus hard to share or communicate.

Nonaka and Takeuchi suggest people create knowledge through four different activities or modes. These are:[13]

1. Socialization, the exchange of tacit knowledge between the different parties in the organization; this requires not only communication between individuals but also an understanding of the values and connections between the individuals.
2. Externalization involves conversion from tacit knowledge to explicit knowledge; this requires communication of individual knowledge to explicit knowledge that others can gather and use.
3. Combination involves the synthesis of explicit knowledge from different areas of the organization. Cross-functional teams are designed to promote this type of knowledge creation.
4. Internalization involves conversion from explicit knowledge to tacit knowledge; explicit knowledge that becomes internalized to the members of the organization becomes tacit knowledge.

Making Knowledge Management a Successful Capability

As discussed in Chapter 9, building capabilities for sustainable competitive advantage is a goal of organizations. Knowledge management is a critical capability for the organization. Knowledge about how to accomplish something may exist in one part of the firm but may be underdeveloped or missing in other areas. For example, there may be processes for experimentation in the R&D lab area that are missing for the operation and production areas. When Apple Computer turned around the first time in 1985, it discovered that multiple units were working on the same product development projects at the same time without knowing that other units were doing the same things.

There are six basic principles that the organization can employ to help ensure success in knowledge management. The first is that managers within the firm need to develop a system/process for knowledge management that will allow it to move toward its strategic goals. The heart of this system is a culture that encourages the sharing of information. An organization greatly hinders its innovation potential if the culture is not supportive of information sharing and exchange, or if the knowledge is in functional silos. To take advantage of and to direct the creative energy of members of the organization, it is important that goals about knowledge systems and processes be well known and communicated. This sharing will serve as an example of knowledge management for the organization.

The second principle is that knowledge management requires, as does organization learning, good information. Too often, there are "secrets" in the organization, which means that people who need certain information cannot find it. Studies indicate that the further one moves from the center of the organization, the harder it is to find information. This can cause the company to miss opportunities or threats that could have a profound effect on performance. As firms grow to be international, the desire to better serve local markets increasingly results in dispersing staff to those markets. Therefore, good knowledge management means that the information is readily available throughout the organization even as the firm internationalizes. Firms such as Microsoft ensure that this occurs through their information system, which actively and rapidly tracks information from around the world.

The third principle of knowledge management is: The system created must make information accessible in a timely fashion. If the firm is not providing information in a timely manner then the system is not working correctly. In many firms, processes for knowledge management are developed and implemented without looking at the needs of all the users. If there are problems with knowledge management in the organization, managers need to examine the system and potentially change it. The fourth principle is that there must be individuals in the system that understand how to use the information. The information may be good and timely, but there can only be knowledge management if people know how to use it. It is up to the organization to develop a process and system that ensures individuals who know how to use the KM system exist in the organization. If the organization has a limited number of individuals who understand how to use the information, then training and development for the rest of the organization needs to occur. If no one knows how to use the information, then the organization needs to hire or develop such individuals or determine that it has the wrong strategic goals or KM system. The information itself is not right or wrong but it may not be the "right" information for the organization. The collection of information is not an organizational goal; using information effectively is.

Information and knowledge exist in a variety of places in the organization. Information and knowledge are often not with the manager or the well-educated members but with the person who is experienced in one area of the firm. This person may not know how to share the information, may not have been asked to share, or may not believe the information is important. Managers often squelch sources of information by "managing" when they should be listening. The knowledge-management system should create efficiencies seeking inputs from the most experienced employees in the firm.

Knowledge management should not only make sure that there are individuals who can use the information, but it must also ensure that the information can be used. Knowledge management requires that information is employed to be successful. Therefore, the fifth principle of knowledge management is that the system needs to be usable and there must be the ability to develop ideas from it. To do this, the system must be people friendly, and the people must perceive that it is friendly. Most knowledge systems have a technical base—decision support systems, information technology, databases, and so on—but if the systems are not understood and used, the knowledge-management system will not help the organization achieve its strategic objectives. In fact, it is quite likely that the knowledge-management system will drain resources without adding value.

Individuals who have high technical skills often develop information within the organization. Thus, an engineering department in a firm like Dell may develop the information. However, a department that principally conducts a business function such as marketing will employ the information that is developed. Such information can accomplish little if the data are not understandable and usable. Therefore, organizations need to ensure that the data are usable by those who will actually employ the knowledge.

The sixth and last principle is perhaps the most important: The knowledge-management system should help the organization develop **organizational wisdom**. Wisdom for the organization is an understanding that goes beyond data and information manipulation. If the knowledge system truly disseminates lessons learned and information for decision making, the wisdom of experience will become part of the organization. This wisdom manifests itself in strategic decisions and organizational effectiveness. The organization that reaches wisdom is more likely to lead the way in the management of technology and innovation. Established firms like Oracle and Microsoft have developed over their history an understanding of the market and various products. Such an organizational memory can sometimes make shifts in paradigms difficult because it shapes a firm's perspective. However, organizational memory can also be a valuable resource if the wisdom of experience emerges. The organization uses the wisdom not to limit its perspective but to help ground its understanding of products and developments in the light of the industry's and firm's histories. To illustrate, the Internet bubble left many firms that had not seen a decline in the market overexposed when individuals acted like the bubble would never end. In contrast, firms with the wisdom that comes from experience, like Microsoft, were aggressive in the boom market but not overly aggressive or unrealistic in their expectations.

10.2 REAL WORLD LENS

Accenture

Accenture is a global management consulting, technology services, and outsourcing company whose mission is to help clients improve performance. At the end of 2008, Accenture had more than 186,000 employees in 52 countries and annual net revenues of $23.39 billion. The company's business model requires attention to KM. The need to access and leverage the experience and knowledge of employees globally, was recognized by Accenture's chairman and CEO, William D. Green. Green stated that for Accenture, the goal is to be organized in such a way that they are constantly learning and bringing new ideas, innovations, and expertise to clients—providing new knowledge and ways to add value.

When Accenture spun off from Arthur Anderson in 2001, major reorganization was required—both in structure and in knowledge systems and processes. The KM reorganization included consolidating over 1,000 separate databases into a central technical infrastructure, as well as creating a global KM strategy and planning group. After realigning the KM architecture, measurement and evaluation of the value added by the KM processes

REAL WORLD LENS *(continued)*

became important. As Accenture moved KM into a centrally managed program, an evaluation process was undertaken. The goal was to develop a measurement framework and reporting process for KM. Accenture wanted to know the value added through its KM systems and processes.

Among the processes that Accenture used to access the value of the KM system were:

- A strategic random sampling plan that helped minimize survey fatigue
- The use of penetration statistics to provide a reasonable basis for ROI calculations
- The use of well-designed electronic surveys that take advantage of branching logic to minimize the number of questions presented—surveys that do not ask irrelevant questions to certain individuals
- The support of KM governance and working groups
- A continuous measurement model, with a consistently updated ROI and other impact metrics that could be measured using scorecards and trend analysis

1. What other benefits could Accenture garner from this knowledge-evaluation system?
2. How can this evaluation system inform future development of the KM systems, structures, and processes at Accenture?

Reference

Aaron, B. 2009. Determining the business impact of knowledge management. *Performance Improvement*, 48(4), 35–45.
www.accenture.com

There is one key characteristic that has to be present for the knowledge-management system to flourish: The knowledge memory must be open to change. A knowledge-management system that does not itself innovate and change is doomed to become an albatross for the organization. It will lose its edge in providing timely information to members of the organization.

Knowledge Management in R&D Departments and Teams

We can see the essence of knowledge management in the examination of R&D departments. Historically, R&D involved a number of researchers located and working together in a single location. One great historical example is Bell Laboratories, which laid the groundwork for most of today's telecommunications industry. However, R&D today is commonly not a team of individuals established in one location that work together over a long period. Instead, R&D today is collaboration that occurs among individuals who work on a given problem over some fixed period. These individuals today conduct R&D from a variety of locations connected by concerns with a given product or process until meeting the given need. These teams of individuals communicate through tools such as the Internet in real time. But, the key is building a team that brings multiple insights from different functional areas and different regions of the world. These multifunctional and cross-cultural teams can be difficult to manage, but they are more likely to develop innovative ideas.

These teams require strong social skills among the team members as well as appropriate communication processes. The ability to interact with diverse individuals and exchange knowledge is critical. Much of the information that needs to be communicated is tacit, or qualitative, in nature. The communication of tacit information can be more difficult than explicit knowledge. This difficulty can become even more severe as teams are internationalized and individuals from different cultures and backgrounds need to communicate with each other.[14] However, there are means that organizations can employ, whether the concern is management of the R&D team or the organization as whole, to ensure that knowledge management is successful.

Just as with so many issues in the management of technology and innovation, the team must be built for success and fit the type of issues it is expected to address. If the team is an R&D team, obviously knowledge about the scientific area and potential market characteristics can be critical. However, if the team in question is in charge of the acquisition of technology, then the financial personnel increase in importance as they try to analyze and assess value. No matter what the goal, there are certain characteristics each team member should have to enhance the quality of knowledge management within the organization: an understanding that people are assets and are the reasons organizations can be innovative and apply technology in ways that help the organization succeed. These characteristics include:

1. A vision for where the organization and the team are going.
2. An openness to new ideas and the ability to display that openness. Listening skills as well as the ability to frame and reframe ideas are important parts of this openness.
3. A strong sense of what the organization is and is not and whether the organization needs to undergo radical or incremental change.
4. A person who is passionate about the exchange of knowledge and the opportunity to learn.
5. An ability to champion ideas and energize the organization toward the accomplishment of goals.

The personnel in R&D can share their knowledge to enhance risk taking and trial-and-error processes in redesigning production processes. There are certain abilities and processes that enable knowledge management in three significant ways within organizations. Figure 10.3 delineates these abilities and processes and how they impact knowledge management. The purpose of these is to help the firm:[15]

1. Explore and discover through internal innovation.
2. Choose direction through exploring partnering options inside the firm among different business units and outside the firm through alliances and mergers/acquisitions.

Knowledge Management Goals	Supporting Competencies	Smoothing Processes	Goal Capability
Encapsulate/Store	Effective questioning—who, what, why, when, where, how Environmental scanning Common language throughout the organization	Cross-functional socialization Experimentation Planned dialogue	Identification and acquisition of knowledge sources
Search/Sort	Option identification—looking for multiple right answers Using tools such as scenario building	Evaluation of progress Supportive systems for decision making Consistent and complete auditing techniques	Dissemination of knowledge
Reuse/Leverage	Negotiation skills Build network interactions Communication enhancement	Managing changing processes Monitoring and adjusting progress	Using knowledge in unique, innovative combinations

FIGURE **10.3** Abilities and Processes That Impact Knowledge Management

3. Act on knowledge managed to meet expectations by transferring learning and skills into other areas. Using products and processes in new ways to gain new benefits is an important goal of knowledge management.

The Fit in Knowledge-Management Systems

When the firm is examining knowledge management it is important that the type of knowledge-management process and systems match its goals and needs. Whether undertaking radical innovation or making slight adjustments in the technology of a single process, knowledge management is a key concern.

Figure 10.4 delineates some of the factors when developing and implementing a knowledge system to meet the demands for information, learning, and knowledge. The information focus will be either primarily internal or external to the organization. Obviously, the more radical the change, the more the firm needs to focus on external information sources for guidance. This does not mean that internal information becomes unnecessary; it means the focus is primarily on external information. Likewise, an incremental adjustment in a manufacturing process would rely more on internal information sources.

The other axis is the level of complexity of the type of innovation or change in technology. The higher the complexity, the more important the knowledge-management process becomes. Because complexity increases the number of issues to consider, complexity increases the need for more communication and more levels of understanding within the firm. In addition, complexity usually adds to the need for more cross-functional, cross-level communication.

INFORMATION FOCUS

	More Internal	More External
High	Complex tasks with local implications; Explicit knowledge exchange across functions Example: Product improvement team—cross-functional	Highly complex, uncertain innovative task; Explicit knowledge exchange and integration of diverse sources Example: Radical new product introduced to the market
Low	Routine, low complexity, transfer of incrementally derived knowledge Example: Change in production line	A task with nonroutine demands; Some degree of uncertainty; Explicit transfer of external information to team Example: Restructuring, especially if using a consultant

(COMPLEXITY — vertical axis label)

FIGURE **10.4** Knowledge Demands

Each of the four blocks in Figure 10.4 illustrates some of the issues and the types of knowledge sharing required. This list is not all inclusive, but it does clearly illustrate that knowledge management is more than telling your boss and colleagues what you did today. For example, Accenture, discussed earlier, brings external information to its clients. That information for Accenture is relatively low in complexity. However, when Accenture developed its KM evaluation system, it found itself facing a complex, internal knowledge demand. Paradoxically, the effective evaluation of KM is difficult and requires excellent KM.

USING ORGANIZATIONAL LEARNING AND KNOWLEDGE MANAGEMENT

Organizational learning and knowledge-management systems can be critical success factors for innovative firms. The management of such a process is complex. It requires that the organization seek to integrate flexibility into its planning processes and a system to ensure that the full range of information is gathered and that the organization learns from that information.[16] Figure 10.5 summarizes some of the issues in trying to facilitate the development and implementation of a knowledge-management system. The keys, like with any development and implementation of new technology, are answering the following questions:[17]

1. What is the relative advantage of undertaking the new approach? What will we gain?
2. What is the relative disadvantage? What will we lose? Any time you implement a new way of managing knowledge, there are some in the organization who will lose power because their knowledge will become diffused or less useful.
3. How complex is the change? Can it be broken into manageable parts?
4. Can we test the new system of managing knowledge? There is more trepidation if there have not been smaller trials.

	Traditional	KM System
People Issues	Control behaviors; Set up predefined goals and procedures	Community of capable individuals with diverse thinking
Company View	Adherence to "how we do things" view	Continuous, multiple perspective assessment; Looking for alignment and fit
Structural Orientation	Top down, trickle down strategies; Hierarchical structure	Proactive involvement at all levels; Flexibility; Network structure
Orientation	Follow processes and procedures; Unit isolation	Community of practice; Cross-functionality
Formality	Higher	Lower

FIGURE **10.5** Facilitating Knowledge-Management Systems

5. What are the "fuzzy" boundaries? How adjustable is the new KM system once it is put into place?
6. What are the risks—of implementing a new way of managing knowledge and of not implementing it?
7. What knowledge is required?

In the knowledge-management system, the community of capable individuals is a key concern. After all, the knowledge of individuals in the organization forms the base for learning activities in the firm. Learning is the key to the knowledge available. Knowledge management is central to making individuals' knowledge available so people can learn from the experiences of others. By combining the knowledge bases of individuals, the organization can leverage knowledge to develop new and better products and processes.

How the organization implements its strategic technology and innovation processes to create value and competitive advantage has been the focus of this book. The firm needs to make sure that the processes of gathering information and sharing knowledge accomplish the desired goals of enhancing its ability to sustain its competitive advantage. Evaluation of what the future holds and control of organizational processes are the keys to maintaining the knowledge system that will aid in creating value for the future. Appendix 4 at the end of the text will address how to look to the future.

SUMMARY

Organizational learning and knowledge management build on each other and are critical to the success of the technology-focused firm. The processes for both activities require extensive efforts to ensure that a wide range of concerns from culture, organizational structure, and management of people encourages both learning and knowledge management.

MANAGERIAL GUIDELINES

We have discussed how to build a learning organization and the capabilities that knowledge management can help the firm build. However, just as in other areas of technology and innovation management, there are certain cautions to keep in mind. When building a knowledge-management system, there are certain mistakes that can hurt the credibility and the outcomes derived from the system.[18]

1. Failure to coordinate efforts among information systems, technical processes and procedures, and human resources is a key potential mistake. Knowledge management is an integration between technology and people. Too often, the technical core begins the process of building a knowledge-management system, and then administration finds itself trying to catch up or developing a system that is not compatible with the technical core of the firm.

2. Starting too small is another common failure. When implementing a knowledge-management system, the initial project needs to be significant. If it is not, then the value of the system is not recognized. Furthermore, the capabilities of the knowledge processes are not well developed. If an organization starts off too small, the benefits of knowledge management are lost because the capabilities are not built in from the beginning.

3. Not changing the reward structures and culture of the firm to recognize excellence in teamwork is another common failure. If you want information and learning shared, then you must reward it. The traditional reward systems are based on individual accomplishments exclusively. In knowledge organization, these systems need to be based on how well information is gathered and shared, how much learning occurs, and how knowledge is used.

4. Building a great database is not knowledge management, but many firms seem to think it is. For a firm to be truly successful in leveraging its knowledge resources, it must build "communities of practice" among those who regularly work together or who have common tasks or goals. Having a great database does not mean it will be used. Knowledge management requires utilization.

5. Assuming that knowledge management will occur because it is important is another common mistake. Organizations do not accomplish goals without someone taking charge. Knowledge management requires change-management skills, and change needs a champion. Knowledge, like learning, does not automatically lead to beneficial growth. It has to be managed.

Guiding Questions

The process of building the knowledge-management system involves systematically mining and interpreting individual knowledge. This is a complex process but can be aided if the organization asks key questions that will reveal the network of information gathering and processing.

1. To accomplish your work, what do you have to know?
2. Where did you get the knowledge on how to do these things?
3. What information or data do you need to accomplish your work?

4. When you are seeking this information or data, where do you look? How easy is it to find the information or data you need?
5. What would make it easier?
6. What information do you feel would make your work easier if you had it?
7. What information do you have or regularly get that seems irrelevant to your work?
8. What information/knowledge systems are available to you?
9. What do you like about the knowledge systems that you have been exposed to? What do you dislike about them?
10. How can we develop a knowledge-management system that you would use?
11. What are the keys to developing a knowledge-management system that will add value to your work and to the organization?
12. How flexible is your knowledge-management system?

CRITICAL THINKING

Relating to Your World

1. This chapter discusses how organizations need to promote organizational learning and knowledge management. One of the points emphasized was that organizational learning and knowledge management are based on individual organizational members sharing what they know with others. Using the following four areas, list specific actions that a manager can implement to encourage such sharing.

   a. Planning
   b. Implementation
   c. Evaluation
   d. Control

2. We discussed a firm developing "wisdom." What do you think are the key steps to such development? Are there certain cultural aspects of a firm that would encourage the development of wisdom? If you were a manager, what would you try to do to encourage the development of organizational wisdom?
3. What advice would you give a manager who is charged with developing a knowledge-management system? What do you believe are the key ingredients of such a system? What approach would you use? Why?

WWW EXERCISES

1. Identify a knowledge-system design process used by a company or organization. What were the goals of the system? What problems did the company encounter? What were the goals and benefits of developing the knowledge system? Did it help the organization's performance? How?
2. Organizational learning has been discussed a great deal in the management literature. Find a website that discusses the process of managing organizational learning and knowledge systems. Does this process seem doable? If so, why do organizations struggle with such development? If not, what advice would you give?
3. Find an article or website that provides guidelines for developing a knowledge-management system. What do you think of the advice? Compare the advice you find to the advice your classmates find.

CASE **10.1** THE REAL WORLD
University of Pretoria—Faculty of Veterinary Science

The University of Pretoria, Faculty of Veterinary Science has a very rich and diverse heritage. Some of their resources are unique, and employed widely by others in Africa and around the world. Therefore, it is critical to safeguard them while enhancing their accessibility. The physical deterioration as well as the loss of information regarding these resources was accelerating, as there were no clear guidelines for the maintenance and preservation of the items at the University.

The faculty established an open access repository to digitize these unique materials. The repository uses DSpace(TM) software, developed jointly by the Massachusetts Institute of Technology (MIT) Libraries and Hewlett-Packard Labs. The software complies with the Open Archives Initiative (OAI) which allows items to be easily discovered by web search engines, and indexing tools. A path for access to the physical veterinary paper-based resources was now defined and could be implemented by the digitization of the collections.

By using defined selection criteria, the Library could identify several valuable collections to be digitized. For the faculty, there are several key areas where value will be added.

- Providing open access to valuable resources that otherwise would be hard to study, especially by outside users.
- Supporting lifelong learning through technology and improved learning initiatives.
- Possibly providing a third stream income by selling of high resolution digital images.
- Capturing of tacit knowledge about resources that might otherwise be lost.

Through the interaction and collaboration with the Department, the Library showed that it could play an important role as a facilitator in lifelong learning and conservation of Faculty output. Through digitization of the elephant collection, the library ensured the preservation of this scarce and unique resource for future use. Marketing of the collection nationally as well as internationally enhances public knowledge about and information source might otherwise have been lost. The collection also showcases the research output of the faculty of Veterinary Science and creates a general awareness of their own heritage and indigenous knowledge.

1. What were the key ingredients to the successful development and implementation of this knowledge system? What type of learning had to take place? What type of knowledge was shared?
2. What other uses could the system have for professors and students besides the dissemination of documents within the repository? How can these uses be fostered and developed?

CASE **10.1** *(continued)*

Reference

Breytenbach, A., and R. Groenewald. 2008. The African elephant: A digital collection of anatomical sketches as part of the University of Pretoria's Institutional Repository—A Case Study. *OCLC Systems and Services,* 24(4), 240–251.

AUDIT EXERCISE

According to research by Zarraga and Garcia-Falcon,[19] the creation, transfer, and integration of knowledge requires several determining factors. These factors are:

- Multifaceted dialogue in the work team
- Existence and use of organizational memory
- Individual autonomy in the daily work
- Common language in the organization

- High care in the work team
- Clarity of organizational intent

If you were a manager in a high-tech company, how would you monitor the presence of these factors to ensure that knowledge in your area of the organization was created, transferred, and integrated as appropriate?

DISCUSSION QUESTIONS

1. From the chapter, it is clear that knowledge management is critical. What steps would encourage knowledge management?
2. How do culture and learning interact?
3. Discuss the interaction between information gathering and knowledge management.
4. If you were charged with developing the specifications for a knowledge-management system in your organization, what key information would you need and how would you go about gathering and organizing it?
5. Compare and contrast data, information, learning, knowledge, and knowledge management. How important are these distinctions to everyday management practice?
6. What is the role of a knowledge-management system in innovation efforts? In obtaining technology from external sources?

PART FOUR OPENING CASE: GOOGLE

1. From the Google case, what elements of information management and learning were present?
2. How will knowledge management in Google have to evolve as the firm evolves?

KEY TERMS

communication 296	knowledge management 303	organizational wisdom 306
explicit knowledge 293	learning 293	tacit knowledge 293
knowledge 293	organizational learning 294	

NOTES

1. Lubit, R. 2001. Tacit knowledge and knowledge management: The keys to sustainable competitive advantage. *Organizational Dynamics,* 29 (4): 164–178.

2. Barney, J., with W. Ouchi. 1986. *Organizational Economics: Toward a New Paradigm for Studying and Understanding Organizations.* San Francisco: Jossey-Bass.

3. Maijoor, S., and A. Witteloostuijn. 1996. An empirical test of the resource-based theory: Strategic regulation in the Dutch audit industry. *Strategic Management Journal,* 17 (7): 549–570.

4. Howard, E. 2000. The dot.com effect. *Strategic Communication Management,* 4 (2): 16–22.

5. Duffy, J. 2008. Bank shaves as much as 40% of telecom costs using UC. *Network World,* 25(22): 32–33.

6. Huber, G., and R. Daft. 1987. The information environments of organizations. In F. M. Jablin, L. L. Putnam, K. H. Roberts, and L. W. Porter (eds.), *Handbook of Organizational Communication: An Interdisciplinary Perspective* (pp. 130–164). Beverly Hills, CA: Sage.

7. Ibid.

8. _____. 2008. Johnson & Johnson CEO William Weldon: Leadership in a Decentralized Company: Knowledge@Wharton (http://knowledge.wharton.upenn.edu/article.cfm?articleid=2003), June 25.

9. Lei, D., J. Slocum, and R. Pitts. 1999. Designing organizations for competitive advantage: The power of unlearning and learning. *Organizational Dynamics* (Winter): 24–38.

10. Davenport, T., and L. Prusak. 1998. Know what you know. *CIO,* 11 (9): 58–62.

11. Malhotra, Y. 1998. Deciphering the knowledge management hype. *The Journal for Quality and Participation,* 21 (4): 58–60.

12. Dingsoyr, T., F. Bjornson, and F. Shull. 2009. What do we know about knowledge management? Practical implications for software engineering. *IEEE Software,* (May/June): 100–103.

13. Nonaka, I., and H. Takeuchi. 1995. *The Knowledge Creating Company: How Japanese Companies Create the Dynasties of Innovation.* Oxford: Oxford University Press.

14. Gassmann, O., and M. von Zedtwits. 2003. Trends and determinants of managing virtual R&D teams. *R&D Management,* 33 (3): 243–262.

15. Beckett, R. 2004. Stimulating and evolving knowledge-oriented improvement processes in a business enterprise. *Journal of Manufacturing Technology Management,* 15 (4): 325–334.

16. Malhotra, Y. 1998. Deciphering the knowledge management hype. *The Journal for Quality and Participation* (Jul.–Aug.): 58–60.

17. Fahey, D., and G. Burbridge. 2008. Application of diffusion of innovations models in hospital knowledge management systems: Lessons to be learned in complex organizations. *Hospital Topics: Research and Perspectives on Healthcare,* 86(2), 21–31.

18. Ambrosio, J. 2000. Knowledge management mistakes. *Computerworld,* 34 (27): 44.

19. Zarraga, C., and J. Garcia-Falcon. 2003. Factors favoring knowledge management in work teams. *Journal of Knowledge Management,* 7 (2): 81–96.

Social Responsibility

This appendix discusses social responsibility and managing technology and innovation. In recent years, society's expectations of business have changed. Society expects that firms will act in the public interest rather than focus on maximizing profits at any cost. The expectations that firms will act to benefit society will continue in the future and in fact will be expected to become even stronger.[1] As a result the social issues surrounding either internal innovation or externally obtaining technology will increase both the complexity of technology management and the impact on firm performance. Thus, by considering social issues, managers may not only impact the firm's value creation in the short term, but also over the long term as they will create value by increasing the firm's reputation and limiting potential future legal liabilities.

Initially, this appendix discusses three broad social issues that impact the firm—sustainability of the environment, corporate social responsibility, and the ethics of leaders and individuals. These are all critical to the firm's ability to add value for its shareholders as well as contribute value to other stakeholders. The appendix then concludes with a discussion of the use of social issues analysis in key managerial decisions, including choosing which technology-based projects to undertake.

THREE BROAD SOCIAL RESPONSIBILITY ISSUES

It is noted above there are three issues that broadly impact a firm's social responsibility—sustainability of the environment, corporate social responsibility, and ethics of leaders and individuals. The purpose of the strategic process view of MTI is to add value to the firm—in both the short-term and in the long-term. Part of adding value is developing processes and products that are sustainable not only for the organization but also for society. Consideration of such issues is critical for technology-focused firms because there is often a regulatory void for such firms. Laws and rules typically lag innovation. Right now (2009) the explosion in nanotechnology applications is outpacing the regulatory process. For nanotechnology firms, it is important to understand the ethical issues that surround that technology if such firms are

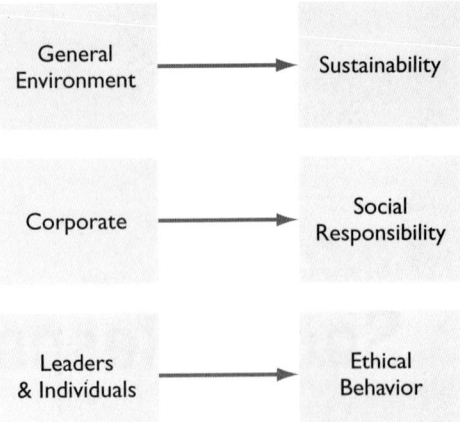

FIGURE **A1.1** Social Issues in Management of Technology and Innovation

to avoid potentially negative impacts later. For such firms, if each firm acts in a socially responsible manner the firm will be less likely to find it has created some legal liability or acted in a way that hampers the firm's growth as the regulatory environment begins to be established.[2] Figure A1.1 shows the three levels at which ethical issues must be considered—society, corporate, individual. It also shows the three areas that most dramatically impact technology-focused firms—sustainability, corporate social responsibility, and ethics. We will next examine these three issues in more depth.

Sustainability

Sustainability for a firm is a pattern of resource use that is designed to meet organizational and human needs while preserving the environment so that these needs can be met in the present, as well as in future generations. Technology has been used to solve a myriad of problems for society, but it has also caused some problems. Some of the problems now being faced by the world include increasing energy needs as more technology becomes more widely available throughout the world, transportation overload with the increase of cars and other vehicles, plus increased demands on natural resources. Each of these areas has implications for the environment. The ability of the ecosystem of the Earth to sustain the changes in carbon emissions (from fossil fuels), the chemical waste from production of computer parts, and the stripping of old growth rainforests are all part of the sustainability question that faces business managers as members of the more global community. More and more countries and individuals are asking for and even demanding accountability for such outcomes.

Environmental sustainability is, as a result, becoming a key social concern for business. In the past, many business firms viewed their social responsibility as supporting economic growth. However, today technology foresight is also becoming an expected part of innovation and technology management.

Costs for Firms	Benefits for Firm	Benefits for Society
Resources for innovations focused on sustainability	Better reputation Decreased liability New products & processes	Decreased environmental damage Better health Resources for future generations
Resources for innovations focused on social responsibility issues	Better employee relations New products & processes New market development Less liability	Better employment opportunities More community development Improved quality of life
Resources to meet demands of the environment for responsible technology and innovation	Cost savings, revenue generation, economic prosperity for future generations, better quality of life	

FIGURE **A1.2** Costs and benefits from sustainability and CSR

Technology foresight demands that the firm not only understand what the new product or process can do for the firm immediately or for the society's immediate employment, but also what the product or process will do to the environment over time. As a result investing resources in sustainability can create value for the firm. This value includes not only a better reputation which attracts customers but also decreased legal liability. (See Figure A1.2.) Thus, a technology- and innovation-focused firm must not only create value for shareholders and economic growth but also consider a richer set of social concerns than ever before.

Therefore, organizations that are trying to manage technology and innovation, sustainability requires that the managers of the firm have a strategy for continuing to be a positive influence in the general environment. This strategy should include five basic tenets:

- *Make use of the best available science.* Because of secrecy issues and fear of industrial spying, some organizations handcuff their scientists by limiting their interaction with other scientists. In areas such as biotechnology and other science-related areas of new product development, such secrecy has a downside: failure to learn from others or to share with others in important realms. This failure to learn from others can prevent the understanding of potential negative impacts on society. As a result there is a balance that must be found between secrecy and helping the entire society move forward with new innovations.
- *Protect, maintain, and rehabilitate ecosystems.* A firm's carbon footprint is the amount of carbon dioxide that is associated with the production and distribution of its product or service. It is possible for a firm to reduce its carbon footprint in a number of ways some of which are very simple, for example, changing light bulbs to more energy efficient

lighting, and reducing paper use by relying more on electronic communication. New products should be developed while evaluating the impact the product will have on resources of society.

- *Base use of resources on strategic plans that are well thought out and implemented in a responsible way.* Sustainability should be part of the overall strategic management process of the organization. If the goal of sustainability is actively stated and pursued, then it is more likely that processes and products will be developed that reflect that goal. The strategic focus on sustainability should be integrated into the very first strategic steps of the firm.
- *Control for new processes and products that emerge during testing phases to ensure sustainability.* Firms should not start worrying about sustainability issues after the introduction of the process or product. Too often technology gets implemented without testing for or thinking about its environmental impact.
- *Make trade-offs as necessary.* Such trade-offs should reflect societal values and should be made in an open, transparent manner. If the trade-offs are made in such a manner, it is likely that the first four tenets are being considered also.

To illustrate such trade-offs, H&M is a Swedish clothing and cosmetics retailer that has a wide ranging code of conduct that conforms to the various conventions of the United Nations. H&M requires that the actions of its suppliers conform to this same code. Included in this code are issues that could reasonably be expected for a retailer like H&M such as a ban on child labor, discrimination, and forced labor. However, the firm also requires compliance with environmental legislation. Even though with the code H&M's efforts there have been increased efforts in the environment, the firm has not been able to make the impact it or others desire in the worldwide cotton-production industry. H&M does not deal with these cotton growers. Cotton uses high levels of water consumption and is heavily fertilized which can damage many sensitive environments. H&M has made a trade-off—it seeks to improve this industry as best it can remotely but at the same time the firm cannot walk away from buying cotton at a reasonable price.[3] Thus, the firm must make trade-offs to move sustainability forward while still making a profit for its shareholders.

As the firm evaluates technology for sustainability, there are indicators that can be considered to help direct decision making. Sustainability indicators provide knowledge on the interplay between the environment, society, and economic activities. Building strategic indicator sets around sustainability typically examines questions such as:

- What is it we are doing here? (descriptive indicators of sustainability)
- How does what we are doing matter to the firm, shareholders, and stakeholders? Are we reaching, or at least moving toward, desired sustainability goals? (performance indicators around sustainability)
- Are we getting better at what we are doing? (efficiency indicators that indicate less energy or input usage)

- Do we understand what is happening in the industry overall and how does this affect our sustainability efforts? (policy effectiveness indicators)
- Is the society better off because of our sustainability activities? (total welfare indicators)
- What can we do to improve what we are doing and what should we do next? (looking to the future for sustainability)

At a more global level there are several key areas that the responsible firm needs to examine to determine if its management of technology and innovation is reflective of environmental, social, and economic sustainability: [4]

- Intergenerational equity – providing future stakeholders with potential that matches or exceeds today's stakeholders.
- Decoupling economic growth from environmental degradation – managing economic growth to be more resource responsible and less toxic to the environment as a whole.
- Integrating environmental, social, and economic concerns when developing policies about sustainability.
- Maintaining and enhancing the adaptive capacity of the environmental system through good corporate citizenship and by avoiding irreversible damage to future generations.
- Avoiding unfair costs on vulnerable populations by actions such as exploiting natural resources in economically underdeveloped areas.
- Accepting global responsibility for environmental effects that occur because of decisions and actions within the organization.

The management of a sustainable course of action requires that the firm accept the notion of corporate social responsibility in which the firm recognizes it has a responsibility to the broader society. The next section will address this critical issue.

Corporate Social Responsibility

Corporate social responsibility (CSR) is where an organization has a built-in, self-regulating mechanism that monitors and ensures its adherence to law, ethical standards, and positive behavioral norms. However, it is important to recognize that CSR involves more than just ensuring that the law is followed. The law is the minimum but insufficient standard of corporate behavior. The socially responsible firm embraces positive behavioral norms. Thus, they need to recognize the impact of their activities on the environment, consumers, employees, communities, stakeholders, and all other members of the public sphere that matter and should be managed. This is especially true in the introduction of new technologies (both processes and products) as well as the acquisition of technology from external sources. Furthermore, CSR promotes that the proactive position of the public interest is important to business. In addition, firms should be good corporate citizens by encouraging community growth and development, and by eliminating practices that harm the public, regardless of legality. Essentially, CSR is the inclusion of public interest into corporate decision-making. It is how the firm addresses the general environmental sustainability by practices and processes within the firm.

If the organization puts resources to work in a socially responsible way, then it should expect several positive outcomes for the firm and society. These outcomes include that the firm can expect better employee relations; new product, process, and market development; and less liability. For society, an excellent corporate citizen can enhance the quality of life for the community and the firm's employees. (See Figure A1.2.)

However, CSR is not without its critics. The critics say that CSR is a distraction from the fundamental economic role of business and as such has no place in the firm's strategic-action portfolio. Milton Friedman, the famous Nobel economics laurate, is the best-known proponent of this view. His arguments on focusing on economic factors are similar to that we saw in sustainability when individuals argue that providing jobs should be the focus. Other critics imply that firms who tout their good citizenship are trying to make themselves look good and do not add value with their CSR activities. These criticisms are not unfounded but the belief here is that over the long term CSR will provide value to the firm, its shareholders, and the society.

The concerns for the firm in the management of technology and innovation in a corporate social responsibility manner revolve around four areas of interest: employees, suppliers, customers, and the community. (See Figure A1.3.)

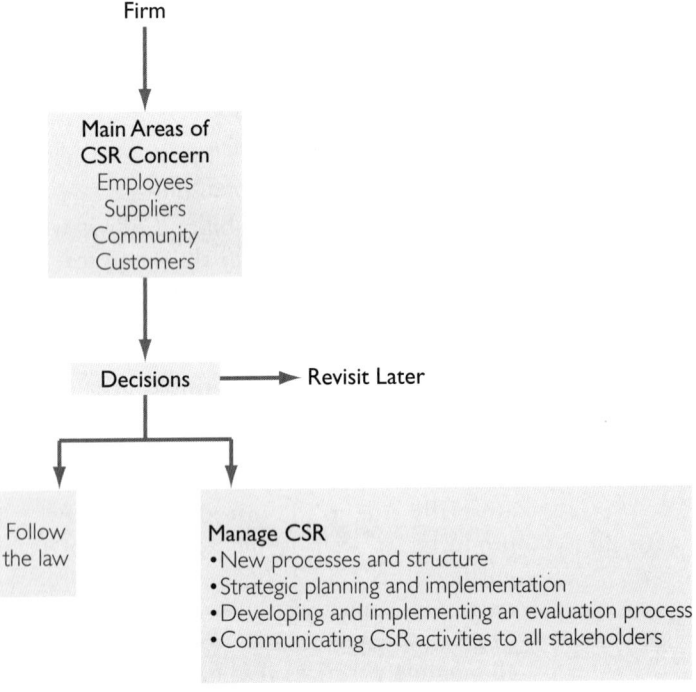

FIGURE **A1.3** CSR in the Firm

The firm has many decision points that can be very complex. For example, if a firm wants to lower the costs of manufacturing, it may consider moving production facilities to a lower labor-cost country. The CSR issues are multiple:

- What happens to the current employees?
- What are the labor laws of the other country?
- What will be the reaction of customers?
- How will the community in which the factory is located react to the shutdown of the current factory?
- Will supplies be available in the new locale? How does the moving of the factory affect the suppliers' employees?
- Is part of the potential cost savings in the new location due to lower environmental standards?
- If so, what standards should the firm follow?
- A myriad of questions that are relevant to these areas can be generated related to the specifics of any firm.

To understand the variety of issues a firm should account for how its actions affect the internal and external social environment. In Chapter 2, we discussed financial analysis using accounting data; here we expand these measures to include social accounting. **Social accounting** is a concept that describes the how, what, and why of social and environmental effects of a firm's actions on internal and external stakeholders. This is a relatively new concept and requires innovation and thought on the part of the corporation as it pursues its economic activities. Thus, the firm seeks to place measures on the costs and benefits of its actions to a wide range of stakeholders in social accounting.

Most companies that want to be seen as socially responsible and have undertaken social accounting have discovered the following:

- There is a need to link CSR reporting to financial and strategic reporting.
- Much of the information needed is readily available, but processes need to be modified to develop and present a social accounting report.
- Surprisingly, cost savings usually emerge. Also, linking CSR to strategic reporting heightened awareness of other potential cost savings.

The benefits of launching a corporate social responsibility program include more directed strategic thinking for the future, building reputation and trust with stakeholders, as well as managing product and market development. The steps for launching a CSR program are:

- Clearly state a set of principles so that employees understand the values, goals, and what are desired outcomes including performance.
- Identify core competencies for corporate social responsibility needs that are facing the firm.
- Organize task forces and teams with CSR responsibilities for each of the identified key areas.
- Include CSR concerns in the decision-making process along with variables such as price, quality, and delivery. This helps to create social accountability for actions.

- Monitor and report on actions and achievements so that patterns of success and failure can be tracked. The firm will need this feedback if it is to improve.
- Be transparent and obtain stakeholder feedback.
- Evaluate CSR as an ongoing strategic initiative in the firm.

One firm that has made a firm commitment to CSR is Ben & Jerry's. In fact, they file a Social and Environmental Assessment Report just as they file SEC required documents. They have a Social Mission Statement as well as product and economic missions. Their Social Mission is:[5]

> To operate the company in a way that actively recognizes the central role that business plays in society by initiating innovative ways to improve the quality of life locally, nationally and internationally.

It is enacted with three ongoing goals:[6]

1. Use our Company to further the cause of Peace and Justice.
2. Harmonize our global supply chain and ensure its alignment with our Company values.
3. Take the lead promoting global sustainable dairy practices.

Obviously, for Ben & Jerry's, CSR is a basic corporate value that they have embraced and maintained throughout their 30-year history. While we talk about corporate social responsibility, CSR has little effect on strategy if individuals and leaders within the organization are not committed to ethical behavior. The next section addresses the issues of ethical considerations in the management of technology and innovation.

Ethics

The third domain of social responsibility is ethics. Arthur Freedman, PhD, a consulting psychologist who specializes in organization development, believes ethics is informed by values. He outlines a hierarchy in which an understanding of morality (right or wrong) informs values (preferences), which in turn inform ethics. Thus, he sees ethics as part of the entire system of the person. Freedman goes on to state that **ethics** are a set of standards that govern behavior. In the management of technology and innovation, there are many potential ethical concerns. The obvious concerns deal with the impact of new products and processes on others, but within the organization there are several important issues that need to be considered.[7]

As we look at ethics here, we will discuss first causes of unethical behaviors in general. Then, we will examine how innovation and creativity can lead to ethical conflict and how changes in stakeholder expectations can influence the processes within organizations. Finally, we will discuss how the firm can help individuals act more ethically.

The reasons given for unethical behavior within organizations are many. The ones that are most related to management of technology and innovation includes:

- The evaluation systems that are in place for firms and individuals are typically focused on "short-termism" or the results needed now.

- The difficulty to translate strategic goals into operational reality.
- Rationalization of individuals that things will be fine and that their actions have no impact.
- The pressures from external stakeholders.

What is ethical in one part of the world may not be ethical in another part of the world. As markets expand and companies move operations to other countries, managers may find themselves in conflict with personally held value systems as well as images of what is socially responsible. It is up to the firm to communicate what it views as ethical. If the firm establishes its foundation for ethical standards no matter where the firm operates then it is those values that should be upheld. For example, it may be acceptable in many parts of the world to discriminate against women. However, a firm can establish a standard that its firm does not discriminate against women. As a result the local standard may allow something that the firm does not.

Organizational practices need to establish and support compliance goals around ethical processes, including such processes as selection, training, and rewards and recognition. Here are several practical ideas to help leaders who wish to build an ethical organization: [8]

- *Implement training programs.* Most organizations require that every employee attend ethics training of some sort because firms want employees who understand that ethics is an important business aspect. The unethical behavior of one employee can cost the firm dearly.
- *Make ethics a part of business strategy.* Firms with ethics programs have a code of conduct that supports employees in making ethical decisions. In addition, questions, policies, and practices related to ethics and compliance need to tie to the business plan and organizational vision.
- *Measure ethics performance.* Make ethics a leadership competency that is measured as part of the company's performance management system. It is well known that the biggest influence on ethical behavior is the behavior of the supervisor. Therefore, for a firm that wants its employees to act ethically,
  - Supervisors must model integrity, which means doing the right thing versus doing the most expedient thing.
  - Managers should lead by example.
  - Ethical issues should be investigated and resolved swiftly.
  - Officers of the company should advance ethical behavior by making sure that policies and practices are aligned to support an ethical culture.
- *Invest resources.* Nothing speaks louder about management's concern for ethical issues than dedicating resources to ethics training and oversight.
- *Communicate regularly.* Another way that firms make ethics tangible is to use technology such as: anonymous, 1-800 phone lines to report questionable practices; "whistleblower" hotlines; published codes of conduct; and cards that display company values. It is important to share ethical practices with stakeholders such as customers, investors, suppliers, and community members.

- *Tap into your company's grapevine.* The grapevine can be used effectively without getting into specifics of a situation. When rules are broken, and there are appropriate consequences, it is important that they be communicated to the company at large. This will reinforce the importance of ethical conduct throughout the firm.

Of these general guidelines, the most influential is the behavior of the leader. The leader of the organization sets the tone. It is true for Ben & Jerry's and it is true for TOMS Shoes. Founder Blake Mycoskie started TOMS Shoes after spending time in Argentina. Many of the children he saw had no shoes. Because they had bare feet they were denied schooling and were susceptible to "Mossy Foot." After some investigation, Mycoskie found that this was a common problem in many countries of the world. He started a shoe company to make simple shoes. When a pair is purchased by a customer, another pair is given to a needy child. TOMS Shoes has made shoe drop-offs in the United States, Argentina, Ethiopia, South Africa, Haiti, as well as other countries. In addition, Mycoskie has developed processes to make shoes with renewable or recycled materials. As a result of Mycoskie's focus on "doing things right," TOMS has been recognized by many as a successful model of socially responsible entrepreneurship. TOMS plans to give away 1 million pairs of shoes by the end of 2012.[9]

Unique Issues for MTI. These general guidelines for ethical considerations in the firm are important, but the management of technology and innovation has ethical concerns that can be unique. When developing technology internally, the firm is encouraging creativity among its employees. There are four problematic areas of creativity/innovation that can raise serious ethical issues.[10] These are:

- *Breaking the rules and ignoring standard operating procedures.* Often innovation will involve the breaking of rules and operating procedures but it must be asked which rules are acceptable to ignore, when, how far can the deviation go, and who can initiate such actions.
- *Challenging the way things are done and those in authority.* Challenging the way things are is also often associated with innovation but in the process other issues should be raised. For example, is the new way at least as "ethical" relative to all stakeholders as the old process?
- *Creating conflict and competition.* Conflict and competition are often associated with innovation but firms need to examine how they create such conflict and what are the trade-offs in the standards of the company that it may create.
- *Taking risks.* Risk taking is also associated with innovation but what types of risks are acceptable and how does the firm ensure that it does not lead to actions that violate the firm's ethical standards is important for the firm to clearly establish.

Because creativity requires "thinking in different ways," those in creative areas of the firm, like R&D, may find themselves in conflict with the norms of the firm. When individuals are expected to be creative, the firm needs to encourage a strong sense of individual values that are not out of line with the organizational value system. Managing innovative people requires the firm to develop a shared vision, clearly communicate expectations, and develop leaders/managers that are able to foster understanding about ethical issues, conflicts of interest, and the values of the firm.

For those firms involved in obtaining technology from external sources through alliances, mergers, acquisitions, and joint ventures, the ethical issues are also critical. The same behaviors should be expected from leaders of any firm in the allied or acquired firm. As noted earlier, ethical behavior involves more than obeying the letter of the law. Thus, the two organizations may both meet the law but have very different ethical standards. This can cause potential for conflict between the two parties. There is a rich set of similar ethical issues that firms must consider as they pursue the external acquisition of technology. For example, if a firm acquires another firm and wishes to obtain new technology, it may be that the firm focuses so much on the technology it forgets the social impact of the merger. The ethical issues associated with the external acquisition of technology through such mergers will be discussed more in Chapter 7, however, it suffices to say now that there are a rich set of issues that are involved in such a complex activity as an alliance or a merger.

The areas where ethics most often involve the management of technology and innovation are product development and intellectual property. Ethical concerns in product development or continuation deals with the expectation that a company will ensure that its products and production processes do not cause harm. Some of the more acute dilemmas in this area arise because there is usually a degree of danger in any product or production process and it is difficult to define a degree of permissibility. The degree of permissibility may depend on the changing state of preventative technologies or changing social perceptions of acceptable risk. Examples of such changes that developed as technology helped us gain a better understanding of risks include, tobacco products, mobile phone radiation, and pollution. Knowledge and skills are valuable but not easily "possessed." It is sometimes difficult to determine who owns intellectual property—the company or the employee themselves. As a result, attempts to assert ownership and ethical disputes over ownership arise. Examples include patent misuse, employee raiding, and industrial espionage.

Strategies for Ethically Building Value-Added

Firms increasingly are under pressure to act in a transparent, responsible manner while pursuing profitability and innovation. The management of technology and innovation is an area where these demands are particularly intense within the firm. Figure A1.4 illustrates the progression of a firm from mere compliance with what society demands—obeying the law—to truly creating new value through balancing social, environmental, and economic sustainability. The creation of new value is characterized by "newness" of products, services, processes, alliances, markets, and business models. But the creation of value in this manner has not changed for the organization in

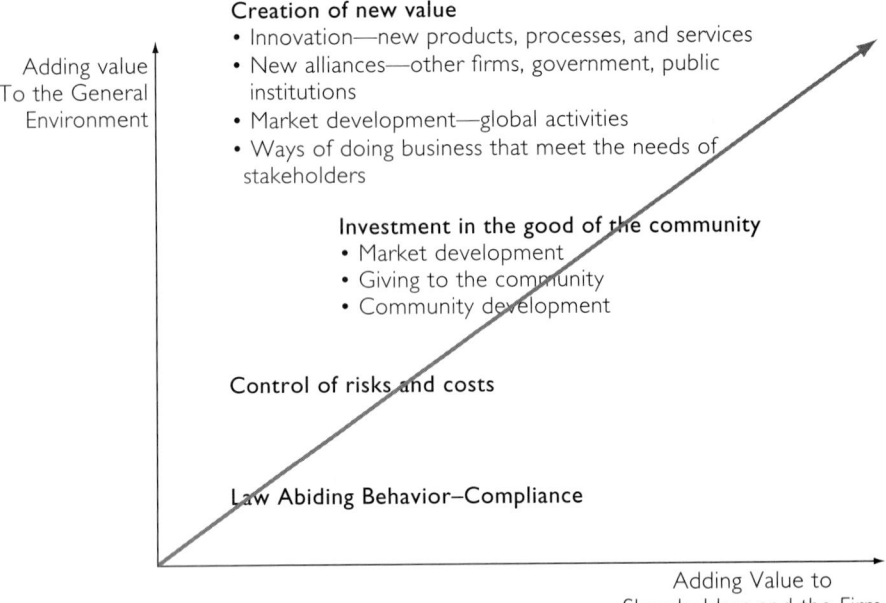

Adding value To the General Environment

Creation of new value
- Innovation—new products, processes, and services
- New alliances—other firms, government, public institutions
- Market development—global activities
- Ways of doing business that meet the needs of stakeholders

Investment in the good of the community
- Market development
- Giving to the community
- Community development

Control of risks and costs

Law Abiding Behavior–Compliance

Adding Value to Shareholders and the Firm

FIGURE **A1.4** Adding value through responsible corporate behaviors.

recent years although the rules on how to do it have. To insure sustainable value and long-term growth and survivability, the firm should:

- Harness innovation for good of society.
- Put human resources, human capital, and customers at the center of consideration.
- Spread economic growth and opportunity; not oppression.
- Engage in new ways of doing business.
- Be performance driven in social, environmental, and economic aspects.
- Develop managers/leaders who are ethical and socially responsible.
- Pursue purpose beyond profit.

When a firm partners its technology with its social venturing like Ben & Jerry's and TOMS, it finds itself doing things in a new way. If done well, the result is cost savings, revenue generation, economic prosperity for future generations, and a better quality of life for the firm and its stakeholders.

SUMMARY

This appendix has examined the social issues that are part of the management of technology and innovation within the firm. Such information is critical to the application of MTI. Without a clear understanding of the firm's values in sustainability and ethical considerations, management may be pursing goals that are either unrealistic or contrary to the needs and wants of society in its technology and innovation efforts. This can be damaging to the firm. This text is unique in that it stresses a strong social understanding as a key part of the strategic management of technology and innovation. Too often, texts and MTI professionals ignore the social aspects of technology and innovation. However, without a strong understanding of these issues, not only can the MTI effort fail but so, too, can the firm as a whole.

EXERCISES

Audit Exercise

In Chapter 2, we discussed assessments of the external and internal environments as well as the strategic process. If you were the CEO of a firm pursuing an internal innovation strategy, how would you assess the areas of sustainability, CSR, and ethics within your organization? Some can be assessed using financial data, some can be assessed using other information that should be present in the organization, and some assessments can be based on experience.

What did you learn about trying to develop this type of an assessment for determining a firm's strategic direction in the MTI arena? Are there other areas of assessment that should be considered?

Case Study

Find a large company in which you are interested. Once you have identified the company, find its industry and its major competitors. Then find the following:

1. Where the focal company ranks in responsibility toward its various stakeholders?
2. How does the firm's level of social responsible actions affect its "bottom" line?
3. What innovative actions have been taken in the industry/organizations to improve its reputation?

All of this information should be available online.

What would you say about the technology and innovation responsibilities of the focus company and its managers in its environment? Explain your answer with information from your analysis.

KEY TERMS

corporate social
responsibility 321
ethics 324

social accounting 323
sustainability 318

technology foresight 319

NOTES

1. Bielak, D. S. Bonini and J. Oppenheim. 2007. CEOs on strategy and social issues. *McKinsey Quarterly*, December, 8–12.
2. Lee, R. and P. Jose. 2008. Self-interest, self-restraint and corporate responsibility for nano-technologies: Emergine dilemmas for modern managers. *Technology Analysis & Strategic Management*, 21(1), 113–125.

3. Svensson, G. 2009. The transparency of SCM ethics: Conceptual framework and empirical illustrations. *Supply Chain Management: An International Journal*, 14(4), 259–269.
4. Stanners, D., P. Bosch, A. Dom, P. Gabrielsen, D. Gee, J. Martin, L. Rickard, and J. Weber. 2007. Frameworks for Policy Integration Indicators, for Sustainable Development, and for Evaluating Complex Scientific Evidence. In *Sustainability Indicators: A Scientific Assessment*, edited by T. Hak, Be. Moldan, and A. Dahl, Island Press: 2007.

5. http://www.benjerry.com/company/sear.

6. *ibid.*

7. Kubal, D., M. Baker, and K. Coleman. 2006. Doing the right thing: How today's leading companies are becoming more ethical. *Performance Improvement*, 45(3), 5–8.

8. *Ibid.*

9. http://www.tomsshoes.com/

10. Baucus, M., W. Norton, D. Baucus, and S. Human. 2008. Fostering creativity and innovation without encouraging unethical behavior. *Journal of Business Ethics*, 81, 97–115.

New Product Development and Project Management in Innovation

To be successful with an innovation-based strategy, a firm must make and implement a number of key decisions in its strategic process. As highlighted in the previous three chapters, the firm must have a plan for undertaking the strategic activities. In developing such a plan, there needs to be a fit among the different organizational factors and characteristics as the firm seeks to implement its strategy. The effort to build fit requires that the firm pay attention to fit with the macroeconomic environment and to fit at the operations level if the strategy is to be successful.

For most firms, innovation is in a project-based framework. This is especially true for new product development. The degree of uniqueness and the processes required to get to the desired innovative outcome means the firm needs to develop a systematic process. Because each innovation effort is unique, the use of project management tools is appropriate to plan, implement, and evaluate innovation and technology.

This appendix discusses innovation project management. First, it presents the basics of a project. Then it introduces a process for managing innovative projects. It presents tools for analyzing various steps during that process. Finally, this appendix discusses the continuous process of innovation development, and the balancing of a portfolio of innovation projects.

INNOVATION PROJECTS

We defined innovation earlier as having some aspect of "newness." Like innovation, projects are unique. A project is an undertaking designed to accomplish a specific objective through a set of interrelated, specialized tasks. The characteristics of a project are:

- Well-defined objective: An expected result or product is the objective of a project. If the innovation is process oriented, then the expected result is some type of increased efficiency. If it is product oriented, a new or improved product is the desired objective.
- Series of interdependent tasks: A project requires a number of non-repetitive sequential tasks.
- Resources: There must be resources available to carry out the tasks. Although not all of the resources need to be dedicated (allocated to only one project), there does need to be a clear understanding of who and what will be available for a project at designated times within the process.
- Specific time frame: A project has a finite life span—a start time and a designated time for accomplishing the objective. In his famous 1961 speech, President John F. Kennedy stated that the United States would have a man walk on the moon by the end of the decade. In July 1969, the United States met that goal. Thus, the timeframe was designated by Kennedy and that shaped many of the resulting decisions.
- Customer orientation: The customer of a project may be internal or external to the organization. For example, an engineering department may do a design project for operations (internal focus). However, in a consumer products company, many projects are devoted to improving or creating products for the marketplace (external focus).
- Degree of uncertainty: At the beginning of the project the firm uses the best knowledge and information available. As a base starting point certain assumptions and estimates are made for the project budget, schedule, scope, and availability of resources. Because these are assumptions and estimates, there is a degree of uncertainty.

The reasons an organization would want to use a project-based approach to innovate should be relatively clear from the definition of the characteristics of a project. The biggest advantage to using an innovation project approach is that it can help *bring order out of chaos* by creating a "road map" for change. A project-based approach shows who, what, when, where, and how things should happen.

INNOVATION PROJECT MANAGEMENT

Figure A2.1 presents a seven-step framework for innovation project management. The firm may have in place systems, policies, and procedures to enhance innovation, but each new product or process implementation is unique. In an innovation project, the commitment of resources will increase over time. As a result, it is critical to develop early planning and understanding of the project. If there are problems in the project, the early identification of these problems will help limit the expenditure of resources on unfruitful ventures.

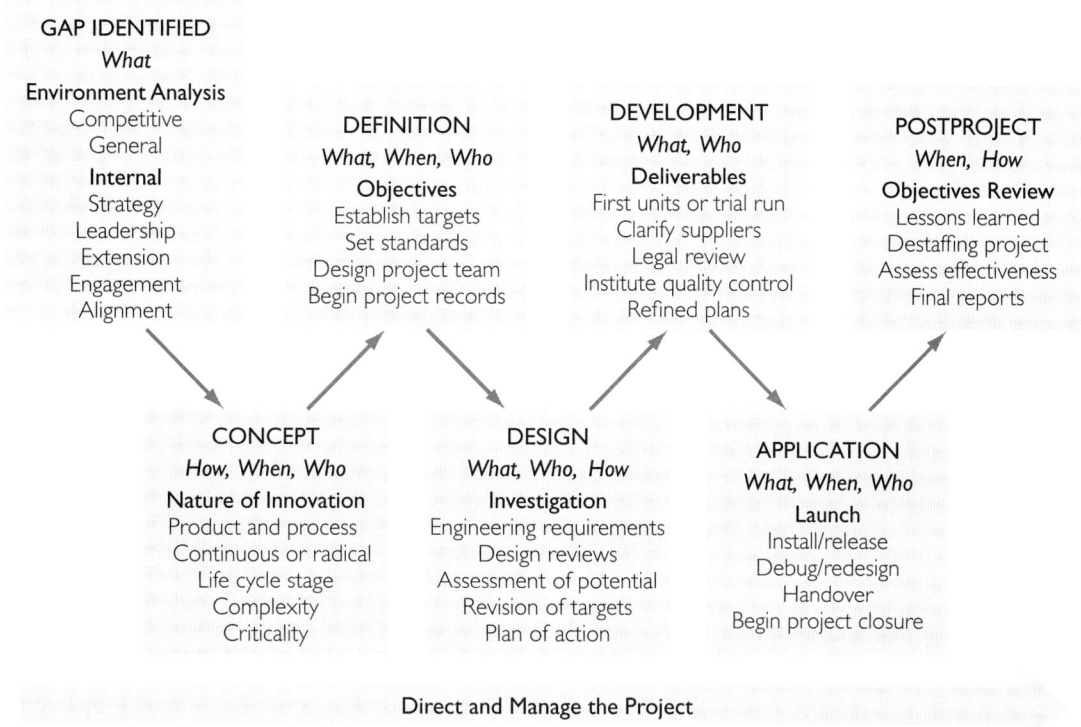

FIGURE **A2.1** Innovation Project Framework

Once the firm makes a decision to undertake an innovation, the resources required to fulfill that decision escalate and the risks for the firm increase. Therefore, the identification of where the organization is, plus the conceptualization and definition of what is to be done are critical to setting the direction for the innovation project. For example, Sun Microsystems initially determines what its strategies and goals are, and then a committee meets weekly to prioritize the projects in its portfolio of projects. The committee uses about 20 variables to rank projects in its rating matrix.[1] This is how Sun begins and moves through the steps in innovation project management. We discuss each of the seven phases of this framework in more detail next. This framework is a good format for any project management effort.

Gap Identification Phase

As stated in Chapter 5, gap analysis is an important evaluation tool for those undertaking an innovation strategy. The gap identification process compares where the organization is to where it wants to be, as well as where the firm's competitors are. The gap can be process oriented or product oriented; however, once a gap is identified, the firm needs to address it or face the risk of being at a competitive disadvantage.

The environmental and internal systems analysis should help the organization identify a number of potential gaps. However, in managing innovation

Gap Analysis Results	Characteristics
Must be addressed	Failure to address would lead to competitive disadvantage; Loss of strategic position in the industry; Risk from not addressing is much larger than potential costs.
Should be addressed	Failure to address might lead to loss of market share or position in the industry; Risks of not addressing are moderately high. Survival of firm may or may not be threatened by ignoring this gap.
Would be nice to address	Addressing likely to enhance firm's competitive position; Potential benefit probably outweighs risks and costs associated with undertaking a project to address. Good place for future-oriented activities.
Does not need to be addressed	The gap is a result of competitive position. The threat of competitors' actions causing this gap to widen or gain in importance is low. May require the firm to move in directions not needed.

FIGURE **A2.2** Gap Analysis Results

projects, it is important to remember that there are always more gaps and ideas in a firm than there are time and resources to address them. Therefore, it is important to separate the gaps into several categories (Figure A2.2):

1. Those that must be addressed
2. Those that should be addressed
3. Those that it would be nice to address
4. Those that do not need to be addressed

The manager should classify the gaps according to the level of risk in addressing and not addressing, potential payoff or cost, organizational strengths and weaknesses, and environmental threats and opportunities.

The firm should undertake a six-step process to analyze the significance of the gaps and whether a gap needs to be addressed or not:

1. List the gaps and potential gaps for the firm. These should be the result of a well-executed gap analysis (described in Chapter 5).
2. Determine the threats and/or opportunities from each gap.
3. Judge the internal and external impacts for addressing the gaps.
4. Judge the feasibility of addressing the gaps. This takes some educated guesswork. However, if the resources are not in the firm or available to the firm, then addressing the gap with an internally oriented strategy may not be realistic, and instead other strategies might be needed.
5. Review the list, feasibilities, risks, and other factors about the gaps and eliminate those that require an inappropriate use of resources and time, or that are not feasible for the organization.

6. Select the most important gaps to address and act on them as soon as possible.

Concept Phase

The second phase in the innovation project framework is the definition of relevant concepts. *The Project Management Institute's Body of Knowledge* defines conceptualization as the "process of choosing/documenting the best approach to achieving the project objectives."[2] To prepare the conceptualization of the innovation project, managers need to:

1. State clearly the gap the project is going to address. This statement should include what the problem consists of, the need to find a solution, and what the intention of the firm is.
2. Gather information about the nature of the innovation to be undertaken and its place in the gap analysis results. As can be seen in Figure A2.1, this would include type of innovation, life cycle stage, complexity, and criticality of the project to be undertaken.
3. Identify constraints. There are constraints in the environment and in the firm. The gap analysis should help identify some of the resource and time constraints. However, customer input as well as the input of other stakeholders should be sought.
4. Develop alternatives to address the issues identified by the gap analysis. By developing alternatives, managers can develop a clearer understanding of how to solve the gap. In addition, the process of developing alternatives may give the management other insights into ways to solve the gap.
5. Record the project objectives that emerge. If the previous four steps have been done correctly, the objectives for the project should be clear. The objectives follow logically from the analysis and lay the groundwork for defining the project parameters.

The concept phase should also define the complexity of the project. The more complex the project, the more work needs to be done during this stage. There are five basic types of complexity that impact project management. The manager should evaluate these complexities for a firm's innovation projects. The complexities and the key concerns are:[3]

1. Technological complexity deals with the newness of the technology, the number of components to be integrated, and the types and variety of skills needed to develop the innovation.
2. Market complexity addresses the market needs, the amount of change in the market as well as the types of changes in the market, competitor's actions and reactions, and the firm's vulnerability to market changes.
3. Development complexity or the "nuts and bolts" of R&D required for the project. This includes complexity in integrating the different innovation projects, assessing what resources are needed to develop a product that is not yet known, assessing changes in the process, building a supply chain, and developing an understanding of where the innovation is going throughout the process.

4. Marketing complexity examines how to educate potential customers, how to develop the marketing channels, how to promote the innovation, and how to promote compatibility with systems/equipment and other user capabilities.
5. Organizational complexity considers issues such as structure, approval systems, number of groups/teams involved in the innovation, communication processes, and how the organization will keep everyone informed and up-to-date.

The 20 variable rating matrix that Sun Microsystems employs (discussed earlier) is complex and requires a number of assessments. However, it does not include budget considerations. Once the projects are ranked, then budget constraints are considered; however, initially it is the non-budget items such as degree of innovativeness and market demand that drive the process. New concepts must compete with ongoing projects for R&D dollars. Therefore, the conceptualization of the project and its benefits are critical factors in deciding which projects to pursue.

Definition Phase

The what, when, and who of the definition phase set forth the task, timing, and team makeup for the innovation project. For a process innovation, the task would be to plan the conversion, the timing would be a time objective with an endpoint with multiple checkpoints, and the team would include a project manager and team members to coordinate the changeover to the new process. The targets of the innovation project should emerge from the conceptualization of the project. When setting the objectives of the project, the project team should make sure the goals are:

1. Specific and well-defined
2. Realistic and doable
3. Timed for achievability
4. Measurable in a realistic manner
5. Agreed on by the team and the management of the firm
6. The person responsible (the project manager) is identified and known to others

It is also during this stage that the organization is moving beyond looking at what needs to be done, and identifying what will be done. Therefore, the relevant gap analysis, the results that have spurred the development of a project team to undertake an innovative activity, and the list of potential alternatives should be brought together to begin the project records. These records will be instrumental in doing the post-project evaluation. Recall that part of evaluation is "are you where you thought you would be."

The definition stage is also a key place for the organization to ask if this is where it wants to go. After all, it is after this stage that the escalation of commitment of resources really begins. Once again, the firm should look at questions such as:

1. Is there a strategic fit between the project and the firm?
2. Do we understand the scope and implications of the project?

3. Do we know what we are going to do next and what we are building toward?
4. Are we moving on for good reasons? What are they?
5. Does the innovation project we are planning and beginning to implement fit with the organizational strategy?

Sun Microsystems does this by first hiring the best people and letting them lead the firm to new ideas and innovative solutions to old problems. The second key is to create communities inside the firm that encourage collaboration and open innovation. The result is that the firm not only has creative people but communities of creative people whose sum is greater than the individual parts. The goal for the firm in this phase is to define projects that matter to its customers and in turn to the firm.[4]

Design Phase

During the design phase, the firm begins to decide what it needs to meet the strategic goals, who will be responsible for the project, and how the process of innovation will take place. The definition phase has established the targets and standards for the project. The first question in the design phase concerns feasibility: Can the project be done? If so, the design phase can begin. There are two key types of individuals who must be on the team if design is to be viable—**concept generators** and **concept implementers**. Concept generators throw out ideas about how to solve the problems, and concept implementers focus on how to accomplish the ideas. In design, it is important to begin with a concept of the whole and then design components to fit into the whole. The definition phase should give the innovation team the concept of the whole, and the design phase should fill in the parts.

The individual or individuals in the design phase need to possess three talents:

1. The ability to recognize future trends while contributing to the designs for their firm
2. The ability to recognize the potential commercial significance of "aha" moments in their own R&D settings as well as in other interactions in their life
3. The ability to integrate the commercial and technical worlds. This requires knowledge of what is possible and what is wanted

The "how" of design is usually carried out on two levels. The first level is technical and involves the engineers and operations personnel. This level involves the actual design of the product or process innovation. The second "how" is at the project management level and is concerned with the budgeting and scheduling. Budgets are developed and details fleshed out as the design of the product or process emerges. The scheduling process is often more complicated but is just as critical as the budget. Even if the time frames involved are difficult to assess, or competitive forces dictate the schedule, the schedule should be set. There are several common tools for scheduling that managers should know.

Activities	Person Responsible	Duration	Months						
			1	2	3	4	5	6	7

FIGURE **A2.3** Gantt Chart Framework

One of the most common scheduling tools used in project management is **Gantt** charts. A Gantt chart indicates two functions: planning and timing. Activities are listed down the left side, with time required usually along the bottom. The estimated duration for each activity is indicated by a bar spanning the estimated time needed to accomplish the activity. Other columns may be added to indicate departments or individuals that will be responsible for the activity. The Gantt chart is one of the oldest project scheduling techniques, but it has some drawbacks. It is hard to maintain and update when delays occur, and it does not show sequential dependent relationships among activities. Figure A2.3 shows the basic framework for a Gantt chart.

A second scheduling tool is **PERT** (Program Evaluation and Review Technique) analysis. This tool shows the network activities to complete, how these activities interact, and the timing required for them. In addition, PERT indicates three levels of time: optimistic, expected, and pessimistic. Other network techniques have been developed since PERT. These include the Critical Path Method (**CPM**), which tracks the longest path of activities to be completed. The danger of CPM is that the critical path may change if activities on other paths are delayed so careful monitoring of activities is required with the changes in the CPM identified as necessary. Precedence Diagramming Method (PDM) and Graphical Evaluation and Review Technique (GERT) are techniques that make the sequencing and the interrelationships among activities clearer. At this time, none of these techniques is dominant. Instead, the planner should find the technique that works best and provides the greatest insights. Whatever technique is employed the schedule and budget together should clearly present a plan of action for the innovation project.

During the design phase, Sun Microsystems allows and encourages the exchange of ideas. In fact, when there is an innovative idea that overlaps an older product or system, Sun Microsystems will continue to develop the innovative project and add the consideration of EOL (end of life) for the old product/process to the goals of the innovation. In this way, Sun Microsystems can work into the design all of the characteristics of the old product/process that Sun wants to keep. The result is that the staff of the old product will know that the product is at its end of life and the staff will need to work with and aid those that are replacing that existing product. This requires the staff to be both flexible and willing to work in a total team effort to move the firm forward as new products are developed and old ones eliminated.[5]

Development Phase

The development phase begins the actual effort to implement the innovation. Until now, laying the groundwork has been the focus. The abstract phase is completed, and now the innovation team needs to enact the first trial run of the product or process. The more planning and thought that go into design, the fewer problems should arise during this phase. However, that does not mean everything will go smoothly. This phase and the next are led by engineering—design and manufacturing. The steps in development are:

1. Define a method for building the product or implementing the process. A prototype should be built and tested against the design requirements.
2. Evaluate the firm's resources for best practice capabilities. This requires continuous evaluation and iteration of the design. The goal is to maximize the firm's ability to produce a desirable deliverable (product or process).
3. Develop a list of materials needed and a design for routing those materials to determine the actual cost. Until a prototype is built, tested, redesigned, and rebuilt that meets acceptable criteria, the costs are estimated. Only when there is a clear set of inputs should the vendor-supply chain be determined and final costs calculated.
4. Determine the ability of the firm to introduce the innovation along with all of the other products and processes in the firm. Are there synergies with other products and processes? Will the innovation take away resources needed in other parts of the firm? If the capacity to implement the innovation is insufficient, then capital resources need to be committed or the product mix needs to be changed. Newness in one area of the firm can affect a number of other areas. Ideally, managers addressed this issue earlier, but during the development phase, it will become clearer what is needed for the innovation.
5. Make sure common sense is still the driving force in decision-making. As the prototype is being developed and tested, the tendency is to become too enthusiastic. This often results in escalation of commitment without solid reasons and analysis for the decisions.
6. Market for when the capacity to produce the product or process is available. Many times, firms will announce an innovation, and then the delays in producing it lead people to wonder if it will ever happen. In the

software development industry such delays in meeting the announced release date occurs often, although it is not a desirable outcome.

7. Determine the required profit margin, market size, and profitability. These need to be part of the decision before full-scale launch. Too often, firms make a great new product, sell a bunch, and then wonder why the bottom line does not grow. This is especially true in small entrepreneurial firms.

If the product or process meets the criteria of the definition and design and the costs and returns look good for launch, then the firm should move to the application phase.

For Sun Microsystems the criteria to move forward on a process innovation to its supply chain management mirror those criteria above. Sun developed the "one-touch supply chain" and worked to overhaul its manufacturing process. Through a closely monitored process, Sun Microsystems's equipment manufacturers now build the firm's hardware, configure relevant software systems, and ship directly to the customer in an integrated manner. The result is Sun Microsystems focuses on information (orders and specifications) but not products.[6] The next step in this evolving process is to create a way for the customer to customize system software in advance of shipment.

Application Phase

The application phase concerns the installing/releasing the new process or products for the whole organization. This is the "do" of the innovation project process. If it is a process innovation, then the installation of the process should be ready for all parts of the organization that will be making the change. If the innovation is a new product then the product (and the associated activities) is ready for full production. As the project team turns over the innovation to the appropriate functions of the firm, some debugging may need to take place. However, as handover begins, it is important to begin project closeout. In preparing for the innovation project closeout, several issues are likely to emerge:

1. Burnout or the loss of interest after working the project is common at the end of a project. Because of burnout, the post-project evaluation is often overlooked. Burnout can also lead team members not to follow through on debugging. The developers of the innovation have the most knowledge of capabilities and potential ways to fix problems, and they should remain involved until the processes are running smoothly.

2. For project team members, concern about what they will be doing next may be distracting. If the team member knows what the next project will be, then excitement over the newness may cause some neglect of the application and post-project review phases. If the team member does not know if there is another assignment or what it will be, anxiety may lead to a failure to focus on finishing the current project. Because project teams are often cross-functional, there is a loss of social network and work group that needs to be addressed as the launch takes place.

3. In most projects, bugs (problems) will still exist. This causes frustrations as the developers and customers try to find solutions. In addition, following the innovation launch there is commonly a reallocation of resources from

the team. If unanticipated bugs arise, then there may be insufficient budget and resources to fix them.

4. Documentation for the product needs to be compiled. This compilation of documents will aid in the evaluation phase. Often, toward the end of the project, the documentation is not as clear because everyone knows what is happening. However, without the documentation, institutional memory can be lost.

5. Comparison of goals, definition, and schedules to actual outcomes as the project ends. How closely does the final product or process match the definition of the project and the design?

6. Contractual commitments finalized with vendors, suppliers, and customers (both internal and external).

7. The last part of application is to transfer responsibilities to those in the organization who will take over the innovation project team's outcome. For a new product, for example, the responsibility will probably move to operations for the manufacturing process. The development of the ongoing manufacturing process becomes a new innovation project. The documentation of the project team should be helpful in making this transfer.

One part of the application phase that is not often recognized is the willingness to shut down a project. Sometimes, a project looks good until the application phase. In the application phase, it becomes clear that the goals of the project are not being met nor carried through to the customer. A common reaction is to try and "fix" the problems. However, one possible solution that is often not considered is to shut down the project. The primary reason for not considering shutting down the project is escalating commitment—the resources and time spent to date make people want to find success rather than recognize the loss and prior mistakes in judgments. At Sun Microsystems, the project must go through yet one more test to make sure that such basic questioning of viability is not ignored. They use the following equation to make this evaluation: $Q \times A = E$ where Q is quality, A is acceptance and E is effectiveness for the customer.[7]

As the transfer of responsibilities takes place, the innovation project team should look to develop a set of lessons learned that can be disseminated throughout the firm.

Post-Project Review Phase

The last phase of the innovation project process is the post-project review. This involves reviewing the objectives of the project and the outcomes in more detail, developing a set of lessons learned, the actual staffing of the project, final assessment, and the delivery of final reports. A mature post-project review process requires a culture that is always looking for ways to improve. The reviews include successes, failures, and surprises. The when, what, why, and how of the project needs to be understood.[8]

The final phase of the innovation project is often overlooked because:

1. The documentation for the project has not been well maintained, and memory of the team members is faulty.

2. The project is over, and everyone is ready to move on to the next project.
3. There are barriers to sharing knowledge. Some cannot see how the problems and solutions in one project can be generalized. In addition, because each project is unique, the managers see the lessons learned as unique.
4. Often, those involved in innovation are forward moving and act individually rather than completing the project-team closure activities. Reflection and consideration of what was and what might have been are not part of what the individual is now doing. However, such reflection allows the development of lessons learned to apply to future projects.
5. Management does not allow the time for post-project review.

If the project team recognizes these potential problems and works to overcome them, the post-project review will be more successful and more useful to the firm. The entire project team should develop the final report. The firm should have a consistent process, but it should generally include most of the following:

- Overview of the innovation project, including the original plan and any major revisions
- Summary of major accomplishments and outcomes
- An evaluation of how outcomes and goals match with an explanation of any differences
- Final accounting for all budgets—financial, time, resources—with an explanation of variances
- Evaluation of team and management performance, individually and collectively
- A list of issues or tasks that should be examined further
- A set of lessons learned for future projects that are similar in nature
- A summary of performance issues, conflicts, and resolutions
- Recommendations for changes to be incorporated in future projects
- Analysis of the innovation project process as a whole

For all of the post-project review, the goal must be how to improve the firm's ability to compete in the changing marketplace.

One project for Sun Microsystems that has been analyzed and reviewed extensively in order to learn more about its failure is Sun Ray. While many products fail, the failure of Sun Ray was particularly difficult for Sun Microsystems. The product is a display device, with applications running on a server elsewhere, and the state of the user's session being independent of the display so that a user can go from one Sun Ray to another and continue their work without closing any programs. The product failure came right after the failure of JavaStation. The Sun Ray project was mismanaged, and there were unrealistic expectations for its adoption. Because of the problems, Sun Microsystems changed its processes and introduced the matrix-based model described earlier in this appendix.[9] The firm today has revised and changed the product based on its initial failure. Today, Sun Microsystems has been able to introduce SunRay 4.0, a better and more useful product. Thus, Sun Microsystems learned from its mistakes and innovatively changed its process of new product development and introduction.

PORTFOLIOS OF PROJECTS

Effective innovation portfolio management typically involves balancing new initiatives with the completion of ongoing projects and using the knowledge of one project to help complete others (Figure A2.4). The more synergies the firm can develop across projects, the more likely it is to be more effective and efficient than competitors are. The process of managing innovation projects can become a competitive position enhancer. As was illustrated in the chapters of this part, 3M has been very successful in gaining synergies across projects. Similarly, Sun Microsystems has been able to manage its portfolios of innovation projects. Those firms that do have success in building portfolios share consistent characteristics that help balance and manage a successful program of innovation projects including:

- A common set of innovation project management structures, information-processing systems, and communication and documentation processes. The 20 variables used by Sun MicroSystems to rank order projects ensures the portfolio of projects is in alignment with the strategy and goals of the firm.
- Monitoring approaches including assessment of progress so there can be intervention if needed. Stopping a project should always be an alternative

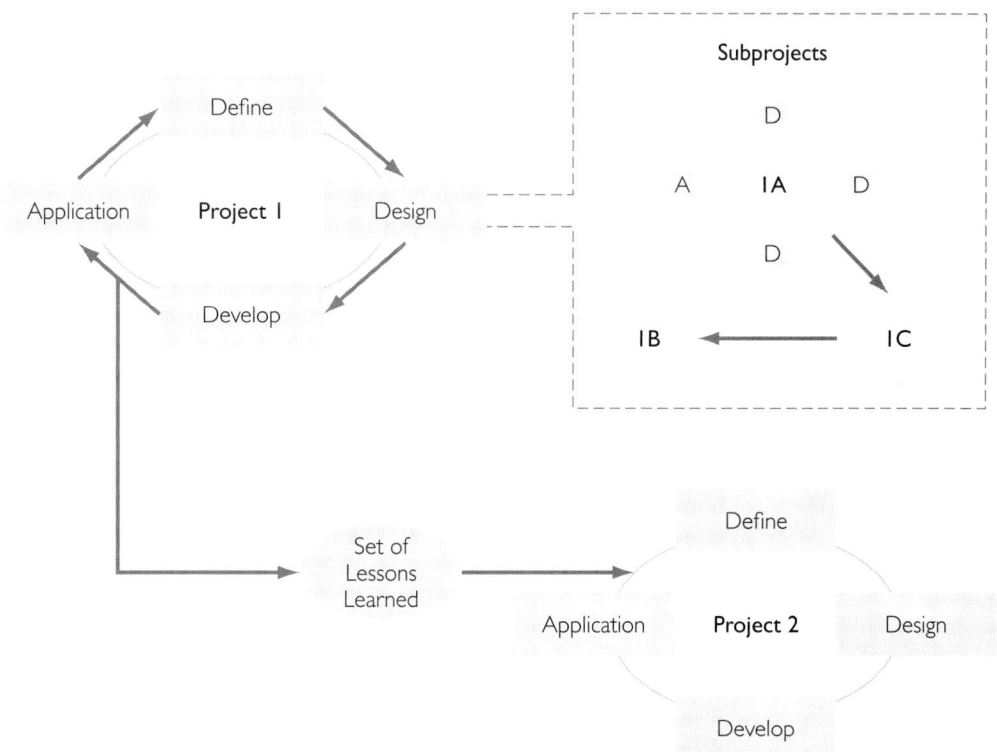

FIGURE **A2.4** Example of Project Interactions

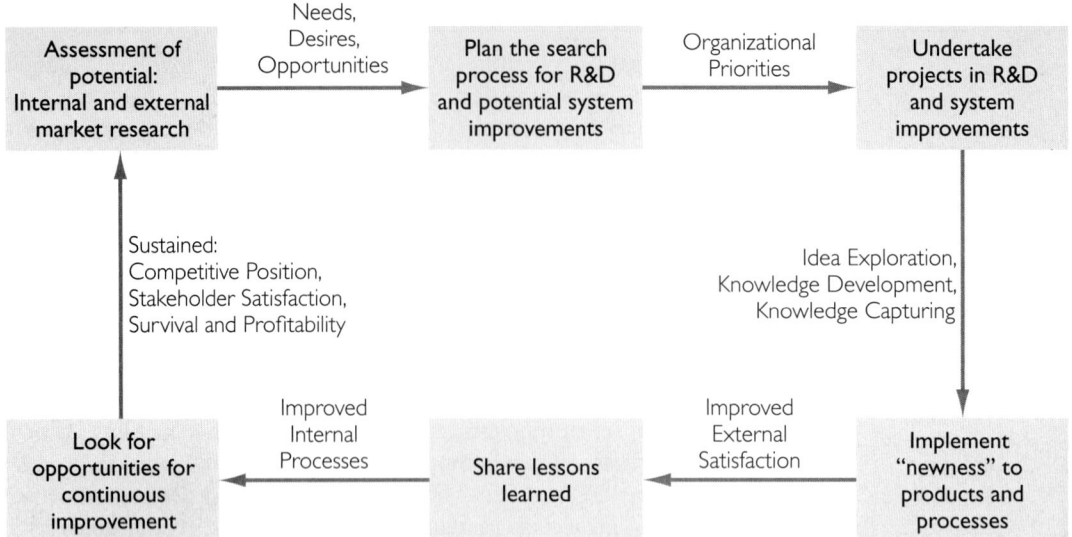

FIGURE **A2.5** Continuous Process of Innovation Development

to consider. As stated by Bill Howard of Sun, "strategy is often about deciding what you are not going to do."[10]

- An innovation project management support system that includes mentoring, planning, auditing, and managing the lessons learned
- Provision of opportunities to share lessons among team members and between teams
- The alignment of reward systems to support efforts in the innovation project teams as well as efforts to improve the management of the total portfolio of projects
- Careful monitoring of financial outcomes to refine, and redirect innovation efforts if necessary. Part of the Sun system is periodic review by senior management to check alignment of the review committee's work.

By having a portfolio of innovation projects that are ongoing in the firm, there is a continuous process of innovation development. Although Figure A2.5 shows what looks neat and orderly, the process is often messy. What is important for the firm is that the steps in innovation development occur and the outcomes and activities are part of the organization's strategic processes for innovation.

SUMMARY

Innovation project management is a critical activity that will impact the implementation of MTI in a firm. Businesses face many unique problems that they can approach through the use of project management techniques. The ability to manage the innovation needs of the firm effectively through projects is a major challenge facing most firms. A few key points that will help managers to better manage these projects include:

- Don't put a lid on the creation of new ideas for projects.
  - Is there a real need for this project and is it viable (on its own)?
  - What is its priority relative to other projects?
  - Do you have the resources to undertake the project?
- As soon as you can, make an informed decision on which projects to continue.
  - Is there a strategic fit?
  - Do you understand the possible scope and implications?
  - Do you know what you are going to do next? What are you building toward?
- Having decided to do a project, do it, but stop if circumstances change.
  - Are you moving on for good reasons? Or is it because of newness factors?
  - What is your organizational strategy? Does the project fit?

EXERCISES

Audit Exercise

When we think of auditing, we often think of financial audits. However, audits can and should be used when examining innovation projects. The difference between financial and innovation project audits can be seen in the following table:

Type of Audit	Financial	Innovation Projects
Status	Confirms status of business performance in relation to some accepted standard	Must ensure goals and objectives are relevant in environment of firm today
Information Gathered	State of firm's economic position and health	Status of the project
Forecast/Prediction	Future stability and financial well-being—ability to cope with crises	Future of the project—likelihood of success and/or continuing the project
Measurement	Financial rations based on balance sheets and income statements as well as other accounting information	Project documentation, including budgets, costs, schedules, logs, resource allocation and usage documents, status of goal accomplishment
Records and Information System	Accounting records are dictated by law and practice; audit can start early on	Must create data bank; develop information-gathering system before the audit can begin
Recommendations	Focused on correctness of information; suggestions aimed at management of accounting system	Part of "excellent" process; suggestions may cover any aspect of the innovation process; should be aimed at continuous improvement

In Chapter 2, we discussed assessments of the external and internal environments as well as the strategic process. In the audit exercise of Appendix 1, we examined the areas listed below to determine how you would assess each of these to better understand social responsibility issues. Some were assessed using financial data, some were assessed using other information that should be present in the organization, and some were assessed based on experience.

1. Timing goals
2. Return on existing assets
3. Investment in new assets
4. Alignment of portfolio with balance of business objectives
5. Need for new business or market areas versus existing business areas
6. New and improved products or processes versus cost reduction
7. Alignment with business risk tolerance
8. Organizational commitment

Just as these can be used to examine how the technology portfolio aligns with the firm's strategy, it can also be used to examine how the innovation project portfolio aligns with the firm's strategy. The greatest value in the analysis is when differences appear then the firm begins to understand where it is not building synergies.

1. Using Figure A2.1 as a base, what information is most critical for the firm to gather and analyze during each step of the innovation project process? Even though innovation is considered an internally focused strategy, are there steps where external information is more important? What characterizes those steps?
2. What did you learn about trying to develop an assessment for determining a firm's success in undertaking an innovation strategy?
3. Are there other areas of assessment that should be considered?

Scheduling Exercise

Find a project you are interested in (it may be innovative or just new to you). Once you have identified the project, develop a set of documents that would be needed to carry the project through. The set of documents should include:

1. A set of goals/objectives
2. A budget
3. A resource list
4. A schedule
5. A plan of action
6. A list of what should be included in your post-project review report

What are the advantages of using this type of documentation for your project? Are there disadvantages?

KEY TERMS

concept generators 337	concept implementers 337	PERT 338
	CPM 338	

NOTES

1. Davidson, A. 2004. Managing a growth culture: How Sun manages its project portfolio. *Strategy & Leadership*, 32(1), 43–46.
2. *Project Management Institute's Body of Knowledge*. 2008. Newtown Square, PA: Project Management Institute.
3. Kim, J., and D. Wilemon. 2003. Sources and assessment of complexity in NPD projects. *R&D Management*, 33 (1): 15–30.
4. Schwartz, J. 2006. The five founding principles that drive innovation. *Financial Times*. London: September 13, 17.

5. Davidson, *op.cit.*
6. McCabe, E. 2007. Technology enables supply chain innovation at Sun. *Industry Week*, 256(1), 6A.
7. Davidson, *op.cit.*
8. Von Zedwitz, M. 2003. Post-project reviews in R&D. *Research Technology Management*, 46 (5): 43–49.
9. Moldenhauer-Salazer, J. and L. Valikangas 2009. The failure of Sun Ray. *Strategic Direction*, 25(6), 36–37.
10. Davidson, *op.cit.*, 45.

Managing Portfolios of Technology and Platforms

In this book, we have discussed the strategic management of technology and how new technology is brought into the organization. We have emphasized two approaches for bringing technology into the organization: internal innovation and external acquisition. However, obtaining the technology is only part of what produces a firm's success. The firm must also have the administrative structures necessary, including the appropriate organizational structure, reward systems, and other managerial systems that support the product and process technologies of the firm. Additionally, the firm must have the right resources available at the right time in the right place to help the firm be successful using the technology. Figure A3.1 summarizes the process leading to a technology plan and the activities necessary to implement the plan.

Thus, a web of choices must be made in addition to how to obtain technology if it is going to be successful. This web of choices is typically more complex because the firm usually focuses on several technologies simultaneously, not a single technology. (Figure A3.2 illustrates how the complexities for a firm increase as the diversity of technologies involved increases.) The success of the firm is greatest if the web of choices connect these technologies into a consistent set of actions. This appendix examines the way the firm should construct its mix of products and processes to be successful in producing a consistent platform or portfolio.

COMPLEXITY AND MTI

The complexity a technology firm must address in its management process will vary widely. For example, Dell Computers builds custom computers for a worldwide market and has a great deal more complexity to address than does a small hometown business that builds custom computers for the community. The more complex the environment, the more complex the processes and structures the organization needs to have in place to be successful.

349

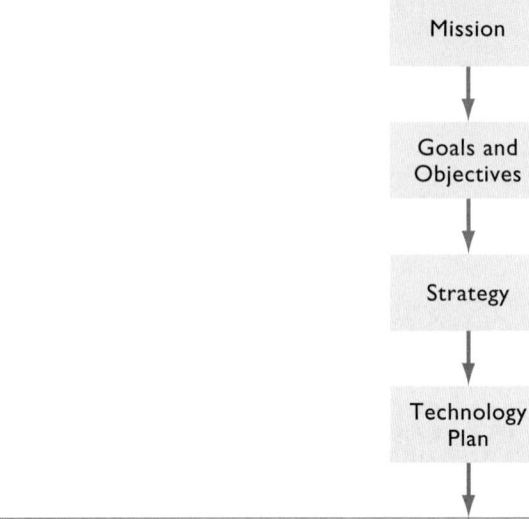

FIGURE **A3.1** Process for Developing the Technology Plan

When the organization starts adapting responses to complexity in the environment, it generally tries to increase the complexity of its technologies and its products and systems. Figure A3.2 indicates that the organization's approach toward technology—single product, platform, or portfolio—is a reflection of the complexity of technology and administration. The decisions about what approach to use depend on the firm's mission, goals/objectives, and strategy. Thus, as detailed in Chapter 1, a firm's approach to technology is a conscious choice by the firm about where it wants to go.

Once the organization selects the strategy to use, it needs to map the complexity it faces and its opportunities relative to the environment so that it develops a technology plan that will enhance its chances for success. Key questions that need to be asked in this process are summarized in Figure A3.3. The answers to these questions depend on the level of complexity that affects the technology strategy of the firm (Figure A3.2). Each of these broad technology strategies we will discuss next.

Single-Product Technology

The single-product technology strategy is the least complex approach a firm can take. Usually, firms using this strategy are small, and the market is local.

FIGURE **A3.2** How Administrative and Technological Complexity Relate

This is the domain of new entrepreneurial firms, and small family businesses. The simplicity of the administrative systems is because the firms are small and the internal processes are well understood by those involved. This does not mean the technology is not sophisticated or advanced; it means the product's technology is well understood by the firm. For example, Site-Specific Technology (SST) Development Group, Inc., is a software development firm creating geographic information systems (GIS) software for the precision farming industry. For example, one of this firm's products uses satellites to provide information processing and data analysis to support decision making in the agriculture industry. SST offers software that performs tasks such as spatial analysis, map generation, fertilizer recommendations, crop records, and field scouting. With this type of information, farmers from around the world can maximize efficiency and yield. While the product line and structure are simple, the technology developed and used by SST is very sophisticated and complex.

There are several ways an organization maintains or enhances its performance using a single-product technology. Specifically, the options for the firm that chooses this approach to its technology strategy include:

1. Product-market exploitation: This is an attempt by the firm to increase its sales of current products through the development of other uses in the market or reaching the market in new ways. The product technology stays the same; however, the administrative support may need to change.

Considerations	Key Questions
Competitive Position	How do we compare to our competitors? Where is the technology life cycle?
Market Appeal of Products	How is the newness factor of our products? Are we in danger of being leapfrogged in our technology or product?
Attractiveness of Technology	Are we cost competitive? Do our products match or exceed the performance of other similarly priced products?
Organizational Capability	Do we have the people, facilities, and other resources to remain competitive? Can we at least match our competitors' potential actions?
Future Orientation of the Firm	Is this the direction we want to go? What is the cost/benefit comparison between our current strategy and a defined new strategy?

FIGURE **A3.3** Key Questions for Determining the Level of Complexity to Pursue

Companies such as L. L. Bean and Barnes & Noble have not changed what they do; however, they have added web-based marketing.

2. Technology enhancement: This strategy aims at improving the systems and processes associated with the production and distribution of the product. Technology enhancement seeks to lower costs through increased efficiency. Typically, the product stays basically the same in this strategy. However, the increased efficiency offers a potential competitive advantage. For example, American Airlines can reduce customer time during ticket purchase and check in with its "remember me" enhancements. By registering with American, the customer can access flight and gate information by entering their phone number.[1]

3. Market enhancement: When the organization sells its current products in new markets, it is pursuing this type of concentration strategy. The market enhancement may be new geographical areas, or it may be new market segments.

4. Product enhancement: This technology strategy is aimed at improving the product—"the new, improved" syndrome. The product remains essentially the same but the product is improved through the adding of new features, options, sizes, colors, and so forth. Apple has been at the cutting-edge of product enhancement in the wireless telephone market since it entered that market with the iPhone.[2] The iPhone has remained basically the same, although it has been enhanced since its introduction.

The advantage of the single-product technology strategy is that the firm becomes proficient at what it does. The major drawback is that the firm is vulnerable to environmental shifts or competitive actions that usurp its competitive position.

Product Platforms

A second more complex strategy occurs when groups of products that are related by the way they are designed, manufactured, branded, distributed, or in some other way are said to be part of the same platform. For a platform to exist among a group of products there needs to be some type of relatedness. Technology is often the basis for the relatedness; both product and process technology within the firm can provide this base. Other factors can be the basis of relatedness and the focus of platform development. These can include resources such as people and location. Alternatively, the relatedness can be based on customers, branding, or global expansion demands. There are a number of potential benefits to platform thinking for a technology strategy. These include:[3]

- Speed: If a firm uses existing technology, components, logistics, or channels, then it is likely the firm can introduce "new" products more quickly. Also, there should be less product development time, and less training or retooling necessary in developing production systems.
- Cost: New products that emerge from existing technologies can be cheaper to design, manufacture, and market. For automobiles, the ability of Honda to use the Accord sedan frame for its new hybrid car allowed it to cut costs.
- Design quality: Because design problems typically are dealt with in earlier versions of the product that forms the foundation for the platform, using a platform for product development should result in better design quality. Any improvements in design should have repercussions in derivative products. For example, Hewlett Packard uses a converged infrastructure that 'converges' virtualized computers, storage, and networks with a firm's given facilities. This single shared-services environment optimizes the workload for the users of the HP products. This approach helps customers make a more efficient use of IT, facility, and staff resources.[4]
- Coherence: Good platform management should lead to better consistency in the makeup of the products and processes of the firm. In addition, where economies of scale and scope can be derived, good platform management will allow the firm to take advantage of such efficiencies. McDonalds has coherence in its products around the world. However, there are differences in the makeup and marketing of the products. For example, in Scandinavia, French fries are sold with mayonnaise, not ketchup, and dessert pie choices in Asia include red bean and corn. The product is essentially the same, but there are slight modifications to adapt to local settings.
- Referenceability: Being platform oriented encourages satisfied customers to try new products the firm develops. Many customers buy their new faster and improved personal computers from the same company as they did their first computer because they know that firm. Other examples include companies like Toyota with its up-market car Lexus and its hybrids or Apple with its iPhones and iPads. In both cases the breadth of the firm's platforms resulted in satisfied customers in one domain

becoming more likely to buy the firm's other products. Brand recognition can be very important in platform development.[5]

- Option value: Investment in the platform can help a firm further develop core technology, cultural understanding in new markets, new manufacturing processes, or build flexibility without committing totally to the newness. This option allows the firm to explore possibilities in a more controlled manner.

Complementary Platform

Another technology strategy is a complementary platform. The benefits listed in the product platform emerge from the operation of the platform for a single related product in which there are efficiencies. However, a platform can also be the foundation for new products and processes. The greatest power comes when the platform is not only the foundation for that firm's new products and processes but also is the foundation for other firms that rely on that product platform. For example, "Intel inside" is a widely recognized technical phrase indicating that Intel's microchips are in the given product that a consumer may buy. Thus, Intel is able not only to push its own products through the distribution channels, but it is also pulling its products through as other firms want to have Intel chips in their products.

You will recall from Chapter 2 the discussion of complementary forces with Porter's five-forces industry analysis model. These complementary forces are reflective of the strength of such platforms in an industry. The power of the platform increases dramatically as more ideas and innovation emerge from the platform. However, if the platform is displaced by another platform or other factors in an industry, it becomes more difficult for the firm to prosper.

There are four factors that influence platform leadership.[6]

1. Scope of the firm: The firm's scope effects both internal innovation and obtaining technology from external sources. A key strategic decision in both cases is: What business are we in? The answer to this question will determine what platform of products and technologies to produce internal to the firm and what product and technological complementors will be encouraged outside the firm.
2. Product technology: This is an internally oriented decision for the firm. The power of a platform comes from the sharing of intellectual property, design of the architecture of the product, and the degree of modularity within the product design and production processes. The more the firm interfaces with others, the more likely unintentional technology transfer may take place. Thus, a firm needs to be alert for potential complementors (recall our discussion of this concept in chapter 2) that may evolve into competitors. For example, Nike designs and markets sports shoes. The firm has a broad platform of different shoes (golf, baseball, running, walking, etc.) that Nike has developed. Independent firms manufacture Nike shoes exactly as directed by Nike. Because the manufacturers know how to make Nike shoes, Nike must be careful to ensure that they do not become competitors.

3. Relationships with external complementors: This reflects the degree of competition or potential competition for platform leadership by a firm that now complements the business. Because of the potential for conflict of interests and the potential desire of a competitor or complementor to leapfrog the product or technology of the platform leader, the relationships need to be managed and monitored. Intel and Microsoft are well-known complementors. They understand each other's strategies, goals, capabilities, etc. However, AMD has emerged as a problem for Intel and Microsoft as it has punctured the powerful hold Intel had on PC chip manufacturing.

4. Internal organization: The administrative support and the technical aspects of the organization allow the platform leader to manage potential conflicts more effectively. The well-managed firm is less susceptible to leapfrogging or to losing platform leadership. It is important that the internal organization allows debate, failure, and questioning of process to stay in a strong position relative to the market, complementors, and potential competitors. Platform leaders have to artfully build alliances and coalitions with internal and external groups with shared interests. It is easier to build these alliances if internal processes are in place to continuously reevaluate and modify the direction of the firm.

DEFINING PLATFORM STRATEGY

Product and complementary platforms can clearly be a successful technology strategy for a firm. However, if the firm is to employ either of these strategies certain steps must be taken to ensure success. These steps are similar to the discussion presented earlier in the text. They include building the appropriate team, understanding the general environment and competitive environment, and having specific operational competitive advantages. The concepts are the same as we have discussed previously, but here we are applying the concepts to this more complex setting. The specific steps to become a platform leader include:[7]

1. Assemble a multidisciplinary team with engineering, marketing, and operations personnel. Too often, decisions about new product development or product renewal are left to one or two of these groups rather than including all three. Because each has a different view, the development of other products and processes is more likely with all three involved. At the very least, failure to recognize major potential problems is less likely.

2. Segment the markets into a grid of niches. These niches may be based on price/performance relationship or some other combination of two important aspects that should be considered. Determine where the firm has coverage in its product line, and where it does not. Then the platform development can be directed toward specific strategic goals to fill gaps or accentuate successes. For example, The Hershey Company began making non-chocolate-based candies when it realized that it was not addressing this attractive niche at all. They had the technology and the ability but no product. They were ignoring a large market segment and limiting their own growth.

3. Identify growth opportunities in those market niches. Hershey identified a range of potential niches in the non-chocolate candy and then determined there was room for it to compete in many of those niches. Today, Hershey has a platform in the confectionary industry that includes both chocolate and non-chocolate-related products.

4. Define and map current product platforms and where they fit in the grid. Defining the product platform is not always easy. For Hershey, if the definition of its product is chocolate-based candies, then it misses the opportunities in the non-chocolate candy industry. Many companies miss opportunities because they do not define the technologies and products they have in ways to open up potential areas of interest.

5. Take a fresh look at the market needs, product technologies, logistics, channel relationships, and manufacturing processes to formulate new ways of viewing the product and process technologies of the firm. This fresh look starts with a clean sheet with all types of interfaces examined. The cross-functional team is looking to leverage new opportunities and new ways of doing things. Hershey found it could use its knowledge of the candy industry to enter the non-chocolate market. This made such entry less costly for Hershey than it would have been for a non-candy manufacturer.

6. Ask customers how to make the product better or to provide more value. If the product is global, then address the differences that exist across cultures. Can the application of a technical change or product change enhance the product? Hershey's approach to its introduction of a premium chocolate product, Bliss, was consumer driven from concept to launch. The consumer sought indulgence. Hershey combined R&D and their understanding of customers' demands to deliver the ultimate personal indulgence—Bliss.[8]

7. Analyze the products of competitors to determine how existing products and processes compare to the products of rivals and to potential substitute products. The analysis should be a breakdown of the other products. This is an area where reverse engineering skills are important to the firm. If the firm understands its products thoroughly, then it has the basis for a systematic, step-by-step comparison. This benchmarking can help the firm improve its products and realize where platform opportunities exist.

8. Examine all processes and distribution channels to be sure they are as good as they can be. Too often, firms assume that processes, suppliers, customers, and distribution systems should be a standard for the new platform products. Such assumptions can limit the ability of the firm to break through to new niches. Hershey began on-line ordering of customized Kisses with messages such as "I Love You" or "Congratulations." As The Hershey Company develops its on-line presence, it will add more products that can be customized.

9. Understand how new product platforms relate to the core competencies of the firm. Before new product platforms can be built, the platform team must determine what needed abilities are found in the firm, which ones can be obtained externally, and which ones must be developed. The assessment of the firm's competencies defines what it can do well

and what it needs to learn or acquire before it can renew a product or develop a new product platform.

10. Plan the project platform, develop a project platform implementation team, and make resources available. The innovation project process outlined in Appendix 2 should serve the firm well in this process. Platform development is directed from inside the organization just as internal innovation is. The introduction of Bliss has been so successful, that Hershey's plans to follow this new process in future new product development efforts and product introductions.[9]

The mindset of product platform development is similar to the mindset of new product development. The process just outlined and the leveraging of resources are keys to success. However, sometimes the firm does not want to build new products around the same technology but instead it wants to diversify its efforts. This diversification requires a portfolio management approach as a part of the portfolio strategy because of the increased complexity in technology and organizational administration.

Portfolio Management

Portfolio management represents the most complex environment in the management of technology and innovation. A portfolio approach by a firm indicates that the firm has different technologies, products, processes, and/or other strategic considerations. Thus, if a platform has some sense of unity to its operations, a portfolio is typically broader. Earlier, we defined a conglomerate as having multiple strategic business units (SBUs). Firms that use a portfolio approach to MTI often have multiple SBUs. At the very least, they have unique product lines that are not related on any of the factors defined as a basis for platforms.

There are several characteristics in the portfolio management of technology and innovation. These include:[10]

- It is dynamic with uncertain information and changing conditions.
- It is ongoing and must be constantly updated.
- It requires evaluation, selection, and prioritization.
- It demands that bad product and process technologies be eliminated.
- It should be designed to review the total portfolio on a regular basis.

Technology management in a portfolio can be viewed along five different processes. These processes are means that a firm can use to manage the complexity that occurs in a setting where the firm must manage a portfolio.[11]

1. *Identification* of opportunities and threats in the external environment and the internal strengths and weaknesses of the firm.
2. *Selection* of technologies that the firm wants to develop and exploit. The firm needs to examine the potential outcomes for each technology area to determine the feasibility of success and development of competitive advantage.
3. *Acquisition* of new knowledge through the development of internal innovation or the obtaining of external technologies.

4. *Exploitation* of opportunities through the development of strong products and processes. Exploitation can also include the development of unrelated platforms in various business segments. For example, 3M is well known for its tape and Post-it Notes, but it also has strong product platforms in safety equipment, replacement joints for hips and knees and medical imaging, to name a few.

5. *Protection* is the last area to consider. Product and process ownership is an important issue. Once a firm decides to add a product or process technology to its portfolio, it must develop ways to protect the market share that is gained. Sometimes the best protection is an aggressive strategy of market penetration or platform leadership. Sometimes the decision is how to keep from being leapfrogged or beaten in the marketplace.

Once the firm understands the portfolio issues and process, it must make decisions about internal and external knowledge development and capturing. Rather than trying to quantify risk, change, and uncertainty, portfolio management allows the organization to identify key elements and spread risk in an ordered, systematic manner.

Portfolio Balancing

As a part of the development of the portfolio strategy it is important to balance the portfolio based on the strategic goals of the firm. Lager defined four categories for portfolio consideration.[12]

1. Optimization opportunities that use a proven technology in an existing environment. This may not be significant in the long run, but for short-term refinement, it can extend the utility of a process or product. This is the lowest risk and lowest cost of the portfolio considerations.

2. Technology transfer of a proven technology in a new environment within the firm. This usually involves obtaining technology from the external environment. The risks of experiencing start-up problems are low; however, the risk of being too far behind the technology curve can be high (i.e., costs cannot be recovered before the technology becomes obsolete). Greeting cards are becoming more virtual; therefore firms like Hallmark need to find new products or ways to use their technology. Hallmark introduced musical cards and cards with small memory chips several years ago. Now, they are introducing recordable storybooks so grandparents or traveling parents can "read" to their kids.[13]

3. Competitive and low-cost technologies developed internally. They require little new investment for the firm, but the newness to the marketplace makes them very attractive with a high potential for profitability. Viagra was an example of such a product when it was introduced as an erectile dysfunction medication. The development and testing of the product as a circulation medication had addressed the safety issues in human subjects. The change in marketing had little production cost or newness for the firm.

4. Radical and risky technologies that are new to the firm and to the marketplace. Usually, these are also internal innovations. Depending on the size and the needed resources, this type of innovation carries the most

risk and the most potential to dominate the technological domain. The Kindle has been successful as a replacement for buying and carrying actual printed books. Because of eyestrain with computers as well as the fact that many readers like "the feel" of a book and the ability to make notes in books, many thought an "electronic book" would not be successful. The attention to detail in the screen clarity has proven otherwise. However, unlike a real book, it is hard to share with friends.

The portfolio of the firm can be analyzed on two levels: technology and product. The technology may be new but the product may not, and vice versa. Once the firm analyzes its portfolio along one of these levels, it needs to break down each of the preceding four areas along the other level. Without conducting analysis along both dimensions, it is possible to analyze the situation inaccurately.

Key to Success in Portfolio Strategy

The key to success in a portfolio strategy is to manage the portfolio; too often, firms keep adding newness, and resources are still being occupied (wasted) on products and processes that are not competitive. The result is the firm gets into a resource scarcity mode and neglects the new development it needs to remain competitive. The underlying causes of this include:[14]

- Preoccupation with short-term financial performance
- Reluctance to kill projects
- Failure to focus efforts
- Desire to get to market too fast—without proper preparation

MANAGEMENT PROCESS CHECKLIST

Figure A3.4 illustrates the process of determining the strategic approach the firm wishes to take in managing technology and innovation considerations for product platforms. The process is based on answering the two questions that we emphasized earlier:

- Where are we now?
- Where do we want to be?

  It is a continuous, ongoing process that involves:

- Monitoring the firm's position relative to current successes and potential threats. The assessment of risk factors lets the firm know what the cost/time factors are as well as other parameters such as technology, competitive environment, and acceptability.
- Examining risk factors relative to characteristics of technology and organizational processes. What alternatives are available, what life cycle stage is the technology/product in, and what resources are available to be allocated to the new product and process technologies?
- Examining strategic options are the same as any evaluation process—stay the same or make a change. The options for change involve the

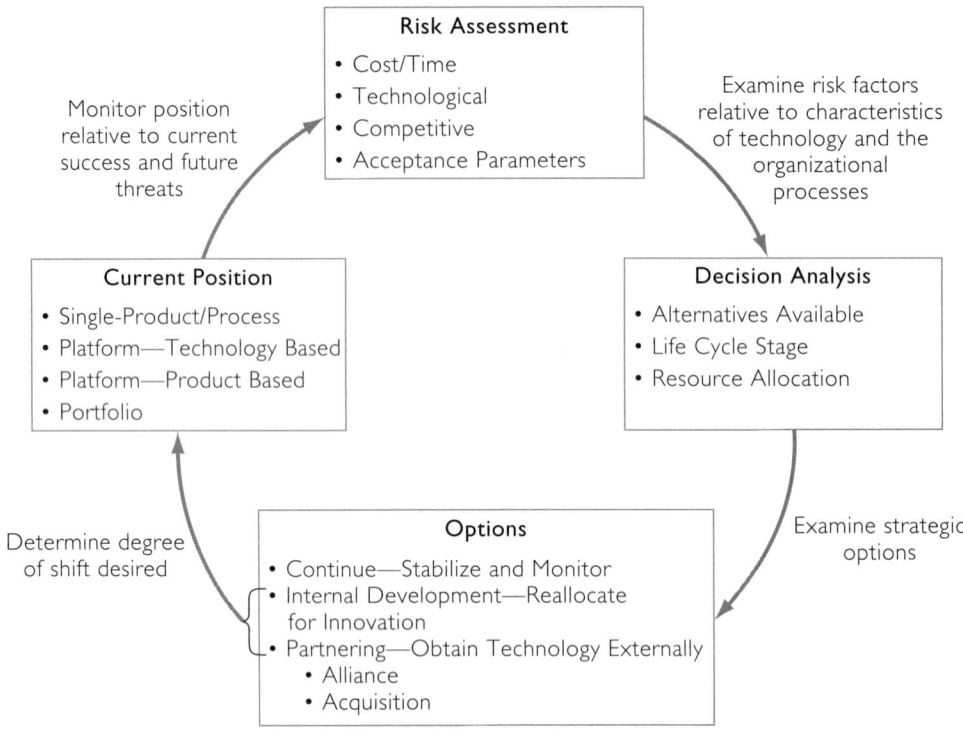

FIGURE **A3.4** Process for Determining the Level of Complexity to Pursue

strategies studied in Part Two and Part Three of the text: internal innovation and obtaining technology externally.

- Determining the degree of change that is needed or desired. The position can shift to single-product/process technology, platform development, or portfolio management.

SUMMARY

This appendix has discussed an important aspect of strategic management of technology and innovation: how to determine the product and market mix the firm will use in meeting its strategic goals. The internal innovation strategy requires determination of where the innovation will take place. The three most common areas are in the product, in processes, and in the market. When obtaining new technology externally, it is more likely the firm will look at related and unrelated products, related and unrelated processes, and related and unrelated markets in building a portfolio. Although both broad strategies are possible for the different levels of complexity, increasing complexity usually indicates the need for more complex activities; alliances and mergers/acquisitions are usually more complex. Too often, the firm finds itself with unplanned complexity in its technologies without the administrative systems to support it. The firm must balance complexity of technology with complexity of administrative systems.

EXERCISES

Audit Exercise

1. To develop a metric for analyzing the type of platform or portfolio needed, it is imperative that there be an understanding of the context of the firm. When we discussed the environment of the firm (Chapter 2), we indicated there were four key areas: economic, political-legal, social-cultural, and technological. As top managers go about determining the strategic direction of the firm's technology:

   a. What key questions should they be asking in each of the four areas? Be specific and be sure the questions relate to MTI.
   b. What type of metrics would be most appropriate for answering the questions you developed?

2. How should the time frame to examine be determined? How does the time frame influence the questions being asked?

Process and Innovation Exercise

Toward the end of the chapter, we presented a process model (see Figure A3.4). How would you relate these processes to internal innovation? To obtaining external technology?

1. What effect do you think each strategy will have on organization structure and processes?
2. How does your view of issues to be addressed in the process differ between being internally or externally focused?

NOTES

1. Anonymous. 2009. AA makes it easier, faster to access flight information and reservations with enhancements to "remember me" technology. *M2 Presswire*. February 5.
2. Hudson, D. 2009. The new iPhone 3GS and You. *Black Enterprise,* 40(2), 42.
3. Sawhney, M. 1998. Leveraged high-variety strategies: From portfolio thinking to platform thinking. *Academy of Marketing Science Journal,* 26 (1): 54–61.
4. Brodkin, J. 2009. HP BladeSystem Matrix takes aim at Cisco. *Network World.* 26(16), 34.
5. Mentzer, T. 2007. Be more than a company; Be a brand. *Landscape Management,* 46(10), 70.
6. Gawar, A., and M. Cusumano. 2002. *Platform Leadership: How Intel, Microsoft, and Cisco Drive Industry Innovation.* Cambridge, MA: Harvard Business School Press.
7. Meyer, M. 1997. Revitalize your product lines through continuous platform renewal. *Research Technology Management,* 40 (2): 17–28.

8. http://seekingalpha.com/article/82178-the-hershey-company-business-update-call-transcript
9. *Ibid.*
10. Cooper, R., S. Edgett, and E. Kleinschmidt. 2001. Portfolio management for new product development: Results of an industry practices study. *R&D Management,* 31 (4): 361–380.
11. Farrukh, C., P. Fraser, D. Hadjidakis, R. Phaal, D. Probert, and D. Tainsh. 2004. Developing an integrated technology management process. *Research Technology Management* (Jul.–Aug.): 39–46.
12. Lager, T. 2002. A structural analysis of process development in process industry. *R&D Management,* 32 (1): 87–95.
13. http://corporate.hallmark.com/Multimedia/Item/Recordable-Storybook-The-Night-Before-Christmas
14. Cooper, R., and S. Edgett. 2003. Overcoming the crunch in resources in new product development. *Research Technology Management,* 46 (3): 48–58.

Predicting the Future and Waves of Innovation

It should be clear from the discussion in this text that technology can have a tremendous impact on both individual firms and entire industries. Therefore, a business needs to understand what might happen in the future. Predictions about the future are not impossible, but they do require insight aided by the use of various tools and methods. Trends observed over time can help shape our understanding of what may occur in the future. There are also specific techniques that a firm or individual can employ to better understand how technology may change in the future.

This appendix examines technological changes and methods for predicting them. It does so first by discussing a historical perspective on innovation that can help us understand how things may change in the future. We then discuss several specific means to predict the future of a given technology. Finally, we briefly examine some of the major changes predicted to occur in technology over the next few years.

LONG-WAVE THEORY OF ECONOMICS—PATTERNS OF CHANGE

Economist Nikolai Kondratieff asserted that capitalistic economies experienced long-wave cycles of forty to sixty years in length.[1] Thus, there is an initial period of innovation and change but ultimately that technology will gain dominance and will be used for an extended time period. However, there will be another major innovation that will displace the existing technology and will gain dominance for another 30–60 years. The innovations commonly happen in multiple industries, not a single industry. Thus, the change is very broad and reflects a wave of change in society. Although Kondratieff identified these cycles, he did not propose a cause for this macroevolutionary process. Figure A4.1 summarizes the waves of technology innovations since the nineteenth century.

Long Wave	Survival	Railroad	Auto	Jet	Internet
Time	Pre-1850	1850–1910	1910–1960	1960–2000	2000–
Travel/ Communication Mode	Horse; Word of mouth	Railroad; Telegraph; Mail	Automobile; Telephone	Jets; Computers	Supersonic; Electronic imaging; Internet
New Inputs	Canals; Water power	Coal; Iron; Steam power	Electricity; Oil	Microprocessors	Biochip; Brain imaging
Driving Sector/ Key Activity	Agriculture; Cotton spinning	Railroads; Tooling machines; Steel	Automobile; Refining oil; Electrification	Microchips; Improving computing speed	Biotechnology; Nanotechnology
Emerging Sectors	Iron tools; Canal transportation	Steam shipping; Construction	Aircraft; Construction services	Networking; Globalization; E-commerce	Neuroceuticals; Bioeducation
Organizational Structure	Simple	Functional	Divisional	Matrix; Network	Satellite

FIGURE **A4.1** Waves of Technological Innovation

Technological Innovation as the Driver of Long Waves

Schumpeter associated Kondratieff's forty- to sixty-year waves with innovations that emerge from entrepreneurial efforts.[2] Distinguishing between invention and application (innovation), Schumpeter argued that innovation was the basic function of the entrepreneur in capitalist societies. The profit decreases and economic recessions that were caused by overcapacity in the industries of one wave would lead to innovations that would drive the next economic boom. These new innovations often provide quantum gains or renewals in productivity by doing fundamentally different things or by doing things in a fundamentally different way.

Van Duijn[3] and others built on Schumpeter's analysis of innovations and business cycles by combining three concepts: innovations, innovation life cycles, and infrastructure investment. Figure A4.2 summarizes Piater's insights on product and process innovation. One insight that can be drawn from Figure A4.2 is that just because a wave declines and is replaced does not imply that the product of that wave disappears.

Although there is much debate about the specifics of long-wave theory, such as exact time frames for waves, there are several points on which techno-economic forecasters agree.[4]

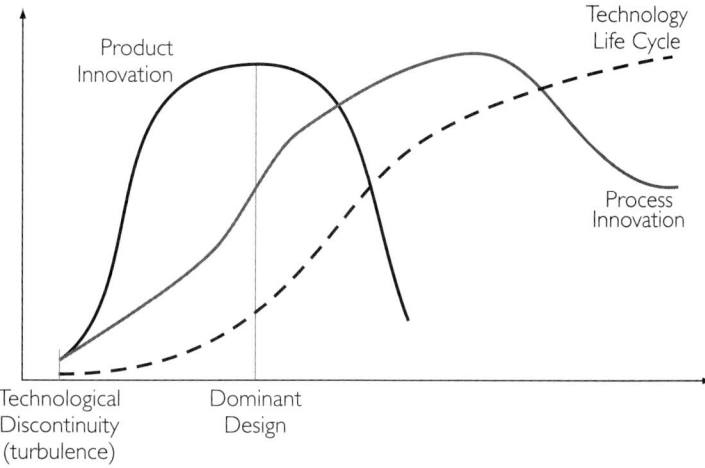

FIGURE **A4.2** Product and Process Innovation

1. Long-wave cycles in economic growth do exist (although some disagreement exists as to the exact timing of the cycles).
2. These cycles are associated with significant fluctuations in the price of important commodities.
3. They are also associated with waves of basic technological innovations that are generally fostered by the decline of the previous cycle.
4. In each wave, dominant technologies can be identified that are associated with primary energy sources.
5. In each wave, organization structure innovations develop that allow organizations to take advantage of the innovations generated.

For managers, the knowledge that innovation waves exist and that they occur in certain patterns means that as they happen, opportunities are created and should be taken advantage of. To illustrate, the development of the Internet created tremendous opportunities and change. Managers at the time should have seen a wave developing. However, the history of waves of innovation also suggests that there will be a substantial decline in Internet innovation in time. The decline, however, helps to generate the next wave of innovation. Managers should take advantage of waves as they develop but also avoid committing too many resources as the innovation wave declines. Each of the waves that have occurred in recent United States history is discussed next.

Survival Wave

Before 1850, most business enterprises had a simple administrative structure, the technology was relatively simple, and the focus of the firm was on survival. The businesses were usually small and family owned and operated. Most of the United States economy was based on agriculture and was localized. Products for sale were those that could not be readily produced on the farm and were manufactured by local craftsmen. The simple administrative structure was appropriate as long as businesses remained localized. However, around 1850, the railroads began to develop as a reliable source of transportation using coal and steam for energy. The railroads by nature were not limited to local areas. Thus, innovations in rail transport led to changes in how business was conducted.

Railroad Wave

During the 1880s, United States railroads were increasing in size, and ownership patterns were changing. Subsequently, railroad companies initiated the adoption of more complex administrative structures. These administrative innovations significantly increased the ability of industrial enterprises and other organizations to grow and deal with the increasing complexity of their internal and external environments. Increased administrative complexity led in turn to the emergence of the functional structure. The late 1800s were a period of mass migration to the West. Trade expanded as transportation that is more reliable became available. The railroads expanded during this time using a functional structure where there were separate departments for each major function in the company such as finance, marketing, land purchasing, and operations. The development of the railroads put businesses such as the pony express and stagecoach companies out of business.

Auto Wave

The early 1900s saw the development of the automobile. This innovation gave people more mobility. No longer did they have to follow the railroad routes or waterways for mechanized travel. Trucking grew as a link between small towns. Airplanes relied on engines that are similar to autos' and are part of this wave too. Barnstormers and trucks moved products faster. Telephones also became more prominent, allowing direct voice contact between people in different locales. As organizations grew larger and more complex, the functional structure began to show one basic weakness that had to be corrected: "a very few men were still entrusted with a great number of complex decisions."[5] The result was the emergence of the divisional or product-focused structure. In this structure, the firm is further compartmentalized. Thus, rather than having a marketing department that deals with everything in the firm, in a division structure, the marketing department would be concerned with only that division of the firm or with that given product.

Jet Wave

Jets and microprocessors were the next wave of innovations. Humans can now travel around the world in hours rather than days or weeks. Microprocessors and satellites allow almost instantaneous information transfer. These innovations have again changed the needs of organizations. A new structure that emerged during this long wave was the matrix structure. This structure allows organizations to have the flexibility to be proactive and reactive to the expanded variety of information available to them.[6] In this structure, individuals do not report to a single authority but report to dual authorities. Thus, an engineer in aircraft manufacturing may report to both a supervisor in engineering and another who manages a specific aircraft project such as a new jet fighter that is under development.

Internet Wave

Most recently, the Internet wave has introduced further change. The innovation of fiber optics and the Internet have led to instant voice, image, and text communication. This change has promoted the development of network organizational structures that are fluid and can change rapidly. The use of titles is minimized within the network organization, and priority is placed on flat structures because the quick response of the organization is paramount. Communication technology has necessitated rapid organizational responsiveness, and the network organization is best suited for this type of rapid environmental response.

Implications of Long Waves

Each innovation wave has led to the emergence of a new organizational structure. The new structure did not completely usurp other structure(s); instead, it gave organizations a new alternative for designing the internal processes, interlocking behaviors, and developing new ways to span new boundaries. For example, prior to 1850, most business enterprises had minimal information-processing requirements. They were usually small and family owned and had simple administrative structures. Information transfer was primarily word of mouth. It took months and weeks to spread the news. For these family-owned businesses, local information was of primary importance and use.

During the railroad wave, regional information became more important. With the advent of the telegraph, information lag was shortened to weeks and days. Railroads made it possible to travel greater distances in a shorter time. The functional structure emerged to meet the organizational demands as railroad companies recognized that the longer roads were not as efficient to manage as those under 50 or 100 miles in length.[7]

When nationwide information became necessary to compete in the marketplace effectively, the divisional or product structure emerged. Information time lags were reduced to days as mail services became faster because of the introduction of trucks and airplanes. The telephone was the first innovation that allowed immediate person-to-person information transfer.

With today's computers and satellites, the information-transfer process has been reduced to minutes and seconds. Satellites allow the organization to process international information immediately. The information systems have become integrating tools in the decision-making process. The matrix or network organization structure developed to facilitate cross-functional information transfers.

In summary, each of the successive waves of innovation has increased the amount of information that is relevant to the organization and decreased the time lags involved in processing the information and making decisions. In addition, after each long wave, the organizational structures that have emerged have decentralized decision-making and spread the organization geographically.

Managing Through the Waves

Firms that expect to survive changing waves of technology need to find ways to move into new technologies. Figure A4.3 shows how two well-known

	General Electric	Nokia
Railroad Wave	First Lamp Factory; Light Bulbs; Locomotives	Paper Mills; Wood Pulp Mills
Auto Wave	Airplane Engines; First Television Network; Electrical Appliances	Electricity Generation Equipment; Cable and Electronics; Automobile tires
Jet Wave	Laser light; Jet Engines; MRI; Fiber optics	Electronics Department Formed; Combined Multiple Businesses under Nokia name
Internet Wave	Better light bulb; 4D imaging; Digital Hospital Miniature Ultra-sound; Holographic Data Storage	Divested all businesses except telecommunications
Primary MTI Process	Internal innovation with some acquisition and divestment	Acquisition until last decade— now more internal innovation oriented

FIGURE **A4.3** How Firms Change Strategically

firms have done just that through their 100+ years of existence. In Part One, we opened with a brief history of General Electric. GE was one of the original firms listed on the Dow Jones Index. Through the years, GE has had a history of developing products and then improving them as technology advances. For example, GE recently unveiled a new locomotive design that has a "greener" engine as well as a jet engine that burns less fuel and is "greener." However, GE realized that while it was introducing new products, it was not making great leaps in technology at the rate strategically desired. This recognition has pushed GE to look outside the firm for acquisitions of more radically innovative products.

Nokia, the Finnish telecommunications company, has followed a very different path, but has also ridden the waves of technological change. It was founded in 1865 as a wood pulp mill. In this business, wood is crushed and wood pulp is created so that paper can be manufactured. The technological foundation for many businesses was shifting in the late 1800s as new technologies emerged. In part, business firms taking full advantage of railroads that had increased the efficiency of product transportation drove this new technological trend. In addition, new chemical processes were beginning to take hold. Nokia integrated some of those new chemical measures in its wood-processing efforts. The firm then built on its chemical knowledge and began to diversify into a new emerging chemical process, the vulcanization of rubber. One outcome of this adaptation to the wave of innovation was Nokia next began to focus on manufacturing rubber boots. When the automobile emerged, Nokia then moved to take advantage of this new technology by making tires for automobiles.

In the 1960s, informational equipment began to take hold. Again, Nokia was at the forefront of these changes. The firm made one of the first mobile telephones, a phone that employed radio technology. This telephone was large, cumbersome, and expensive, but it was a predecessor of today's mobile phones. Nokia in the 1970s and 1980s decided to concentrate on one business—telecommunications. As a result, Nokia divested its other businesses at the beginning of the 21st century. Today, Nokia's strategy is to focus on one industry; whereas GE has decided to continue its focus on multiple product lines in multiple industries.

METHODS FOR VIEWING THE FUTURE—LOOKING OUT 5 TO 10 YEARS

Although the firm should recognize that long waves of innovation occur, there are also more immediate concerns about the future that the firm needs to monitor. For example, while 40 to 60 year waves are important, businesses need to look 5 to 10 years out to see what might be more immediate threats, even if those threats are not as substantive of change. If they do not they will not be in business to ride the next wave of innovation. There are a number of ways to look at what future trends may be. All of these methods have their advantages and disadvantages. Vanston proposed five different ways in which a firm can gain insights to the future. These five methods are very useful to predict events 5 to 10 years in the future.[8]

1. Extrapolation is based on the premise that the future will represent a logical extension of the past. This process assumes that the patterns of the past will continue into the future in a reasonably predictable manner. Thus, extrapolation of past information into the future will give a strong indication of what the future holds.
2. Pattern analysis is based on the belief that the future will replicate past events. The saying that "history repeats itself" is the basis of this view of the future. Identifiable patterns do exist, but each one is different from the others. This seeking of patterns has a place in viewing the future, but caution should be taken not to take the analogy too far.
3. Goal analysis is based on a self-determination type of model. The general belief is that actions taken by relevant stakeholders will determine the future direction of technology and the organization. While evaluating the importance and role of various stakeholders, there is always the danger of not recognizing some critical groups or overestimating the importance of others. In addition, some group may emerge as overwhelmingly powerful because of changes in the environment. When safety issues emerge relative to a new technology, all of the other stakeholders may lose influence until those issues are addressed.
4. Counterpunching involves the belief that the future will result from events and actions that cannot be predicted and may even be random. This view of the future recognizes the complexity of events and decisions that organizations make to add value, but it minimizes the planning for contingencies or most likely scenarios.
5. Intuition involves "gut feelings" about the forces that will shape the future. Random events and the actions of individuals and social institutions so determine the future that rational, analytical methods of trying to determine

the future may not be possible. Therefore, one method is to gather individuals with high levels of expertise and let them discuss the future. The decisions based on this method are only as good as the ideas generated and the ability of the group to recognize the value of the ideas. Politics, the desire to maintain power, and blind spots are major problems with this view.

These five methods on how to gather and analyze information each have advantages and disadvantages when looking 5 to 10 years ahead. The conclusions that are drawn from these techniques determine the new products, markets, and processes the firm will pursue in the future and the amount and type of knowledge sharing and generation. What is important for the organization is to gather relevant data, transform them into information, learn from the information, and process the newly acquired knowledge in a way that leads to competitive advantage. The ability of an organization to forecast successfully can be a competitive advantage. The longer the time frame for prediction, the more important it is that the organization pursues multiple techniques and methods to understand what that future may look like. We will next look at those methods to predict the future less than 5 years into the future.

FORECASTING TECHNIQUES—LOOKING OUT 5 YEARS OR LESS

The historical waves occur over long periods of time (forty to sixty years). The preceding section discussed methods that provide insights on shorter-term trends but still are relatively long-term perspectives (five to ten years). But, there are also methods to make forecasts that are more immediate (less than five years). These forecasting methods are driven by more specific methods than the five methods cited in the prior section and are particularly useful for forecasting specific technologies to pursue.

Several assumptions about these forecasting methods need to be made clear before examining specific techniques. First, there is no way to forecast with certainty. By definition, forecasting is a "crystal-ball" proposition. The purpose of forecasting is to provide a basis for future decision making. However, as pointed out in evaluation and control (Chapters 5 and 8), the firm needs to constantly monitor the direction it is going to see if goals are being met. Second, no forecast is perfect. Each method or technique has weaknesses. In addition, information gathering and interpretation are arts, not sciences. Finally, forecasts can help in formulating strategic direction for the firm. It is the best guess of what changes will be occurring. Managers need to be constantly aware that it is a best guess. However, there are techniques that can improve the odds of the guess. Six specific methods will be discussed next. Some of these methods have similarities with the five methods cited earlier; however, the underlying methods here rely on mathematical support to a greater degree than do the prior methods.

Trend Extrapolations

These methods examine trends and cycles based on historical data. The trends are then extended based on mathematical techniques such as regression, weighted smoothing, decomposition, and flexpoint analysis methods. Each of these methods has its advantages and disadvantages. More complex modeling does not necessarily mean better forecasts; however, the use of computer databases and computer modeling techniques gives the forecaster the opportunity to examine the results of a number of different techniques for extrapolation.

The basic assumption of extrapolation is that the past is the best predictor of the future. Therefore, the only criterion for producing a forecast is the availability of historical data. It should be noted that two forecasters using the same data can develop very different forecasts based on method used, interpretation of data, and decisions about the variables and parameters used. These techniques can be quite successful in the short term, but the longer the time frame, the more inaccurate the extrapolation is likely to be. Thus, the model is most useful looking forward a year or two, and its predictive ability drops as the time frame gets longer. Although mathematical models are straightforward, judgment-based models are generally superior.

Expert Consensus

Another group of techniques that is used in forecasting involves the judgment and consensus of experts. Sometimes the group of experts meets and, through a series of discussions, finds some level of consensus. Often, a vote is taken among participants with consensus being chosen as some arbitrary number of the group voting for some given view of the future—75 percent or so. One of the best-known methods is the Delphi technique. In this method, responses and those making the analysis remain anonymous. The technique involves a set number of experts writing their predictions, the responses being collated, and then everyone responding to the list. The process continues with the narrowing of opinions through each round until consensus is reached. Generally, the Delphi technique is more accurate than face-to-face discussions. This is because power, politics, position, and force of personality are eliminated from the discussion.

Simulation Methods

With the development of various technologies, simulation methods have become more popular in trying to predict future events. Simulations rely on analogs to model complex systems. For example, an airplane simulator can be designed using a mechanical analog and mathematical analogs. The mechanical analog is the simulator itself, and the mathematical analog is the probability of different events occurring. When there is an airplane crash due to mechanical failure or pilot error, there are studies done in simulators to see if such crashes can be prevented in the future.

For predicting the future, mathematical and gaming analogs are most commonly used. The S-curve is a common mathematical analog for predicting the life cycle of a particular technology. Gaming analogs involve creating an artificial environment or situation and then examining what happens during the manipulation of different components of the system. It will be interesting to see how the generation that grows up with computer games and computer-based learning will approach gaming analogs in the future.

Scenario Building

This method is based on the development of a worst-case scenario, a best-case scenario, and a most likely scenario. In this method, experts consider the impact of shifting components on the system as a whole. The outcome of scenario building is to provoke thought about how the firm would handle each of the scenarios. Most companies plan for the most likely scenario and hope to come fairly close. Sometimes the best case can become a nightmare without planning. The small manufacturer who has demand for its product exploding in a short time frame is faced with a multitude of problems that must be addressed. This often happens with the "hot" new toy item or other fad items. However, it can also happen with new technologies that discover a new market.

Decision Trees

Decision trees were originally developed as graphical representations of alternative choices. The original decision trees were based on yes/no questions, and the next question was based on the answer. Computer technology has allowed the development of much more complex trees with feedback loops and multiple alternatives. In addition, decision support systems have emerged that allow mathematical probabilities, risk factors, utility measures, and expected value to be calculated and factored into the forecasts.

Hybrid Methods

These methods combine several approaches to gain different views of what the future might hold. Hybrid methods have developed for several reasons. The need to view the future from several vantage points is the strongest reason for using a hybrid method. Each forecasting technique has its own strengths and weaknesses. By using several methods, the firm can potentially improve the chances for successful forecasting. A second reason for hybrid methods is that the goal is accuracy. Some areas of the firm have better quantitative data for analysis, and some must rely more on expert judgment. By using the best methods for the data available, the firm should have better results.

PREDICTED FUTURE TECHNOLOGIES

New waves of technology have already begun to develop. These technologies will impact the economy. A few of the more interesting or promising new technologies are briefly reviewed here.

Biosensors

Biosensors already are used in the medical field. For example, when people enter an emergency room, they will have small sensors attached to constantly monitor their heart rate, blood pressure, and so forth. If any substantial changes occur, there is almost instant feedback on those changes to staff in the emergency room. However, in the future, these sensors will be further developed so that the clothes you wear may react to potential illness before you can recognize it. For example, Nicholas Kotov, a chemical engineer has transformed fabric into a biosensor and an electrical conductor simply by dipping it into a solution of carbon nanotubes, antibodies, and a polymer. Thus, a service or your doctor's office would receive almost constant feedback on your health and could alert you to any abnormalities that may arise. As a result, the *biosensors* would keep people safe from disease or chemical poisons that may arise in the workplace.[9]

Fuel Cell Vehicles (FCV)

One of the most heavily researched new technologies is fuel cells that are powerful and more environmentally friendly than current technology. The desire is that these cells will be smaller than those that the current technology as well as more powerful. Although they are not expected to reach the mass market before 2012, fuel cell vehicles (FCVs) may someday revolutionize on-road transportation. This emerging technology has the potential to significantly reduce energy use and harmful emissions, as well as dependence on oil. FCVs will have other benefits as well. Like battery-electric vehicles, FCVs are propelled by electric motors. But while battery electric vehicles use electricity from an external source (and store it in a battery), FCVs create their own electricity. Fuel cells onboard the vehicles create electricity through a chemical process using hydrogen fuel and oxygen from the air. The emissions from these types of vehicles will be water and heat.[10] If hydrogen-based fuel cells are deployed in autos, trucks, and buses, the pollution associated with the internal combustion engine would be eliminated.

Smart Grids/Smart Meters

Between 2010 and 2030 more that $10 trillion will be spent on designing, developing, and installing smart grids for the transmission of electricity. The primary objective of smart grids is to overcome the endemic problems common to current electrical grid systems. In other words, smart grids will make the distribution and consumption of energy more efficient and cheaper. Smart grids will combine alternative energy sources, including wind and solar, as well as lead to the installation of an advanced metering structure. This system will help energy companies identify peaks and lags in consumption, limit electricity loss, and enable them to distribute loads more effectively and efficiently. The smart meters and smart grids will work in concert to optimize the power flows using renewable sources. These should be more cost efficient and reduce carbon dioxide emissions substantially.[11]

Solar Arrays

One new technology that is still in the developmental stage is orbiting solar arrays. Scientists are currently investigating the feasibility of this potential renewable energy source. It is thought that an array of solar panels could be placed into orbit. These panels would then beam the solar power back to a receptor for conversion into electricity. The primary attraction of this would be the ability to tap into an energy source that is larger than all other known sources combined. Building the array and positioning it in orbit around the Earth means that there would be a continuous flow of solar energy without any interruptions from weather or nighttime. There are a number of technical challenges remaining; however, those involved in the research are enthusiastic about the potential to generate significant power in the future.[12]

APPLICATIONS TO MTI

The record for accurate prediction of future trends in society and business is uneven at best. The goal of organizations is to evaluate trends and events in the hope of identifying opportunities in the marketplace and developing a competitive advantage. There are, however, several keys for making forecasting a useful tool to an organization.

1. Watch developments in other industries and fields of study. Look for opportunities and threats from other areas. Too often, companies are focused on their own industry and current competitive environment and miss the threats from other industries. The same is true for opportunities. The laser technology that is used in printers and pointers came from the photocopy industry.
2. Watch out for the "vested interests" that provide information. Enthusiasm and optimism may be the result of wanting something to happen rather than the reality that it will happen.
3. Remember that people generally do not like change. New processes and technologies have to add value for individuals and groups before they will be accepted. Just because it is an improvement does not mean that the new technology or process will be adopted.
4. Predictions should come from multiple views. Earlier in this chapter, we presented five views for predicting future trends. Potential mistakes are easier to avoid if multiple perspectives are used in the analysis.
5. Remember that it takes time for new technologies to realize their potential value. It may be ten or fifteen years before the value of a new technology is realized. Time for diffusion must be recognized and allowed. There may be a number of reasons for delays: poor organizational processes, failure to apply the technology in a way that excites consumers, or poor infrastructure that will support the technology. The original iPod by Apple revolutionized music availability. However, Apple was not a phone company. When Apple combined the iPod with the telephone, it revolutionized what expectations were for cell phones. Apple continues to develop the iPhone and its myriad of applications. Others are following rapidly. The telephone, the digital camera, and the music player are fast becoming one combined product. The firm that makes the next leap will realize new potentials that were not imagined in the 1970s when portable telephones were introduced.

SUMMARY

This appendix has detailed forecasting techniques and methods. In addition, it has demonstrated the continuing cycles of innovation that society and business are exposed to. In this text, we have emphasized that it is the prepared, forward-looking organization, and manager who have the best potential for finding competitive advantage. Whether the firm chooses internal innovation or external technology acquisition, thought and knowledge about potential future trends are a key to long-term success.

EXERCISES

Audit Exercise

1. To develop a judgment-based tool for forecasting, it is imperative that there be an understanding of the context in which the forecast is going to be used. When we discussed the environment of the firm (Chapter 2), we indicated that there were four key areas: economic, political-legal, social-cultural, and technological. As top managers go about the visioning of the future process:

    a. What key questions should they be asking in each of the four areas? Be specific and be sure the questions relate to MTI.
    b. What type of forecasting technique would be most appropriate for answering the questions you developed?

2. How should managers determine the time frame to examine? How does the time frame influence the questions being asked?

Trends for the Future Exercise

Toward the end of the appendix, we presented some technologies that appear to be ready to support the next wave of innovations.

1. How would you describe the drivers of the next wave of technological change?

2. What effect do you think this wave will have on organization structure and processes?
3. What environmental changes will we see in technology, social-cultural issues, political-legal activities, and in the global economy?

Justify your descriptions based on what you learned from this text and your other readings in MTI.

NOTES

1. Kondratieff, N. 1935. The long waves in economic life. *Review of Economic Statistics*, 17: 105–115.
2. Schumpeter, J. 1935. *Business Cycles: A Theoretical, Historical and Statistical Analysis of the Capitalist Process.* New York: McGraw-Hill.
3. Van Duijn, J. 1983. *The Long Wave in Economic Life.* Boston: Allen & Unwin.
4. Volland, C. 1987. A comprehensive theory of long-wave cycles. *Technological Forecasting and Social Change*, 32 (2): 123–145.
5. Chandler, A. 1962. *Strategy & Structure: Chapters in the History of the Industrial Enterprise.* Cambridge, MA: MIT Press.
6. Miles, R., and C. Snow. 1986. Organizations: New concepts for new forms. *California Management Review*, 28 (3): 62–73.
7. Chandler, A. 1962. *Strategy & Structure: Chapters in the History of the Industrial Enterprise.* Cambridge, MA: MIT Press.
8. Vanston, J. 2003. Better forecasts, better plans, better results. *Research Technology Management*, 46 (1): 47–58.
9. Bourzac, K. 2009. Nanotubes come into fashion. *Technology Review.* 112(3); 80–82.
10. http://www.fueleconomy.gov/feg/fuelcell.shtml
11. http://www.environmentalgraffiti.com/featured/green-technologies-future/7228 and http://money.cnn.com/2009/10/27/news/economy/smart_grid/index.htm
12. http://www.environmentalgraffiti.com/featured/green-technologies-future/7228

GLOSSARY

A

acquisition: The outright purchase of a firm or some part of a firm. The takeover might be by agreement or hostile. An MTI acquisition should be designed to obtain technology that the acquired firm has and the acquiring firm wants or needs to build and maintain a competitive advantage.

agency theory: A theory concerning the relationship between a principal and an agent. There are costs associated with resolving conflicts and aligning interest. Agents (company managers) may act in their own best interest rather than that of the firm and thus need to be actively monitored.

alignment: A fit among the systems within the firm as they support the firm's strategy. It involves monitoring and adjusting processes and structures to environmental changes and organizational outcomes.

analysis paralysis: Not getting anything done by focusing solely on analysis.

autoadjudication: Computerized systems that automatically determine the outcome of a query, such as claim approval for insurance companies.

applied research: Research that utilizes the new knowledge developed by basic research to develop new products or processes.

B

balance sheet: An internal accounting document that all firms generate and that provide information on the item.

barriers to entry: Structural factors within an industry or a firm that discourage potential new rivals. These may include government regulations, economic factors (large capital investments), or market conditions.

basic research: Research that focuses on the creation of new knowledge.

benchmarking: A systematic comparison of processes and performance to create new standards or to improve processes. It should help build competitive advantage. The firm seeks out the "best" products or processes in other units or firms. It then imitates or adapts the "best" to produce a better product or to improve processes.

buyers: Individuals who actually buy the output of the industry being analyzed.

C

capabilities: The set of organizational characteristics that facilitate and support its strategies. These are the building blocks for the firm's strategies and include the skills and abilities the firm possesses.

cognitive institutions: Institutions that shape the individual's behavior that come from the broader society. Most commonly, this is viewed principally as the culture of the country.

competitive advantage: The condition that enables a firm to operate more efficiently and/or effectively than the companies it competes with. This results in benefits accruing to the firm. It is something that the firm does better than any of its competitors.

complementors: Products or processes that interact so that the moves by one fit the moves by the other. Complementors are often seen as a sixth force in Porter's five-forces model. Product complementors are products that sell well with another product such as computer peripherals, which are complementary products for computers.

communication: The transfer of meaning from one source to another.

concept generators: Individuals who throw out ideas about how to solve the problems.

concept implementers: Individuals who focus on how to accomplish the ideas of the concept generators.

continuous technology: Changes in technology that occur over relatively short periods of time and tend to be incremental.

consortia: Characterized by several organizations joining together to share expertise and funding for developing, gathering, and distributing new knowledge.

corporate social responsibility: (CSR) is where an organization has a built-in, self-regulating mechanism that monitors and ensures its adherence to law, ethical standards, and positive behavioral norms.

CPM: The network technique that tracks the longest path of activities to be completed.

cybernetic control: A control concept that comes from the biological sciences and deals with the behavior of dynamic systems such as innovation and change over time. When a system needs to show a certain behavior over time, the inputs of the system are changed to realize this desired output of the system.

D

defensive technology: Technology the firm obtains that competitors currently use to gain an inroad to the firm's customer base. A defensive technology strategy would most likely involve acquiring the technology from an outside source.

delegation: The authorization of someone (usually subordinates) to make designated decisions about various aspects of the implementation process.

diffusion: It is a special type of communication that is concerned with the spread of new ideas.

disruptive technology: This type of technology changes the industry in such a way that previous competitive and business rules no longer apply. The new technology replaces the established thinking in a given domain. It is similar to a radical technology because both change how an industry competes. However, a technology does not always have to be radical to be disruptive.

divisional structure: A structure in which a firm has multiple business units organized by some competitive aspect (product, market, customer, technology, etc.) with separate function-based groupings of employees.

downsizing: When a firm either sells some of its units or lays off some employees to decrease the size of the firm.

due diligence: An investigation of important aspects of a potential acquisition target or alliance partner to ensure that the target or partner is as the acquiring firm believes and to better understand how value will be created.

E

explicit knowledge: Knowledge that concerns "knowing that" which may be shared by several individuals. Explicit knowledge is that which can be expressed clearly, fully, and leaves nothing implied. An example is knowledge that can be formally expressed and transmitted to others through manuals, specifications, regulations, rules, or procedures.

F

fast follower: A firm that quickly follows the first mover into the market.

first mover: The competitor that is able to enter into a product domain or develop a process improvement first. The firm may be first to market with a given product, first into a given market area, or first to use a given a technology.

financial fitness: The difference between the desired financial outcomes (objectives) and those actually produced indicates the level of financial fitness. The closer actual and desired outcomes match or the more desired outcomes exceed those desired, the higher the level of fitness.

formal leadership: Leadership that occurs because of a person's official position in the organization.

franchise agreement: A contract between the company (franchisor) and the person who buys an individual business unit (franchisee) to sell a given product or conduct business under the franchisor's trademark.

G

gap analysis: A performance measurement and analysis technique that searches for the difference between what a firm wants to occur, what actually has occurred, and what is likely to occur. The purpose is to uncover where strategic successes and failures have occurred or might occur.

H

horizontal merger or acquisition: Merger/acquisition in which the acquired and acquiring firms are in the same (or a very similar) industry.

I

innovation: The process whereby new and improved products, processes, materials, and services are developed and utilized.

income statement: An internal accounting document that all firms generate and that provide information on the item.

informal leadership: Leadership activities that occur in the everyday activities of an individual that are not related to the individual's formal position in the firm. Knowledge and charisma are two common bases for such leadership.

inertia of success: The complacency experienced by some firms that have had long-term success. It may cause the firm to miss new opportunities or threats in the environment.

Intrapreneurial: The term used to describe entrepreneurial activities that occur within organizations.

J

joint venture: Two or more firms combine their knowledge, capital, and so on to form a new third entity with a specific goal or objective in mind.

just-in-time (JIT) inventory management: A process innovation that ensures the inputs for the production process are there just as they are needed for the process.

K

knowledge: Familiarity, awareness, and/or understanding gained through the process of using information, studying events, and experience.

knowledge management: The ability to acquire, integrate, store, and share knowledge using human and technical systems. It includes the organization of intellectual resources and information systems within a business environment.

L

leapfrog: When a new technology skips over the existing generation of products to introduce a product with significant new qualities.

learning: The gathering and sharing of existing knowledge, which can come from internal or external sources. An increase in knowledge or skill.

licensing arrangement: When a firm agrees to pay another firm for the right either to manufacture or sell a product.

liquidity ratios: Used to judge how well the firm can repay its debt. These rations examine both short and long-term debt.

low-end disruption: A technology that enters the market with lower performance than the incumbent but exceeds the requirements of certain segments of that market. A radical technology does not fit this definition, but a disruptive technology can. The purpose of such entry into the market is to gain a foothold with lower cost.

M

management of innovation: A comprehensive approach to managerial problem solving and action based on an integrative problem-solving framework and an understanding of the linkages among innovation streams, organizational teams, and organization evolution.

management of technology: The linking of different disciplines to plan, develop, implement, monitor, and control technological capabilities to shape and accomplish the strategic objectives of an organization.

market power: When a firm has enough market share to shape that market's actions. Often gained through merger or acquisition, it is the power held by a firm over price and the power to subdue competitors.

mentoring: One-on-one activity between and among employees in the organization or a system that is designed to allow two parties to learn from each other.

merger: When two firms combine as relative equals. The joining of two firms where one transfers all of its assets to the other. In effect, one corporation "swallows" the other, but the shareholders of the swallowed company receive shares of the surviving corporation.

metrics: The measurements that the organization uses in its evaluation and control processes. It is a system of related measures that facilitates the quantification of some particular characteristic in which the firm has interest.

mission: A brief statement, usually fewer than sixty words, that builds on the firm's vision of itself to specify what it does and how.

N

normative institutions: The norms of the industry and profession. For example, the values of an accountant or a doctor are very similar around the world.

next-generation technologies: The change in the technology and its impact on society that is more than the small step experienced in continuous change but is not revolutionary either.

O

offensive technology: The use of technology in a manner that is not used by competitors, which gives a firm a competitive advantage.

operational fitness: The difference between the desired and actual operational performance. It is reflected in measures that examine the efficiencies that emerge from the combined activities in areas such as sales and manufacturing.

organizational learning: The acquisition of knowledge through the application and mastery of new information, tools, and methods.

organizational wisdom: Wisdom for the organization is an understanding that goes beyond data and information manipulators.

P

PERT: A scheduling tool that shows the network activities to complete, how these activities interact, and the timing required for them.

platform: Products and processes whose technology complement and interconnect to support each other.

pulling: When society makes demands for changes in product or process that lead to new developments or products being brought to market.

pushing: When new technologies that are not known or anticipated by society are brought to, promoted in, and adopted by society because a "need" is created.

process innovation: New ways of accomplishing tasks in the organization that are designed to increase efficiencies or effectiveness in that organization.

product platform: Groups of products that are related by the way they are designed, manufactured, branded, distributed, or in some other way are said to be part of the same platform. There needs to be some type of relatedness—technology, both product and process, people, location, customers, branding, or global expansion demands.

R

radical technology: Technology that causes a dramatic change in the way things are done in a society.

radio frequency identification technology (RFID): This technology places a small tag on each item at the manufacturer. This tag allows the product to be actively tracked from the time it leaves the manufacturer until it leaves the store.

ratio: Where the number of interest is divided by some relevant measure, such as total assets, sales, or equity, should be employed because it controls for issues such as size.

reengineering: A process that involves fundamental rethinking and redesign of work processes in a firm. The purpose is to discover new and better ways of accomplishing the tasks necessary for competitive advantage.

regulatory institutions: Laws and regulations in a given country.

relationship fitness: The difference between the desired and actual relationships within the combined firm. Such fitness is reflected in a number of issues within the organization such as: Are

decisions being made in a timely fashion? Is the proper information getting to the proper place within the organization? Are roles clearly defined? Is senior management involved? Are the cultures at least compatible? Are projects being properly monitored?

retained earnings: Net profits retained in a business after dividends are paid.

retrenchment: The process of reducing expenditures to become financially stable. Retrenchment may involve a number of activities—downsizing, divestment, and so on—but the fundamental purpose is to get to the core activities that the firm does well.

S

S-curve: A curve that graphs the four phases of the technology life cycle: embryonic, growth, maturity, and aging.

second movers: Not first into a market, but companies that move into it quickly after the first mover.

social accounting: A concept that describes the how, what, and why or social and environmental effects of a firm's actions on internal and external stakeholders.

strategic alliance: A partnership of two or more corporations or business units to achieve strategically significant objectives that are mutually beneficial.

strategic business units (SBUs): An organizational structure in which independent business units within the firm market their own products. The head of each SBU reports directly to the CEO.

strategic fitness: The degree to which the firm has the ability to align its strategic goals to its strategic outcomes.

strategic group: A group of firms that competes in a similar manner (i.e., customer, product, geography).

strategic group map: A tool to segment an industry into relevant groups so the business can identify which firms are the most direct competitors.

strategic management: The effort by a firm to analyze its environment and its own strengths and weaknesses and then consciously choose the competitive path it wants to follow. On that path, the firm will seek to build upon its strengths and address its weaknesses.

strategic planning: The firm should gather information on these various elements and understand which forces are strong and why the industry profitability is where it is. This external analysis of industry can aid the firm in understanding where it, as an individual firm, needs to act in the future to gain a competitive advantage.

strategy: A coordinated set of actions that fulfills the firm's objectives, purposes, and goals. It is a long-term action plan for achieving a goal or set of goals.

subcontracting: A type of alliance that is intermediate in formality.

sustainability: A pattern of resource use that is designed to meet organizational and human needs while preserving the environment so that these needs can be met in the present, as well as in future generations.

sustainable competitive advantage: A competitive advantage that can be maintained by the business over a significant period of time.

switching costs: The costs incurred when a customer changes from one supplier or marketplace to another. The higher these costs are, the more difficult it is to justify a switch in suppliers.

systems integration: Integration aimed at supporting existing businesses' improvements in existing products or opening of new markets.

systems view: A view of the firm as an association of interrelated and interdependent parts. It is an interdisciplinary field that studies relationships of systems as a whole—inputs, throughputs, outputs, and feedback.

T

tacit knowledge: Knowledge that concerns "knowing how." It is knowledge that is not easily shared. Tacit knowledge often consists of habits and culture that we do not recognize in ourselves and is embedded in group and organizational relationships. It is hard to identify, locate, quantify, map, or value.

tactics: The planned activities of the functional areas.

technology: The practical implementation of learning and knowledge by individuals and organizations to aid human endeavor. Technology is the knowledge, products, processes, tools, and systems used in the creation of goods or in the provision of services.

technology foresight: Demands that the firm not only understand what the new product or process can do for the firm immediately or for the society's immediate employment, but also what the product or process will do to the environment over time.

transaction costs: The costs of conducting and maintaining the alliance.

turnaround: The positive reversal in the fortune of a firm that is a significant change from the previous direction.

tweaking: Adjusting the ways the firm organizes its existing knowledge to increase its leverage.

V

vertical merger or acquisition: The merger/ acquisition of a firm that is either a supplier or customer of the acquiring firm in the value chain.

vision: Summarizes where a business wants to go and includes an understanding of how technology supports the firm's vision. The vision helps the firm focus its efforts more clearly on what the innovation plan wishes to accomplish.

INDEX

Page numbers followed by f indicate a figure

Case Name	Case Authors	Source	Part Number
Gold Peak Electronics: R&D Globalization from East to West	Kumar and Kumar	Harvard Case	1, Strategic Foundation
Pharma Technologies Inc.	Nicholls-Nixon, White and Herbert	Harvard Case	1, Strategic Foundation
GPS-to-GO Takes on Garmin	Pillittee	Richard Ivey School of Business Case	1, Strategic Foundation
Amyris Biotechnologies	Chess and Raffaelli	Harvard Case	2, Internal Strategy
Bell Canada: The VoIP Challenge	White and Day	Richard Ivey School of Business Case	2, Internal Strategy
For the Love of Good Food: The pLate Trace Project	Haggerty, Jang and Liu	Richard Ivey School of Business Case	2, Internal Strategy
Post-Merger Integration at Northrop Gruman Information Technology	Freccia and Bourgeois	Harvard Case	3, External Strategy
Shilling & Smith Acquisition of Xteria Inc.: Data Center Technology Leasing	Jeffery, Shield, Ekici and Conley	Harvard Case	3, External Strategy
Skype	Eisenmann and Coles	Harvard Case	3, External Strategy
CQUAY Technologies Corp.	Beamish and Boeh	Harvard Case	4, Strategic Success
GE's Growth Strategy: The Immelt Initiative	Bartlett	Harvard Case	4, Strategic Success
HCL Technologies: Employee First, Customer Second	Ramdas, and Gajulapalli	Darden Business Publishing Case	4, Strategic Success

MAKE IT YOURS

Make It Yours, add business cases to your text. Your course is unique, create a text that reflects it. Let us help you put together a quality Strategic Management of Technology and Innovation casebook simply, quickly, and affordably. We have aligned best-selling business cases from leading case providers such as Harvard and Ivey at the part level for this text. Create a comprehensive learning solution by adding cases into your text, or simply select cases to create a casebook. Contact your local Cengage Learning Representative for details.

AVAILABILITY OF RESOURCES MAY DIFFER BY REGION. Check with your local Cengage Learning representative for details.